Springer Proceedings in Advanced Robotics 26

The Springer Proceedings in Advanced Robotics (SPAR) publishes new developments and advances in the fields of robotics research, rapidly and informally but with a high quality.

The intent is to cover all the technical contents, applications, and multidisciplinary aspects of robotics, embedded in the fields of Mechanical Engineering, Computer Science, Electrical Engineering, Mechatronics, Control, and Life Sciences, as well as the methodologies behind them.

The publications within the "Springer Proceedings in Advanced Robotics" are primarily proceedings and post-proceedings of important conferences, symposia and congresses. They cover significant recent developments in the field, both of a foundational and applicable character. Also considered for publication are edited monographs, contributed volumes and lecture notes of exceptionally high quality and interest.

An important characteristic feature of the series is the short publication time and world-wide distribution. This permits a rapid and broad dissemination of research results.

Indexed by SCOPUS, SCIMAGO, WTI Frankfurt eG, zbMATH.

All books published in the series are submitted for consideration in Web of Science.

Pablo Borja · Cosimo Della Santina ·
Luka Peternel · Elena Torta
Editors

Human-Friendly Robotics 2022

HFR: 15th International Workshop on Human-Friendly Robotics

Springer

Editors
Pablo Borja
Faculty of Mechanical, Maritime and Materials Engineering
Delft University of Technology
Delft, The Netherlands

Cosimo Della Santina
Faculty of Mechanical, Maritime and Materials Engineering
Delft University of Technology
Delft, The Netherlands

Luka Peternel
Faculty of Mechanical, Maritime and Materials Engineering
Delft University of Technology
Delft, The Netherlands

Elena Torta
Department of Mechanical Engineering
Eindhoven University of Technology
Eindhoven, The Netherlands

ISSN 2511-1256 ISSN 2511-1264 (electronic)
Springer Proceedings in Advanced Robotics
ISBN 978-3-031-22730-1 ISBN 978-3-031-22731-8 (eBook)
https://doi.org/10.1007/978-3-031-22731-8

This Springer imprint is published by the registered company Springer Nature Switzerland AG
The registered company address is: Gewerbestrasse 11, 6330 Cham, Switzerland

Organization

Organizing Chairs

Pablo Borja	Delft University of Technology, the Netherlands
Cosimo Della Santina	Delft University of Technology, the Netherlands
Luka Peternel	Delft University of Technology, the Netherlands
Elena Torta	Eindhoven University of Technology, the Netherlands

Program Committee

Marcello Bonfé	University of Ferrara, Italy
Gianni Borghesan	KU Leuven, Belgium
Jonathan Cacace	Università di Napoli Federico II, Italy
Federico Califano	University of Twente, the Netherlands
Raffaella Carloni	University of Groningen, the Netherlands
Fei Chen	Istituto Italiano di Tecnologia, Italy
Nikhil Chopra	University of Marylandy, USA
Francesca Cordella	Università Campus Bio-Medico di Roma, Italy
Raymond Cuijpers	Eindhoven University of Technology, the Netherlands
Wilm Decré	KU Leuven, Belgium
Fanny Ficuciello	Università di Napoli Federico II, Italy
Giovanni Franzese	Delft University of Technology, the Netherlands
Nico Hübel	University of Antwerp, Belgium
Cristina Iacono	Università di Napoli Federico II, Italy
Sagar Joshi	Delft University of Technology, the Netherlands
Marta Lorenzini	Istituto Italiano di Tecnologia, Italy
Pauline Maurice	LORIA-CNRS, France
Sara Moccia	Scuola Superiore Sant'Anna, Italy
Rocco Moccia	Università di Napoli Federico II, Italy

Gianluca Palli	Università di Bologna, Italy
Salvatore Pirozzi	Università degli Studi della Campania Luigi Vanvitelli, Italy
Rodolfo Reyes-Báez	University of Groningen, the Netherlands
Peter Ruijten	Eindhoven University of Technology, the Netherlands
Matteo Saveriano	University of Trento, Italy
Büsra Sen	Eindhoven University of Technology, the Netherlands
Daniel Tozadore	École Polytechnique Fédérale de Lausanne, Switzerland
Elise D. T. Verhees	Eindhoven University of Technology, the Netherlands
Valeria Villani	University of Modena and Reggio Emilia, Italy
Jelena Vuletić	University of Zagreb, Croatia
Chenguang Yang	University of the West of England, UK

Series Editor Foreword

At the dawn of the century's third decade, robotics is reaching an elevated level of maturity and continues to benefit from the advances and innovations in its enabling technologies. These all are contributing to an unprecedented effort to bringing robots to human environment in hospitals and homes, factories and schools; in the field for robots fighting fires, making goods and products, picking fruits and watering the farmland, saving time and lives. Robots today hold the promise for making a considerable impact in a wide range of real-world applications from industrial manufacturing to healthcare, transportation and exploration of the deep space and sea. Tomorrow, robots will become pervasive and touch upon many aspects of modern life.

The *Springer Tracts in Advanced Robotics (STAR)* was launched in 2002 with the goal of bringing to the research community the latest advances in the robotics field based on their significance and quality. During the latest fifteen years, the STAR series has featured publication of both monographs and edited collections. Among the latter, the proceedings of thematic symposia devoted to excellence in robotics research, such as ISRR, ISER, FSR and WAFR, have been regularly included in STAR.

The expansion of our field as well as the emergence of new research areas has motivated us to enlarge the pool of proceedings in the STAR series in the past few years. This has ultimately led to launching a sister series in parallel to STAR. The *Springer Proceedings in Advanced Robotics (SPAR)* is dedicated to the timely dissemination of the latest research results presented in selected symposia and workshops.

This volume of the SPAR series brings a selection of the papers presented at the 15th edition of the International Workshop on Human-Friendly Robotics (HFR), which took place in Delft, the Netherlands, from September 22 to 23, 2022. The volume edited by Pablo Borja, Cosimo Della Santina, Luka Peternel and Elena Torta is a collection of 17 contributions on human-robot co-existence including theories, methodologies, technologies and experimental studies.

From its classical program with presentations by young scholars, the 15th edition of HFR culminates with this valuable reference on the current developments and new directions of human-friendly robotics—a genuine tribute to its contributors and organizers!

Naples, Italy
Stanford, USA
October 2022

Bruno Siciliano
Oussama Khatib
SPAR Editors

Preface

We are witnessing a rapidly growing number of applications carried out by robots in close collaboration with humans, either through direct interaction or by operating in their immediate vicinity. This new generation of human-friendly robots needs to be safe and trustworthy in a physical and cognitive sense. Furthermore, new paradigms for the holistic development of collaborative cognitive robots must be envisioned and experimentally validated. The benefit of humans and society is the central objective of such a research effort, and existing opportunities arise from different application domains such as healthcare and high-tech systems.

The International Workshop on Human-Friendly Robotics (HFR) is an annual meeting devoted to covering a broad range of topics related to human-robot interaction at physical and cognitive levels, where the participants discuss theories, methodologies, technologies and empirical and experimental studies. Additionally, HFR brings together academics, researchers, industry partners and scholars to exchange and share their experiences and results on all aspects of the role of robots in our daily life.

The 15th edition of HFR followed a single-track format and took place in Delft, the Netherlands, on the 22 and 23 of September 2022. The conference consisted of 17 talks and 17 posters, three plenary talks and several demonstrations from academic and industrial partners. This book contains the 17 full papers corresponding to the talks presented during the conference. Each article was evaluated through a single-blinded peer-reviewed process carried out by at least two program committee members.

We are incredibly grateful to the authors and reviewers for their invaluable contribution to the conference. We are also indebted to professors Jens Kober, Dongheui Lee and Maarten Steinbuch for the great plenary talks. Additionally, this endeavor would not have been possible without the support of our sponsors, Franka Emika, Fizyr, RoboHouse and TU Delft Robotics institute.

September 2022

Pablo Borja
Cosimo Della Santina
Luka Peternel
Elena Torta

Contents

A Literature-Based Perspective on Human-Centered Design and Evaluation of Interfaces for Virtual Reality in Robotics

Chenxu Hao[1(✉)], Anany Dwivedi[1], and Philipp Beckerle[1,2]

[1] Chair of Autonomous Systems and Mechatronics, Department of Electrical Engineering, Friedrich-Alexander-Universität Erlangen-Nürnberg, Erlangen, Germany
chenxu.hao@fau.de, anany.dwivedi@fau.de, philipp.beckerle@fau.de

[2] Department of Artificial Intelligence in Biomedical Engineering, Friedrich-Alexander-Universität Erlangen-Nürnberg, Erlangen, Germany

Abstract. Over the past decade, the use of virtual reality applications is increasing in the field of robotics, which is set to increase even further with the advent of "Metaverse". Metaverse is considered as an "embodied internet" which will seamlessly integrate machines, robots, and computer applications including those in virtual and augmented reality. In this work, we review the current state of research and discuss the underlying problems with the current design and evaluation methods for the state-of-the-art virtual reality interfaces employed in the field of robotics. To overcome these issues, we present our perspective on the methods for design and evaluation of such interfaces. Our perspective takes into account human-centered design methods along with the functional measures of evaluating the performance of the interfaces. This interdisciplinary perspective emphasizes rigorously designed empirical experiments that require close collaborations among engineers, designers, human-computer interaction researchers and cognitive scientists.

Keywords: Virtual reality · Augmented reality · Human-machine interfaces · Human-robot interaction

1 Introduction

Recently there has been talks for a new means of interaction with the digital world, called the "Metaverse". The term metaverse was first coined in Neal Stephenson's novel "Snow Crash" in 1992 [1]. It is a concept where internet is a combination of persistent and shared 3D virtual spaces linked into a perceived virtual universe [2]. Big tech companies like Meta Platforms, Microsoft, Nvidia etc,. consider it as the future of internet, with Mark Zuckerberg (CEO,

Chenxu Hao and Anany Dwivedi authors contributed equally to this work.

P. Borja et al. (Eds.): HFR 2022, 26, pp. 1–13, 2023.
https://doi.org/10.1007/978-3-031-22731-8_1

Meta Platforms) describing it as an "embodied internet" [3,4]. Such technological improvements will enable a more seamless integration of machines, robots, computer applications including those in virtual reality (VR) and augmented reality (AR) in our everyday life. The use of robot systems has already become prevalent in various complex, repetitive, and dangerous tasks. Such systems can be controlled from a distance by human operators when autonomous operation is infeasible (e.g., tasks in unknown and unstructured environment) or undesired (e.g., when encounter unknown objects).

Virtual immersive environments play a crucial role in providing researchers with attractive and motivating technologies in human-machine interaction research [5,6]. One example is the use of games-based interactions in the virtual environments for immersive training and rehabilitation with exergames [6,7]. Combined with appropriate interfaces, virtual reality devices allow researchers to collect various types of data such as saccades (rapid eye movements) and limb movements. Such data help researchers to investigate the movement accuracy and exercise duration of the participants, allowing researchers to monitor the rehabilitation process better, and design more suitable programs for the participants [5]. The games-based interactions in virtual environments can keep human participants more motivated and engaged, making the training and rehabilitation processes more enjoyable, which can affect the results of training and rehabilitation, as well as the quality of research data. The virtual environments can also provide more ecologically valid scenarios, for investigating situation-dependent behaviors and decisions under realistic conditions, e.g., in driving [8].

Virtual reality-based applications are also widely used in the field of robotics including medical robotics, motion planning and control, and human-robot interaction [9]. In medical robotics, such applications are used in training the medical professionals in robot-assissted surgery [10], treatment of virtual limb pain [11], training the users' of prosthetic devices to efficiently use the device leading to better prosthesis acceptance rates [12]. In motion planning and control, virtual reality applications find use with trajectory generation, robot programming, and simulation [13]. Virtual reality systems are also useful in facilitating safe human-robot interaction during human-robot collaboration tasks [14], teleoperation of robotic systems [15], use of wearable robots [16] etc.

These virtual interactions often involve joint information processing on the cognitive and physical level. Different tasks in the virtual environment demand different types of virtual interfaces (in terms of both hardware and software). One example is testing hazard warning systems for driving in the virtual environment [17]. In this case, the user is required to navigate through an urban environment while keeping an eye on potential hazardous conditions. On cognitive level, the user receives visual and auditory information from the virtual environment that reflects the road and traffic condition, and produces corresponding driving behaviors. On physical level, the user interacts with physical driving controls. To remind the user to pay attention to potentially dangerous situations, the system can provide the user with visual and auditory feedback from the virtual environment, or haptic feedback through physical feedback devices. Several types of feedback may not be required at the same time, and different feedback modalities may lead to varying levels of effectiveness in engaging the user. The requirements for the type of interface can change based on the target application. For an effective interface design, researchers have proposed a human-in-the-loop design methodology that required a collaboration between engineering and psychological sciences [18].

The problem of finding the appropriate interface given a specific type of task in the virtual environment is constrained by several factors including the hardware, the software, and the user themselves. Therefore, it is important to develop good methods to evaluate virtual interfaces. Current design and evaluation methods employed for assessing the performance of virtual interfaces lack formal user studies and empirical experiments analyzing the interactions between humans and virtual interfaces [19–21]. These methods may be adapted from research in human-computer interaction (HCI) and cognitive science [17], where such practices are common.

In this paper, we present our perspective on how we can evaluate and improve virtual interfaces for various types of tasks more effectively by applying rigorously designed empirical experiments—a crucial step that requires effort and close collaborations among engineers, designers, HCI researchers and cognitive scientists. We also discuss how our perspective is related to the recent theoretical work by Oulasvirta et al. [22], which presents computational rationality [23,24] as a theory of human-computer interaction. We briefly illustrate our perspective with a potential example in VR and AR-based robotics research.

The rest of the paper is organized as follows: Sect. 2 briefly reviews the application of virtual reality in the field of robotics, Sect. 3 discusses the current state of research on design process and evaluation methods of interaction interfaces, Sect. 4 presents our perspective on design and evaluation methods for such interfaces and interactions, and Sect. 5 concludes the paper.

2 Virtual Reality Interfaces in Robotics

The virtual and augmented reality systems are gaining increased interest in both commercial and research aspects. Figure 1 shows the number of papers published in the field of robotics relating to the topics of virtual reality, augmented reality, mixed reality, or extended reality. The graph was generated following a search on Scopus with the search string *Robotics AND (Virtual Reality OR Augmented Reality OR Mixed Reality OR Extended Reality).* We see that the interest in these topics is reaching up to ~ 600 publications annually.

Although VR and AR-based systems are mostly marketed for games and entertainment, such systems find themselves very useful in the field of robotics largely due to their capability of facilitating interactions with the robotic systems in an immersive environment without endangering the user [25]. Such applications may also be used for simulating the robotic systems to test functionality of

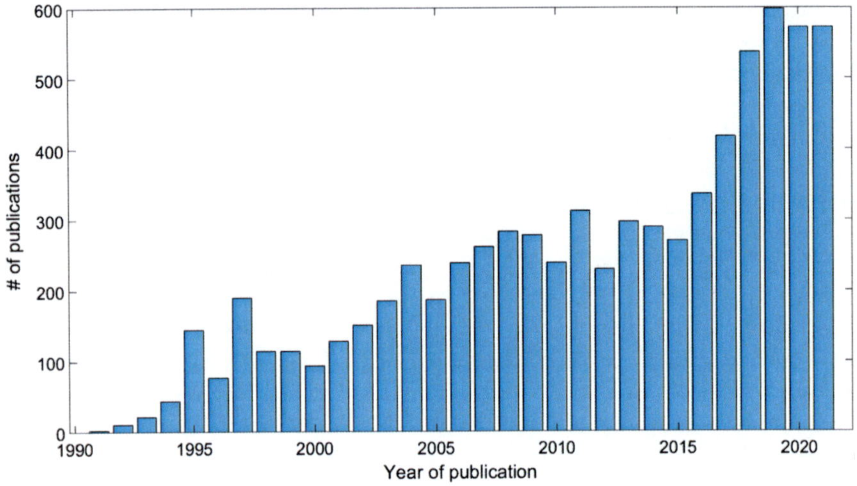

Fig. 1. Number of papers published each year in the field of robotics relating to the topics of virtual reality, augmented reality, mixed reality, or extended reality.

the system without developing a test workspace. This enables designing and testing of human-robot collaboration applications with iterative development process at a reduced cost, since different iterations could be developed and tested before real-world deployment [26]. With improving computational capabilities of computing devices leading to realistic simulations, VR and AR systems are increasingly being used in training user to efficiently use a system, such as, driving and flight simulators, training users of prosthetic devices, or for allowing surgeons to practice surgical techniques before operating on a living patient.

In medical robotics, VR and AR-based systems are employed to augment the capabilities of a surgeon to perceive and act beyond the limitations of physical perception [27]. One such interface that augments the capabilities of surgeons is Immersive Touch platform. It offers training modules for various surgical techniques like ventriculostomy, aneurysm clipping, and percutaneous spinal screw placement [28–30]. A user study with surgeons assessing Immersive Touch platform for its usefulness and resemblance to real surgery shows that such systems would be useful in preoperative surgical rehearsal and neurosurgical training for real-life surgery [30]. AR-based interfaces have also been employed in visualization of synthetic computer-generated images (3D virtual models) obtained from medical imaging overlaid on the real-time patient images to visualize unapparent anatomical details [31].

To ensure safe interactions in close contact human-robot collaboration tasks, researchers have proposed systems that overlay an aura around the robot showing the reachable workspace to inform the user of the area where collisions with the robot are possible [32]. To increase the immersive feeling and the sociability of an interaction with virtual systems, researchers explore wearable haptic interfaces that provide haptic sensations to complement the visual and auditory feedback [16]. For execution of tasks remotely with robot systems (e.g., remote surgeries, tasks in hostile environment, telepresence), researchers have explored use of various methods (e.g., virtual reality environment, augment reality environment, mixed reality environment) for efficient task executions. For assessing the performance of different interaction methods, researchers employ various evaluation metrics, e.g., calculation of the robot pose accuracy, trajectory tracking performance, the usability and workload of the interface, and the users' impressions of the interface during the task execution [15].

The interfaces that facilitate interaction with such applications aim at recognizing user's intentions by acquiring information relating to various aspects of their interaction, such as their motion, emotions, and their gaze, using different sensing modalities (e.g., infra-red cameras, bio-signals, direct physical input using buttons and touch pads). Despite the wide range of use cases for VR and AR-based applications, they suffer with issues and drawbacks that poses a challenge for their widespread adoption outside a laboratory environment. These interfaces usually have a limited field-of-view, poor motion tracking capabilities (exacerbated by occlusions), and poorly designed user interfaces especially to while interacting with 3D content [9]. Generally, these interfaces either require manipulating physical devices such as keyboards, mice, and joysticks etc. or rely on sophisticated motion capture systems (e.g., optical, inertial measurement units (IMUs), magnetic) and voice control modules. The use of button-based interfaces is unsuitable for dynamic and unstructured environments, while the hand-held controllers can be bulky and fatiguing to use. The motion capture systems, such as a the optical systems are prone to occlusions from the surroundings, while magnetic motion capture systems are generally inaccurate around metallic objects. IMU based motion capture systems suffer from sensor drifts and the voice command based systems are dependent on the background noise [33,34]. This requires development of more intuitive interfaces for enhanced feeling of presence and embodiment for the user without compromising the ease of control too much. For assessing the performance of the developed interfaces, it is important to evaluate their performance in a systematic and quantifiable way that enables taking informed decisions while selecting the interfaces for different applications.

3 Discussion of Interface Evaluation Methods

Virtual environment interfaces can be applied to a wide variety of domains. However, application with different tasks imply different requirements for human-virtual environment interactions. This suggests that the design of virtual environment and the selection of interfaces can be task-specific. However, while adapting to the specific task domains, there are some common factors in the evaluation processes of interfaces for virtual interactions. Several previous work have reviewed and investigated the user experience, usability, and immersion with different VR systems [21,35,36].

In the selection and evaluation of interfaces, there are some common trade-offs. A prominent one is the trade-off between the usability of the device and the level of immersion achieved by the system. Usability is often connected to the extent to which a user can achieve the specified goals with effectiveness, efficiency, and satisfaction in a specified context of use, while the level of immersion is connected to the user's sense of presence or embodiment in the virtual environment [37]. A system that has a high level of usability may not be as immersive as desired, and vice versa. For example, among common hand input systems, keyboards and mice have high usability, but they limit natural hand motion of the user, thus reduces the level of immersion. On the other hand, hand-worn devices may have a high level of immersion, but they may cause physical discomfort for the user, making these devices less easier to control [21].

In addition, the level of the trust the user has on the interactions with a virtual environment is important. This requires gaining insights regarding the factors that affect the level of trust. These insights can be used to maximize the trust leading to widespread adoption of the technology and more comfortable interactions with the system [38]. Another factor to consider while evaluating different VR and AR-based interfaces is the level of distraction caused by such interfaces. It is important to reduce the distractions in order to reduce the cognitive load on the user during interactions

The study aiming to reduce distractions while interacting with a system is termed as "calm techonology" [39]. Using human-in-the-loop design methodology, interfaces maybe be designed such that the cognitive and physical load on the users during interactions is reduced [40]. To make interactions with VR and AR environments more convenient, researchers are also exploring context aware rule-based decision making approaches to decide where and when virtual elements should be placed in the environment, and how much information should be provided [41].

Due to these features of the interaction between humans and virtual systems, the development and evaluation of virtual interaction interfaces often requires solving the problem of finding the optimal interface for human-virtual system interaction given a specific task. With such a goal, the development of these interfaces can benefit from applying recent perspectives in human-robot interaction that emphasize the cognitive aspects in human-robot interaction such as embodied cognition [18,42] and design principles from HCI research such as affordance, reducing cognitive load, low physical effort, learnability, flexibility in use, feedback, etc [19,43,44]. However, recent reviews suggest that such principles are not addressed enough in AR systems design [19].

The problem-solving ability of HCI research in VR and AR systems is indeed often limited. This may be partly due to the difficulties in interpreting and following HCI design principles directly. Currently, common methods to evaluate interfaces include evaluations by experts, user experience studies [21], case studies [17], etc. However, there is still a lack of formal user studies [19]. Additionally, current HCI research in virtual environment development are more likely to be constructive rather than conceptual or empirical [20,45]. In other words, there are not enough studies investigating and explaining various existing phenomena in the interactions. Practically, designing empirical experiments to evaluate interaction interfaces could be challenging, as the evaluation of virtual interactions require not only considerations for constraints from the hardware and software, but also considerations for the user, the task, and the interactive environment.

These considerations have been addressed in the domain of cognitive science to some extent (e.g., user's sense of presence, how efficiently can the user learn to use the device, etc.). Oulasvirta et al. [22] present a theoretical framework where computational rationality [23,24] is considered as a theory of human-computer interaction. This theoretical framework assumes that the user behaves in a way that maximizes expected utility, given internal bounds such as cognitive abilities and external, or environmental bounds [22]. We believe that this theoretical framework could bring ideas for evaluating interfaces. More specifically, while focusing on task-specific goals in selecting and developing virtual interfaces, researchers should also take into account the users' information processing bounds, physical limitations, and the interaction environment. Such analyses of virtual interfaces could potentially require both empirical experiments and computational modeling approaches.

4 Perspectives for Interface Design and Evaluation

As our review suggests, the VR and AR-based interfaces play an important role in the field of robotics and human-robot interaction, which is only expected to increase in the future. However, the current interface design methods often prioritize the absolute performance of the interface rather than the ergonomic design, resulting in some potentially powerful interfaces (e.g., hand-worn devices that provide a high-level of immersion) not being used more widely or optimally. Additionally, in the interface selection and evaluation processes, there appears to be an insufficient amount of empirical studies and user studies. This indicates that is a gap between this future perspective of virtual environment interfaces and the lack of human-centered approach in the current development processes for such interfaces.

As an attempt to start bridging this gap and address the issues in the application of virtual interaction interfaces, we propose two research directions for improving the design, selection, and evaluation processes of interfaces for virtual interaction. Our perspective emphasizes the importance of human-centered approach, particularly how empirical experiments could inform the development of interfaces and provide groundwork for further potential computational analyses, leading to better design and use of virtual environments and interfaces.

4.1 Human-Centered Design of Interfaces

The first research direction that we propose is to apply human-centered design principles from robotics [46–48] to interface design, since that virtual interactions and human-robot interactions share a common feature where both types of interactions involve joint information exchanges on the cognitive level and on the physical level.

During the design process of interfaces, it is important to empirically evaluate the interfaces while taking into account the functional and psychological performance of the user during interactions with a system [46]. This emphasizes on a need for a human-centered approach, since consideration of the user's state during interactions will be more likely to result in interfaces that do not cause inconvenience, threat, obstacle, annoyance, or harm to human users during interactions. It is also important to ensure that the interfaces can provide accessibility, functionality, protection, and can minimize the computational demands, complexity, or costs for human users during interactions, except where such support would conflict with users' safety or convenience. Such design principles can better ensure that the final interface has high functional performance while being affordable and maintaining the portability of the device. Essentially, regarding the previously mentioned trade-off between the usability of the device and the level of immersion provided by the system, these principles are aiming at improving the user's experience by increasing the usability of potentially powerful interfaces while preserving the level of immersion as much as possible.

4.2 Empirical Evaluations of Interfaces

The evaluation of interfaces, especially with empirical experiments, is a crucial component of human-centered approach in improving interfaces for virtual interactions. Although researchers have proposed existing frameworks for evaluating the usability of interfaces for virtual interactions [35], currently there is still a lack of evaluations based on formal user studies and empirical experiments. We believe that the problem of interface selection and evaluation is an interdisciplinary research problem by nature. Therefore, we propose the second research direction focusing on the use of well-designed empirical experiments in the evaluation of interfaces. This requires joint efforts and close collaborations from related fields such as engineering and design, HCI, and cognitive science.

Given that there are common principles underlying interface selection and evaluation for different tasks, we believe that a systematic empirical evaluation of interfaces for different types of tasks in the virtual environment will be helpful for making informed decisions for selecting optimal interfaces for the applications at hand.

As the requirements for interfaces are often task-specific, it is crucial to identify or define what does it mean for the interface to be 'optimal' or 'intuitive'. Following the theoretical framework in Oulasvirta et al. [22], we argue that a definition for the goodness of an interface should not only account for the specific task goals, but also the cognitive and physical limitations of the user while considering the interactive environment. For example, in the domain of robotic teleoperation and telemanipulation, the user needs to control a robot remotely by interacting with a virtual system. The user receives information of the robotic agent's working environment virtually, and controls the robot to perform certain tasks [49]. If the human user is remotely controlling a robotic agent to navigate through hostile environments, the desired interface may provide a list of tasks that can be performed with the objects identified by the robotic agent. This list can be displayed to the user in the virtual environment of the user or on a two-dimensional display [49]. Based on the selected task by the user, the robot can assist the user in execution of the task by performing shared local control and thereby reducing the complexity of the task [50]. The robot may adjust the level of assistance based on the cognitive and physical load the user is experiencing by measuring the physiological signals from the user [51]. If the human user is using teleoperation system for surgical training or performing surgery, then it may be important for the system to provide delicate and precise haptic feedback.

With clearly defined task goals, it is possible to evaluate or compare interfaces with rigorously designed empirical experiments that measure different users' task performances, and learning (i.e., how can the users adapt to the interfaces). For example, in teleoperation, researchers can test how well different users perform a variety of decision tasks, design tasks, or motor tasks that share common elements with specific, real-life teleoperation tasks. The objective performance measures could include solutions in real-life scenarios and reaction time. For example, in the study by Bowman et al. [36], participants were given visualizations of internal rock structures in virtual systems and then asked to provide judgments on spatial relationships and make decisions related to designing a tunnel. An example of a motor task in virtual environment could be as simple as remote operating a robotic arm to follow a predefined trajectory to evaluate the executions of precise motions in the virtual space [15]. These examples illustrate that it is possible to design empirical experiments with objective quantitative measures for human performances in virtual environments while using different interfaces.

Such empirical experiments can also potentially provide us with a basis for analyzing interaction interfaces from a computational rationality perspective [22]. By applying analyses of the users' information processing bounds (e.g., visual attention, spatial reasoning ability, emotional states and trust for the system, etc.), physical limitations, and the bounds of the task environments, researchers may ideally be able to identify the (boundedly) optimal virtual interaction setup given the user's characteristics and the task domain in focus. Naturally, it is challenging to identify the internal bounds of the user and external bounds of the interaction environment. However, a variety of modeling work in cognitive science including some on object manipulation [52] and VR illusions [53] could be informative for such analyses for virtual interactions.

Finally, using empirical experiments in the design, selection, and evaluation processes of interfaces under the theoretical framework of computational rationality can potentially provide researchers with guidelines to develop virtual systems that can adapt to the human user's internal bounds, thus facilitating the user's task performance in the virtual environment. We believe that creating virtual systems that can co-adapt to the user is crucial for achieving human-centered design of virtual systems.

5 Conclusion

In this work we present the increasing importance of the virtual reality and augmented reality interfaces in the field of robotics and discuss the underlying problems with the design and evaluation methods of the current state-of-the-art interfaces through a review of the literature. We believe that these issues call for human-centered interface design, selection, and evaluation methods that can lead to more comfortable interactions between human users and robotic agents in virtual environments. To do this, we suggest applying rigorously designed empirical experiments with domain or task-specific measures to assess the human user's level of performance with and level of adaptation to the chosen interface. This is an interdisciplinary problem by nature and requires further collaborations among engineers, designers, human-computer interaction researchers and cognitive scientists.

References

1. Stephenson, N.: Snow Crash: A Aovel. Spectra (2003)
2. Hackl, C.: Defining the metaverse today. [Online]. Available: https://www.forbes.com/sites/cathyhackl/2021/05/02/defining-the-metaverse-today/?sh=433cc19a6448
3. Iannuzzi, J.B, Faulds, Z.: Top companies chasing the metaverse. [Online]. Available: https://www.thestreet.com/technology/companies-making-money-metaverse
4. Newton, C. Mark in the metaverse. [Online]. Available: https://www.theverge.com/22588022/mark-zuckerberg-facebook-ceo-metaverse-interview
5. Alnajjar, M., Abutabikh, A., Issa, M., Debeljak, M., Cikajlo, I..: Development of 3d exergame for upper limbs rehabilitation using leap motion controller and unity. In: 2020 International Conference on Assistive and Rehabilitation Technologies (iCareTech), pp. 24–29. IEEE (2020)
6. Trombetta, M., Henrique, P.P.B., Brum, M.R., Colussi, E.L., De Marchi, A.C.B., Rieder, R.: Motion rehab ave 3d: A vr-based exergame for post-stroke rehabilitation. Comput. Methods Programs Biomed. **151**, 15–20 (2017)
7. Bui, J., Luauté, J., Farnè, A.: Enhancing upper limb rehabilitation of stroke patients with virtual reality: a mini review. Front. Virtual Reality, 146 (2021)
8. Zgonnikov, A., Abbink, D., Markkula, G.: Should i stay or should i go? evidence accumulation drives decision making in human drivers (2020)
9. Makhataeva, Z., Varol, H.A.: Augmented reality for robotics: a review. Robotics **9**(2), 21 (2020)
10. Schmidt, M. Köppinger, K., Fan, C., Kowalewski, K.-F., Schmidt, L., Vey, J., Proctor, T., Probst, P., Bintintan, V., Müller-Stich, B.-P., et al.: Virtual reality simulation in robot-assisted surgery: meta-analysis of skill transfer and predictability of skill. BJS open **5**(2), zraa066 (2021)
11. Ambron, E., Miller, A., Kuchenbecker, K.J., Buxbaum, L.J., Coslett, H.B.: Immersive low-cost virtual reality treatment for phantom limb pain: evidence from two cases. Front. Neurol. **9**, 67 (2018)
12. Boschmann, A., Neuhaus, D., Vogt, S., Kaltschmidt, C., Platzner, M., Dosen, S.: Immersive augmented reality system for the training of pattern classification control with a myoelectric prosthesis. J. Neuroeng. Rehabil. **18**(1), 1–15 (2021)

13. Stadler, S., Kain, K., Giuliani, M., Mirnig, N., Stollnberger, G., Tscheligi, M.: Augmented reality for industrial robot programmers: workload analysis for task-based, augmented reality-supported robot control. In: 25th IEEE International Symposium on Robot and Human Interactive Communication (RO-MAN), pp. 179–184 . IEEE (2016)
14. Liu, O., Rakita, D., Mutlu, B., Gleicher, M.: Understanding human-robot interaction in virtual reality. In: 26th IEEE International Symposium on Robot and Human Interactive Communication (RO-MAN), pp. 751–757. IEEE (2017)
15. De Pace, F., Gorjup, G., Bai, H., Sanna, A., Liarokapis, M., Billinghurst, M.: Leveraging enhanced virtual reality methods and environments for efficient, intuitive, and immersive teleoperation of robots. In: 2021 IEEE International Conference on Robotics and Automation (ICRA), pp. 12 967–12 973. IEEE (2021)
16. Zhou, R., Wu, Y., Sareen, H.: Hextouch: a wearable haptic robot for complementary interactions to companion agents in virtual reality. In: SIGGRAPH Asia. Emerging Technologies, pp. 1–2 (2020)
17. Sutcliffe, A.G., Poullis, C., Gregoriades, A., Katsouri, I., Tzanavari, A., Herakleous, K.: Reflecting on the design process for virtual reality applications. Int. J. Hum. Comput. Interact. **35**(2), 168–179 (2019)
18. Beckerle, P., Castellini, C., Lenggenhager, B.: Robotic interfaces for cognitive psychology and embodiment research: a research roadmap. Wiley Interdisc. Rev.: Cogn. Sci. **10**(2), e1486 (2019)
19. Dünser, A., Grasset, R., Seichter, H., Billinghurst, M.: Applying HCI principles to AR systems design. In: 2nd International Workshop at the IEEE Virtual Reality 2007 Conference, p. 2007. North Carolina, USA, Charlotte (2007)
20. Boletsis, C., Cedergren, J.E., Kongsvik, S.: Hci research in virtual reality: a discussion of problem-solving. In: International Conference on Interfaces and Human Computer Interaction. IHCI 2017, Portugal, 21–23 July 2017 (2017)
21. Kim, Y.M., Rhiu, I., Yun, M.H.: A systematic review of a virtual reality system from the perspective of user experience. Int. J. Hum. Comput. Interact. **36**(10), 893–910 (2020)
22. Oulasvirta, A., Jokinen, J.P., Howes, A.: Computational rationality as a theory of interaction. In: CHI Conference on Human Factors in Computing Systems, pp. 1–14 (2022)
23. Lewis, R.L., Howes, A., Singh, S.: Computational rationality: Linking mechanism and behavior through bounded utility maximization. Top. Cogn. Sci. **6**(2), 279–311 (2014)
24. Gershman, S.J., Horvitz, E.J., Tenenbaum, J.B.: Computational rationality: a converging paradigm for intelligence in brains, minds, and machines. Science **349**(6245), 273–278 (2015)
25. Weistroffer, V., Paljic, A., Callebert, L., Fuchs, P.: A methodology to assess the acceptability of human-robot collaboration using virtual reality. In: Proceedings of the 19th ACM Symposium on Virtual Reality Software and Technology, pp. 39–48 (2013)
26. Badia, S.B.I., Silva, P.A., Branco, D., Pinto, A., Carvalho, C., Menezes, P., Almeida, J., Pilacinski, A.: Virtual reality for safe testing and development in collaborative robotics: challenges and perspectives. Electronics **11**(11), 726 (2022)
27. Madhavan, K., Kolcun, J.P.G., Chieng, L.O., Wang, M.Y.: Augmented-reality integrated robotics in neurosurgery: are we there yet? Neurosurg. Focus **42**(5), E3 (2017)

28. Luciano, C.J., Banerjee, P.P., Sorenson, J.M., Foley, K.T., Ansari, S.A., Rizzi, S., Germanwala, A.V., Kranzler, L., Chittiboina, P., Roitberg, B.Z.: Percutaneous spinal fixation simulation with virtual reality and haptics. Neurosurgery **72**(suppl_1), A89–A96 (2013)
29. Yudkowsky, R., Luciano, C., Banerjee, P., Schwartz, A., Alaraj, A., Lemole, G.M., Jr., Charbel, F., Smith, K., Rizzi, S., Byrne, R., et al.: Practice on an augmented reality/haptic simulator and library of virtual brains improves residents' ability to perform a ventriculostomy. Simul. Healthc. **8**(1), 25–31 (2013)
30. Alaraj, A., Luciano, C.J., Bailey, D.P., Elsenousi, A., Roitberg, B.Z., Bernardo, A., Banerjee, P.P., Charbel, F.T.: Virtual reality cerebral aneurysm clipping simulation with real-time haptic feedback. Operative Neurosurg. **11**(1), 52–58 (2015)
31. Pessaux, P., Diana, M., Soler, L., Piardi, T., Mutter, D., Marescaux, J.: Towards cybernetic surgery: robotic and augmented reality-assisted liver segmentectomy. Langenbeck's Arch. Surg. **400**(3), 381–385 (2015)
32. Makhataeva, Z., Zhakatayev, A., Varol, H.A.: Safety aura visualization for variable impedance actuated robot. In: IEEE/SICE International Symposium on System Integration (SII), 805–810. IEEE (2019)
33. Saponas, Z., Tan, D.S., Morris, D., Balakrishnan, R., Turner, J., Landay, J.A.: Enabling always-available input with muscle-computer interfaces. In: Proceedings of the 22nd annual ACM symposium on User interface software and technology, pp. 167–176 (2009)
34. Chapman, J., Dwivedi, A., Liarokapis, M.: A wearable, open-source, lightweight forcemyography armband: on intuitive, robust muscle-machine interfaces. In: 2021 IEEE/RSJ International Conference on Intelligent Robots and Systems (IROS), pp. 4138–4143. IEEE (2021)
35. Sutcliffe, A.G., Kaur, K.D.: Evaluating the usability of virtual reality user interfaces. Behav. Inf. Technol. **19**(6), 415–426 (2000)
36. Bowman, D.A., McMahan, R.P.: Virtual reality: how much immersion is enough? Computer **40**(7), 36–43 (2007)
37. Nostadt, N., Abbink, D.A., Christ, O., Beckerle, P.: Embodiment, presence, and their intersections: teleoperation and beyond. ACM Transa. Hum. Robot Inter. (THRI) **9**(4), 1–19 (2020)
38. Morra, L., Lamberti, F., Prattickó, F.G., La Rosa, S., Montuschi, P.: Building trust in autonomous vehicles: role of virtual reality driving simulators in HMI design. IEEE Trans. Veh. Technol. **68**(10), 9438–9450 (2019)
39. Moore, B.A., Urakami, J.: The impact of the physical and social embodiment of voice user interfaces on user distraction. Int. J. Hum. Comput. Stud. 102784 (2022)
40. Oviatt, S.: Human-centered design meets cognitive load theory: designing interfaces that help people think. In: Proceedings of the 14th ACM international conference on Multimedia, pp. 871–880 (2006)
41. Lindlbauer, D., Feit, A.M., Hilliges, O.: Context-aware online adaptation of mixed reality interfaces. In: Proceedings of the 32nd Annual ACM Symposium on User Interface Software and Technology, pp. 147–160 (2019)
42. Schürmann, T., Mohler, B.J., Peters, J., Beckerle, P.: How cognitive models of human body experience might push robotics. Front. Neurorobotics **13**, 14 (2019)
43. Rizzo, A.A., Kim, G.J., Yeh, S.-C., Thiebaux, M., Hwang, J., Buckwalter, J.G.: Development of a benchmarking scenario for testing 3d user interface devices and interaction methods. In Proceedings of the 11th International Conference on Human Computer Interaction. Las Vegas, Nevada (2005)

44. Isaacs, E., Walendowski, A.: Designing from both sides of the screen: How designers and engineers can collaborate to build cooperative technology. Sams Publishing (2002)
45. Oulasvirta, A, Hornbæk, K.: Hci research as problem-solving. In: Proceedings of the 2016 CHI Conference on Human Factors in Computing Systems, pp. 4956–4967 (2016)
46. Sosa, R., Montiel, M., Sandoval, E.B., Mohan, R.E.: et al.: Robot ergonomics: Towards human-centred and robot-inclusive design. In: DS 92: Proceedings of the DESIGN 2018 15th International Design Conference, pp. 2323–2334 (2018)
47. Beckerle, P., Salvietti, G., Unal, R., Prattichizzo, D., Rossi, S., Castellini, C., Hirche, S., Endo, S., Amor, H.B., Ciocarlie, M., et al.: A human-robot interaction perspective on assistive and rehabilitation robotics. Front. Neurorobotics **11**, 24 (2017)
48. Beckerle, P., Christ, O., Schürmann, T., Vogt, J., von Stryk, O., Rinderknecht, S.: A human-machine-centered design method for (powered) lower limb prosthetics. Robot. Auton. Syst. **95**, 1–12 (2017)
49. Gorjup, G., Dwivedi, A., Elangovan, N., Liarokapis, M.: An intuitive, affordances oriented telemanipulation framework for a dual robot arm hand system: On the execution of bimanual tasks. In: 2019 IEEE/RSJ International Conference on Intelligent Robots and Systems (IROS), pp. 3611–3616. IEEE (2019)
50. Dwivedi, A., Shieff, D., Turner, A., Gorjup, G., Kwon, Y., Liarokapis, M.: A shared control framework for robotic telemanipulation combining electromyography based motion estimation and compliance control. In: 2021 IEEE International Conference on Robotics and Automation (ICRA), pp. 9467–9473. IEEE (2021)
51. Bianchi, M., Valenza, G., Greco, A., Nardelli, M., Battaglia, E., Bicchi, A., Scilingo, E.P.: Towards a novel generation of haptic and robotic interfaces: integrating affective physiology in human-robot interaction. In: 25th IEEE International Symposium on Robot and Human Interactive Communication (RO-MAN), pp. 125–131. IEEE (2016)
52. Aguilera, F., Alarcón, R.A., Guerrero, L.A., Collazos, A.: A cognitive model of user interaction as a guideline for designing novel interfaces. In: IFIP World Computer Congress, TC 2, pp. 62–76. Springer (2006)
53. Gonzalez-Franco, M., Lanier, J.: Model of illusions and virtual reality. Front. Psychol. **8**, 1125 (2017)

A Multi-Modal Feedback Communication Interface for Human Working Posture Adjustments

Kushal Thirani, David A. Abbink, and Luka Peternel(✉)

Department of Cognitive Robotics, Delft University of Technology, Mekelweg 2, 2628, CD Delft, The Netherlands
l.peternel@tudelft.nl

Abstract. This paper studies non-physical feedback mechanisms to guide human workers toward ergonomic body postures. Specifically, the focus is to solve the tasks that involve no direct physical interaction between the human and the robotic system, therefore tactile guidance by the robot body is not feasible. We propose a multi-modal ergonomic posture guidance system that comprises visual feedback and speech-based audio feedback. We hypothesise that the proposed multi-modal system leads to better performance compared to uni-modal feedback systems when trying to guide users from one pose to another. To test the hypothesis we conducted an experiment that compared conditions with only audio feedback, only visual feedback and multi-modal feedback. In addition, we examined speech-based audio guidance in joint space and in endpoint space. The results showed that the speech-based feedback in joint space came out as the preferred audio feedback due to its ability to allow users to carry out efficient and coordinated inter-joint movements, especially in cases of high redundancy. Furthermore, the proposed multi-modal feedback system was superior compared to the other feedback modalities both in terms of objective measures and subjective measures.

Keywords: Human-machine interaction · Ergonomics · Visual feedback · Audio feedback · Human factors

1 Introduction

The manufacturing industry is one of the key components of modern society. Much of the production process can be automated by robots, such as mechanical arms and non-mechanical smart systems, to improve the production speed and quality of products. Nevertheless, human workers are in many cases still an essential part of the production process due to their superior cognitive capabilities and adaptability. Unlike robots that can perform demanding tasks uninterrupted for prolonged periods of time, human workers are susceptible to cognitive and physical stress caused by various working conditions, such as unsuitable working style [6] or excessive forces on the body [11]. The continuous presence of such physical stress can lead to a dramatic increase in work-related musculoskeletal disorders (WMSD) [2,23].

P. Borja et al. (Eds.): HFR 2022, 26, pp. 14–29, 2023.
https://doi.org/10.1007/978-3-031-22731-8_2

When humans are working in robotised environments, we can exploit smart robotic systems to account for human ergonomics and work-related musculoskeletal risks. These aspects can be monitored by using predefined ad-hoc tables like RULA and REBA [4,7,15,24] or biomechanical models [1,10,16,18]. When the robotic system detects ergonomic risks in the existing working configuration, it can alert the human worker to reconfigure the body posture online through different types of feedback. For example, the approaches [10,18,19,26] used collaborative robotic arms to reconfigure the human co-worker's body into a more ergonomic pose through physical interaction. Similarly, force feedback through haptic interfaces are used to constrain [22] or reconfigure [17] human operator's arm in teleoperation. Nevertheless, not all manufacturing tasks involve physical human-robot interaction or mechanical robotic systems. In such cases, other feedback modalities have to be employed to alert the human co-worker about ergonomic risks during the production process. For instance, wearable vibrotactile devices can be used to alert and guide the human joints to a more ergonomic posture [9]. However, wearable devices can be time-consuming to equip and may irritate the workers with vibrations.

Alternatively, contactless and non-tactile feedback can be provided in a visual or audio form by a smart system (i.e., a non-mechanical robot). For example, the method in [14] used an external camera to track the human body posture and display excessive joint torques to the worker on a monitor in real-time. To a similar end, audio feedback can be used to guide the human posture [13,21]. The study in [20] developed a visual interface to map the shared human-robot workspace in 3D based on multiple ergonomic metrics. The work in [27] exploited latent space to generate a 2D map that can encode a complex multi-degree of freedom (DoF) information about the body. However unimodal visual or audio systems each have their limitations. Visual feedback systems require human workers to look at the monitor, which can draw attention away from the actual task. In addition, small but critical posture corrections might not be easily visible to the user. On the other hand, by using audio feedback the human worker does not need to look away from the actual task. Nevertheless, the transmission of complex multi-dimensional data for pose correction (e.g., name of the joint, direction of movement and magnitude of movement, etc.) through audio techniques is typically slow and can easily overwhelm or confuse the user. Furthermore, audio feedback systems can suffer from problems of habituation, where a continuous presence of sound can get obscured in the background and the user realises this only when the audio feedback either stops or changes [3].

The solution to these limitations is to combine visual feedback and audio feedback into a multi-modal system. Visual feedback offers a fast guidance method for large posture corrections since the human worker can see the actual posture with respect to the desired one in real-time. On the other hand, audio feedback can be used to guide humans for minor pose adjustments once the pose is already close enough to the desired one, or when the actual task requires the full visual attention of the worker. There is plenty of existing literature that has already shown that multi-modal feedback improves motor coordination and learning in humans [25]. Nevertheless, a similar study for methods related to body posture adjustments to improve human worker ergonomics is still missing.

In order to address this, we developed a multi-modal ergonomic guidance system to combine the advantages of visual and audio feedback and hypothesis that *the multi-modal system will lead to better performance compared to either visual feedback modality or audio feedback modality when trying to guide the human users from one pose to another*. To test the hypothesis we conducted an experimental study to compare the three contactless guidance methods: audio-only feedback, visual-only feedback and multi-modal feedback.

2 Methods

2.1 Feedback Mechanism Design

We designed four types of feedback mechanisms in order to carry out a comparative study and test our hypothesis. We examined two speech-based audio feedback methods; one acting in joint space and the other acting in endpoint space. We designed a visual feedback method that can display the current human body posture with respect to the desired one. Finally, we designed a multi-modal feedback method that combines audio and visual feedback modalities.

Audio Feedback in Joint Space Speech-based audio feedback in joint space involves a feedback method where the user adjusts the body posture in a joint-specific manner through verbal commands. The commands were limited to the shoulder and elbow joints for this study. The wrist was not considered since it has less impact on the arm's dynamics, compared to the shoulder and elbow. The commands were in the English language and were output at a rate of 180 words per minute, which is slightly higher than the average of 140–160 words per minute worldwide. The reason for this was to increase the bandwidth and at the same time retain the understandability. The direction and magnitude of the movement were given relative to the current angular position of the joints. The commands were given sequentially for each joint with a short pause in-between. Further corrective commands were given after the user made the movement based on the previous commands and temporarily maintained the arm configuration. Once the user oriented his/her joints in the desired pose (which would be required to perform tasks such as drilling/polishing), the system alerted the user by producing a beeping sound. Table 1 shows the list of commands the user received for this particular modality.

Table 1. List of speech-based audio commands for joint space feedback

Command	Joint	Meaning
"Move arm up by xx degrees"	Shoulder	Shoulder flexion
"Move arm down by xx degrees"	Shoulder	Shoulder extension
"Flex elbow by xx degrees"	Elbow	Elbow flexion
"Extend elbow by xx degrees"	Elbow	Elbow extension

Audio Feedback in Endpoint Space Another speech-based audio feedback mechanism was created which provided verbal commands for endpoint movements in Cartesian space. The wrist joint was assumed to be the endpoint of our study. The commands were in the English language and were output at 180 words per minute. The direction and magnitude of the movement were relative to the current position of the endpoint and the unit of movement was in centimeters. Table 2) shows the list of commands the user received for this particular modality.

Table 2. List of speech-based audio commands for endpoint space feedback

Command	Meaning
"Move arm back by xx"	Move EP towards the body
"Move arm forward by xx"	Move EP away from body
"Move arm down by xx"	Move EP vertically down
"Move arm up by xx"	Move EP vertically up

Visual Feedback The visual feedback system was assumed to be the benchmark in the comparison study because of several reasons. First, concurrent visual feedback is the easiest and the most natural form of feedback modality [25]. Second, visual feedback is superior to other senses when it comes to understanding spatial information [25]. Third, humans usually rely on visual feedback when it comes to following trajectories and interacting with the environment [5]. For providing visual feedback to the user we developed a graphical user interface (GUI) in OpenCV (see Fig. 1). GUI was based on a monitor that presented a graphical representation of the human body (red) and the target pose (white) in real-time. The user had to reach the desired target pose by monitoring GUI, and when it was reached the colour of the screen changed to green.

Multi-Modal Feedback The multi-modal system interactively combines the two modalities of visual feedback and speech-based audio feedback in joint space. The proposed system exploits the advantages of both visual and audio systems. The initial larger pose adjustments can be done by relying on visual feedback. Then, when the body is close enough to the desired pose (± 5 degrees), the final adjustment can be done completely based on the audio feedback. This way the users do not have to constantly look up at the GUI to receive visual feedback about their pose. Instead, they can concentrate on the actual task (e.g., performing drilling/polishing with a tool at the endpoint) while listening to the audio feedback for minor adjustments. The users can still switch to visual feedback at any time if larger corrections are required or when looking at the task is not so important.

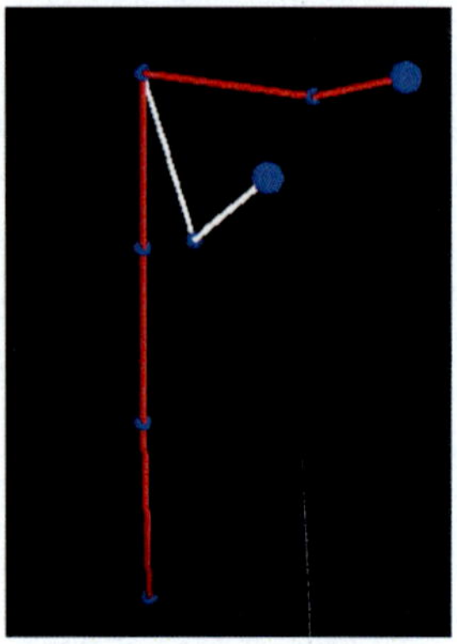

Fig. 1. Graphical user interface of the visual feedback system. The current pose of the user is shown in red and the desired pose is shown in white. The blue circles represent the five major joints of the human body that we considered in this study: ankle, knee, hip, shoulder and elbow. The larger blue circle at the end represents the wrist, which is the endpoint.

2.2 Experimental Design

The study included 14 participants (10 male, 4 female) between 23 and 27 years old (25.14±0.99 years[1]) who were all proficient in the English language. Exclusion criteria were movement, vision and/or hearing impairments. The participants gave informed consent prior to the study. The study was approved by TU Delft Human Research Ethics Committee.

A within-subject experiment was conducted to compare the user performance in three conditions: *audio-only feedback*, *visual-only feedback* and *multi-modal feedback*. The performances for the audio and multi-modal feedback were compared against the visual feedback, which we used as a benchmark. The user performance was evaluated by both objective and subjective measures. An additional experiment was conducted to compare audio feedback acting in joint space to audio feedback acting in endpoint space (see Sect. 5).

Objective Measures

- **Task Completion Time** (TCT) measures the time to complete the pose adjustment, which was defined from start time until the time when a user holds the desired pose for 10 s. A smaller TCT indicates a better performance as it implies that the feedback helped the user to complete the task quicker (See Fig. 2).
- **Total Distance Moved** (TDM) refers to the sum of integrated joint angular movements that are required for the user to successfully orient both joints (shoulder and elbow) within the threshold (±5 degrees) of the desired angular position and any movements while staying inside the threshold for 10 s. A larger TDM indicates

[1] This format of reporting values corresponds to mean±standard deviation and applies for the rest of the paper.

worse performance as it implies that the feedback was not efficient and therefore the user made multiple pose adjustments (e.g., overshoots and undershoots) before reaching the desired pose.(See Fig. 2)

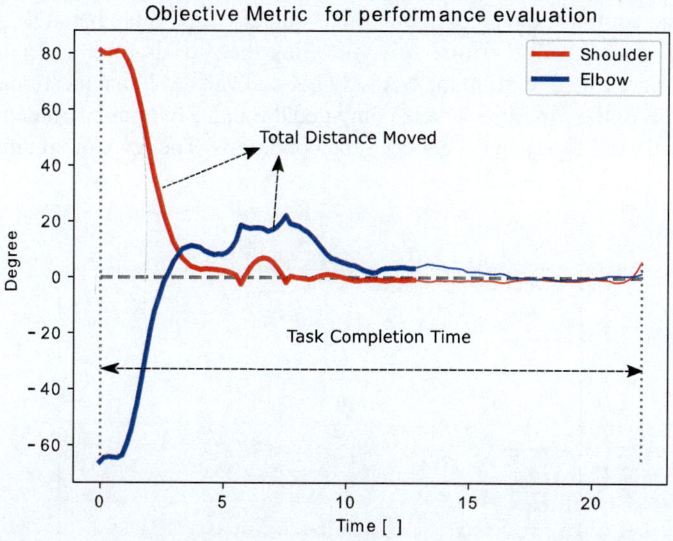

Fig. 2. An example of a pose reconfiguration task with highlighted Task Completion Time (TCT) and Total Distance Moved (TDM) measured. For clarity in this example, the final pose angles have been subtracted from both joints so that they converge to zero.

Subjective Measures

- **NASA-TLX** is a subjective measure used to evaluate the overall workload by calculating a weighted average of 6 different measures: mental demand, physical demand, temporal demand, performance, effort and frustration [8]. A low score indicates that the user-perceived low workload in pose reconfiguration by using a given feedback method.
- **van der Laan** is a subjective questionnaire used to evaluate the acceptability of a given method with usefulness and satisfaction scores [12]. Each score ranges between -2 to +2. The final score can be plotted on a 2D graph, where usefulness is on the x-axis and satisfaction is on the y-axis. The point in the upper right quadrant of the graph indicates good user acceptability.

In the main experiment, the participants underwent 5 trials for each of the three conditions (i.e., audio, visual and multi-modal). The order of the conditions was randomised among the participants. Note that there were two audio conditions (see Sec. 5 for details). To familiarise with the feedback mechanisms and to minimise the learning effect, each subject underwent two or more practice trials before the experiments.

The participants were asked to stand next to a camera system which was responsible for capturing their body posture in real-time. All angular measurements were done in the sagittal plane for the right side of the body. Different starting and ending poses were selected for the 5 repetition trials in each condition. The participants were then asked to follow the feedback commands and to reach the desired pose. Their main objective was to try and complete the reconfiguration as quickly as possible but at the same time avoid unnecessary position errors. After finishing the 5 trials for each condition, the participants were asked to fill in the NASA-TLX and van der Laan questionnaires.

The results of the experiment were compared through a two-tailed Student's t-test to check for statistical significance between the conditions. The level of significance was set to 0.05.

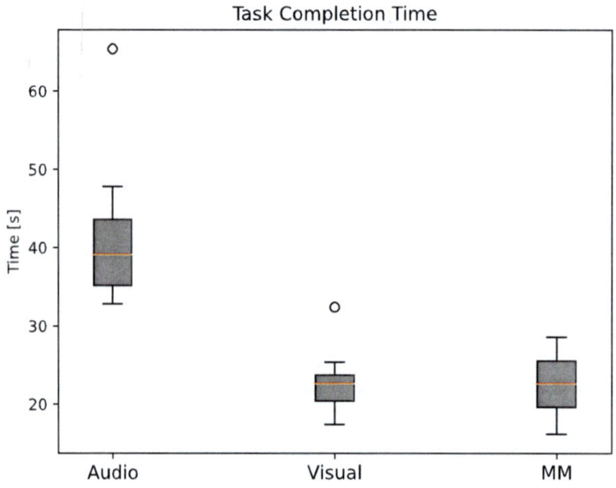

Fig. 3. Box-plots of time to complete the reconfiguration for each condition.

3 Results

3.1 Task Completion Time

Results of TCT comparison are shown in Fig. 3. Audio feedback had the highest TCT (40.97±8.04 s) when compared to visual feedback (22.49±3.62 s). The difference was statistically significant ($p < 0.05$). On the other hand, multi-modal (MM) feedback had a similar task completion time as visual feedback (22.29±3.86 s). The difference was statistically insignificant ($p = 0.89$). The overview of all measures is in Table 3.

3.2 Total Distance Moved

Results of TDM comparison are shown in Fig. 4. Audio feedback had a higher overall movement of the shoulder (134.53±42.46 degrees) and elbow (172.31±54.80 degrees)

Table 3. Table shows the mean results for the objective and subjective metrics.

Metric	Audio	Visual	Multi-modal
Task Completion Time (seconds)	40.97*	22.48	22.28
Total Distance Travelled (degrees)	Shoulder = 134.53*	Shoulder = 108.39	Shoulder = 119.62*
	Elbow = 172.31*	Elbow = 106.31	Elbow = 119.51
NASA - TLX Workload Index	42.16	41.49	28.99*
Van der Laan	Usefulness = 1.27	Usefulness = 1.15	Usefulness = 1.54
	Satisfaction = 0.66	Satisfaction = 1.03	Satisfaction = 1.28

Audio and multi-modal systems are individually compared to the visual system with a two-tailed Student's t-test to check for statistical significance. The values highlighted by * indicate $p < 0.05$.

when compared to visual feedback for the shoulder (108.396±7.85 degrees) and elbow (106.42±14.86 degrees). The differences were statistically significant ($p < 0.05$ for both joints). On the other hand, multi-modal feedback had a slightly larger overall movement for both shoulder (119.63±16.34 degrees) and elbow (119.51±24.21) when compared to visual feedback. The difference was statistically significant for shoulder ($p < 0.05$) and insignificant for elbow ($p = 0.10$)

3.3 Subjective Measures

Results of subjective analysis are shown in Fig. 5. Audio feedback had an average overall workload index of 42.16 and visual feedback of 41.50, making both feedbacks almost equally demanding. There was no significant difference ($p = 0.90$). Multi-modal feedback had an average overall workload index of 29.00, which is significantly lower as compared to visual feedback ($p<0.05$).

According to the van der Laan questionnaire, multi-modal feedback was deemed to be the most acceptable by the participants and it achieved higher scores on both usefulness and satisfaction scales (see the top-right graph in Fig. 5).

We also asked the participants to rank the systems between 1-3 (1 being the best and 3 being the worst) based on their preference to further corroborate the results of the subjective measures. The multi-modal feedback system stood out to be the clear favourite with 9 out of 14 participants voting it to be their preferred form of feedback for pose guidance.

4 Discussion

4.1 Task Completion Time

Audio feedback had on average the highest task completion time among the three methods. This can be attributed to the low information transmission bandwidth of a speech-based format. The participants had to wait and listen to the entire command before could

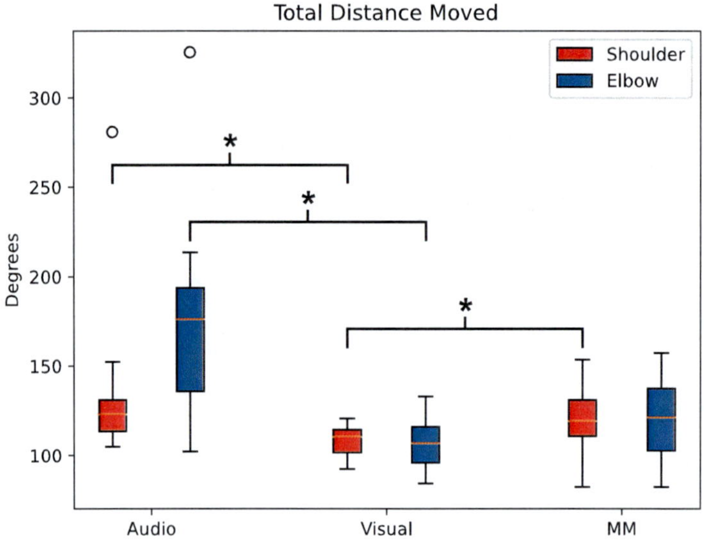

Fig. 4. Box-plots of total distance moved by the individual joints during the reconfiguration for each condition. Statistical significance is indicated by *.

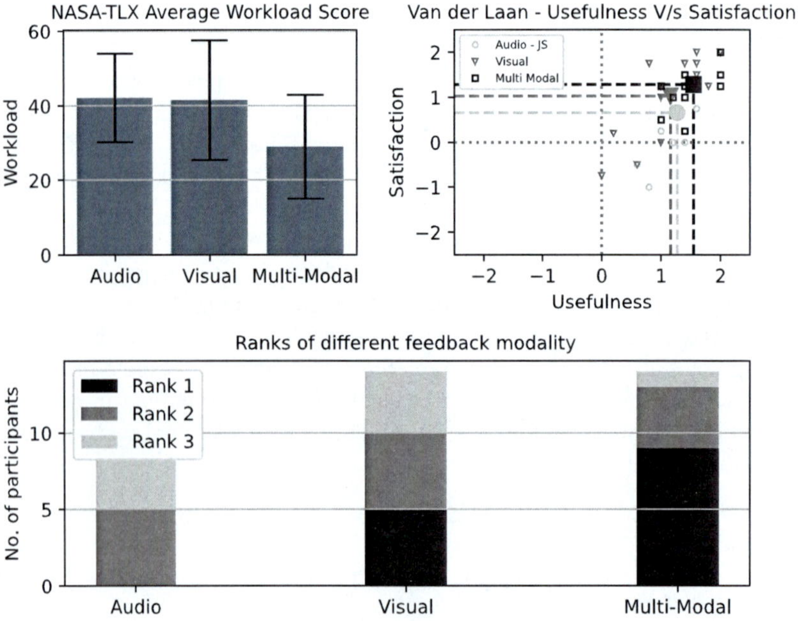

Fig. 5. Results of subjective measures: NASA-TLX (top-left), van der Laan (top-right) and preference (bottom)

they carry out the pose adjustments. In addition, unlike using visual feedback, they did not know how far or close they were to the desired pose, which led to overshooting or undershooting of the target pose. On the other hand, visual feedback allowed the participants to be aware of their current pose with respect to the desired pose. This in general prevented overshooting or undershooting of the desired pose and thus reduced TCT.

Multi-modal feedback had a similar TCT compared to visual feedback, even though the participants had to switch from one modality to another during the pose reconfiguration. The participants relied on visual feedback to carry out larger movements since they could visualise the difference between the current pose and the desired pose, making them move more quicker. Audio feedback was activated only when the participant was ± 5 degrees away from the desired position, thus limiting the comparatively slower audio feedback to only a small portion of the task. The proposed multi-modal system, therefore, achieved the benchmark performance in terms of TCT.

4.2 Total Distance Moved

Audio-based feedback had a significantly larger distance travelled for both the shoulder and the elbow when compared to visual feedback. This can also be attributed to the participants not knowing how far or close they were to the desired pose, which led to overshooting or undershooting of the target pose. Further adjustments for overshoots or undershoots accounted for extra movements that lead to an overall higher TDM. By using visual feedback, the current pose with respect to the target pose is clear at all times and therefore overshoots or undershoots can be minimised.

Multi-modal feedback had slightly larger overall movements for both joints when compared to visual feedback. In the case of multi-modal feedback, the audio modality was activated only when the participant was within ± 5 degrees from the desired position. The additional adjustments during the audio mode could account for the slightly larger TDM. Still, this might be an acceptable tradeoff when visual attention should be given entirely to the actual task of the worker. For example, if the worker should weld or polish a part held by a robot, he/she should visually pay attention to the welding action rather than to the visual feedback related to ergonomics.

Another interesting aspect that can be seen in Fig. 4 is that on average the participants moved their elbows more than their shoulders in the case of audio feedback. This could potentially be attributed to several reasons. Since the elbow constitutes a connection between the shoulder and the hand (arm endpoint), the participants might have unintentionally changed the elbow while changing the shoulder position. Thus, when the users moved the shoulder by the commanded amount, they also moved the elbow for a small amount without having received any command regarding the elbow movement. This would naturally lead to more subsequent corrections of the elbow.

The above-mentioned problem is not present for either multi-modal or visual feedback. This could be because in both these conditions the participants relied more on visual feedback, which would make them quickly aware of any such unintentional movements.

4.3 NASA-TLX

The participants deemed audio feedback and visual feedback to be almost equally demanding. The perceived high workload for audio feedback can be attributed to not seeing the current pose with respect to the target pose, and therefore having to make more adjustments. In addition, constant audio feedback might have been perceived somewhat as an annoyance by the participants. The perceived high workload for visual feedback can be attributed to participants having to constantly pay attention to the GUI while carrying out the task.

The participants deemed multi-modal feedback to be significantly less demanding as compared to the other two. This could be attributed to the method exploiting the advantages of both audio feedback and visual feedback. On one hand, the participants could visually check for the current pose with respect to the target pose at any time. On the other hand, they no longer had to constantly look at the GUI, but could switch to audio feedback and concentrate on the task after reaching a pose close enough to the target pose.

4.4 Van der Laan

All feedback modalities are in the upper right quadrant, which indicates that they are acceptable to the participants. However multi-modal feedback was indicated to be more useful and satisfying as compared to the other two. The usefulness could be higher given that it helped to redistribute the mental workload by allowing participants to concentrate more on the task rather than the feedback, unlike in the case of only visual feedback where the participants could either concentrate on the task or on the feedback. The satisfaction rating was the highest most likely because the system presented audio feedback only when participants were close enough to the desired pose and not throughout the trial, which was the case for the only audio feedback and might have caused annoyance.

5 Supplementary Experiment

The goal of the supplementary experiment was to motivate our choice of using speech-based audio feedback acting in joint space instead of in endpoint space when designing the multi-modal feedback system. We compared the user performance between audio feedback in joint space and audio feedback in endpoint space using the same metrics as mentioned in Sect. 2.2. The same group of participants performed the experiments under these two different audio feedback conditions in order to experience both. The order in which the conditions were presented to the participants was randomised.

5.1 Results

Audio feedback in endpoint space (29.41±3.83 s) had a significantly lower ($p < 0.05$) TCT as compared to joint space feedback (40.97±8.04 s).

Shoulder (shoulder = 126.66±34.13 degrees) and elbow (161.17±64.38 degrees) TDM for endpoint space feedback were slightly lower as compared to shoulder

(134.53±42.46 degrees) and elbow (172.32±54.80 degrees) TDM for joint space feedback. However, there was no significant difference ($p = 0.6$) between either joint in either condition.

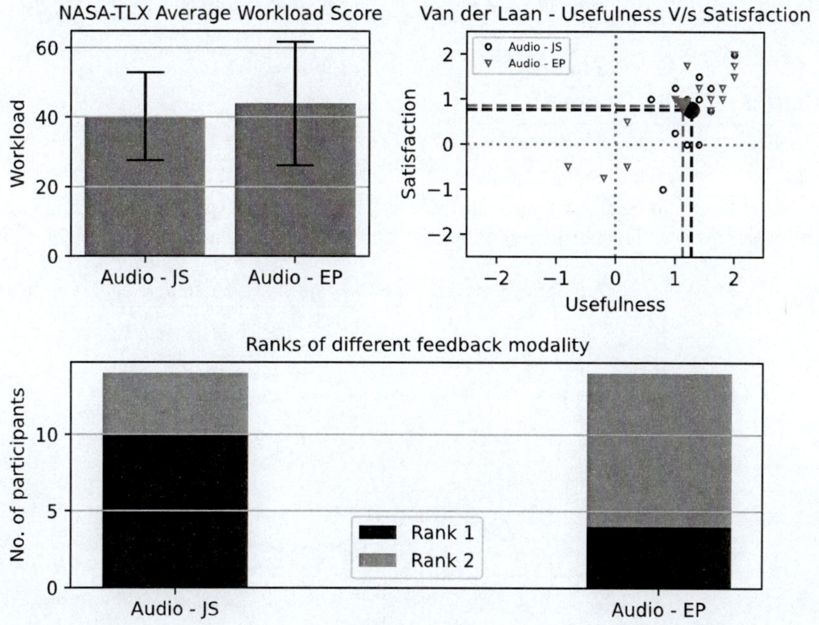

Fig. 6. Results of subjective measures comparing audio feedback acting in joint space and in endpoint space.

The results of subjective evaluation are shown in Fig. 6. Audio feedback in joint space had an overall workload index of 40.24 as compared to 44.00 for endpoint space feedback. The difference was not statistically significant ($p = 0.53$). The results of the van der Laan questionnaire show that the scores for both feedback methods lie in the upper right quadrant (top-left graph in Fig. 6), making them acceptable to the participants. Additionally, we also asked the participants to rank their preferred feedback mechanism to further corroborate the results of the subjective questionnaires. Surprisingly, 10 out of 14 participants preferred the audio commands in joint space (bottom graph in Fig. 6).

5.2 Selection Justification

According to the result, the participants generally took significantly less time to complete the reconfiguration in endpoint space as compared to joint space. In addition, they had an overall slightly less movement of either joint in endpoint space as compared to joint space. On the other hand, subjective scores for workload and acceptability were

Table 4. Table shows the mean results for the objective and subjective metrics.

Metric	Joint Space	EndPoint Space
TCT (seconds)	40.97*	29.41*
TDM (degrees)	Shoulder = 134.53	Shoulder = 126.66
	Elbow = 172.31	Elbow = 161.17
NASA-TLX	40.24	44.0
Van der Laan	Usefulness = 1.27	Usefulness = 1.13
	Satisfaction = 0.77	Satisfaction = 0.86

The results of the audio feedback systems in joint space and endpoint space are compared through Student's t-test to check for statistical significance. The values highlighted by * indicate $p < 0.05$.

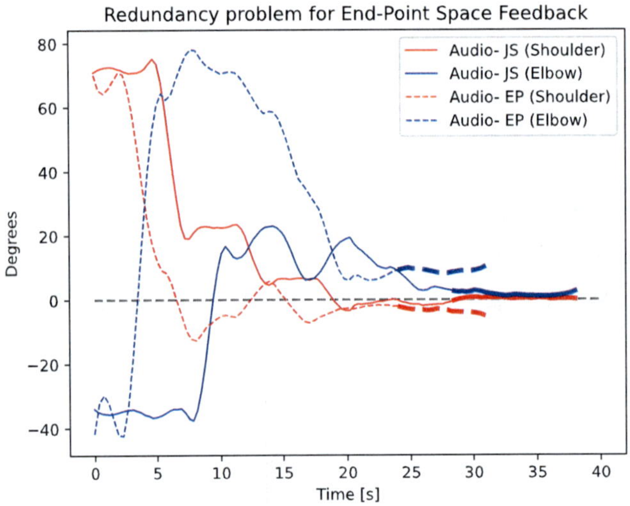

Fig. 7. Demonstration of redundancy problem while using audio feedback in the endpoint. The highlighted portion indicates the last 10 s of the trial where the participant was asked to hold after having reached the desired joint or endpoint positions. By using feedback in joint space, the participant was able to converge his joints to the desired ergonomic pose, whereas by using feedback joint space, the participant is far away from the desired joint positions even though the endpoint (wrist) has successfully reached its desired position.

similar. However, most participants preferred to work with feedback in joint space over endpoint space.

While the evaluation results are generally favourable for the endpoint space method, it has one major conceptual limitation; it cannot be used in cases where there are redundant joint degrees of freedom. The endpoint feedback only gives commands for the positioning of the endpoint and disregards the individual joint positions, therefore the user can orient their joints in multiple ways to reach the same endpoint position (See Fig. 7). This cannot ensure that the user will select an ergonomic configuration among many options, which is the main application of such the proposed feedback modalities. While joint space feedback alone can be quite challenging for the user, it does not suffer from the redundancy problem. Considering the human body has many redundant degrees of freedom and based on the preference results, we, therefore, deem audio feedback in joint space as the preferred method for worker's pose reconfiguration application. Nevertheless, a combination of joint space feedback and endpoint space feedback might be used in some cases.

6 Conclusion

In this study, we have designed a multi-modal feedback system for pose correction intended for improving human ergonomics during task execution. The proposed feedback system was made up of non-physical types of feedback modalities. Visual feedback was used to carry out a large range of body movements, while speech-based audio feedback was used for minor pose adjustments once the user was close enough to the desired pose. Through this feedback system, we were successfully able to combine the advantages of visual feedback and audio feedback.

We hypothesised that the proposed multi-modal feedback system would provide a better performance as compared to the existing non-physical uni-modal feedback systems. While multi-modal feedback generally achieved similar performance as visual feedback in objective measures, it did outperform it in subjective assessment.

References

1. Aldini, S., Lai, Y., Carmichael, M.G., Paul, G., Liu, D.: Real-time estimation of the strength capacity of the upper limb for physical human-robot collaboration. In: 2021 43rd Annual International Conference of the IEEE Engineering in Medicine & Biology Society (EMBC), pp. 4533–4536. IEEE (2021)
2. Barondess, J., Cullen, M., De Lateur, B., Deyo, R., Donaldson, K., Drury, C., et al.: Musculoskeletal disorders and the workplace: low back and upper extremities. Washington, DC: National Academy of Sciences pp. 1–512 (2001)
3. Brewster, S.: The Human-Computer Interaction Handbook, vol. 1. CRC Press (2012)
4. Busch, B., Maeda, G., Mollard, Y., Demangeat, M., Lopes, M.: Postural optimization for an ergonomic human-robot interaction. In: 2017 IEEE/RSJ International Conference on Intelligent Robots and Systems (IROS), pp. 2778–2785. IEEE (2017)
5. Chen, Y., Huang, H., Xu, W., Wallis, R.I., Sundaram, H., Rikakis, T., Ingalls, T., Olson, L., He, J.: The design of a real-time, multimodal biofeedback system for stroke patient rehabilitation. In: Proceedings of the 14th ACM international conference on Multimedia, pp. 763–772 (2006)
6. Feuerstein, M.: Definition, empirical support, and implications for prevention, evaluation, and rehabilitation of occupational upper extremity disorders. In: Beyond Biomechanics: Psychosocial Aspects of Musculoskeletal Disorders in Office Work, pp. 177–206. Taylor & Francis (1996)
7. Figueredo, L.F., Aguiar, R.C., Chen, L., Chakrabarty, S., Dogar, M.R., Cohn, A.G.: Human comfortability: Integrating ergonomics and muscular-informed metrics for manipulability analysis during human-robot collaboration. IEEE Robotics and Automation Letters **6**(2), 351–358 (2020)

8. Hart, S.G.: Nasa-task load index (nasa-tlx); 20 years later. In: Proceedings of the human factors and ergonomics society annual meeting, vol. 50, pp. 904–908. Sage publications Sage CA: Los Angeles, CA (2006)
9. Kim, W., Garate, V.R., Gandarias, J.M., Lorenzini, M., Ajoudani, A.: A directional vibrotactile feedback interface for ergonomic postural adjustment. IEEE Transactions on Haptics **15**(1), 200–211 (2021)
10. Kim, W., Lorenzini, M., Balatti, P., Nguyen, P.D., Pattacini, U., Tikhanoff, V., Peternel, L., Fantacci, C., Natale, L., Metta, G., et al.: Adaptable workstations for human-robot collaboration: A reconfigurable framework for improving worker ergonomics and productivity. IEEE Robotics & Automation Magazine **26**(3), 14–26 (2019)
11. Kumar, S.: Theories of musculoskeletal injury causation. Ergonomics **44**(1), 17–47 (2001). https://doi.org/10.1080/00140130120716
12. van der Laan, J.D., Heino, A., De Waard, D.: A simple procedure for the assessment of acceptance of advanced transport telematics. Transportation Research Part C: Emerging Technologies **5**(1), 1–10 (1997)
13. Lee, J., Cho, E., Kim, M., Yoon, Y., Choi, S.: Preventfhp: Detection and warning system for forward head posture. In: 2014 IEEE Haptics Symposium (HAPTICS), pp. 295–298. IEEE (2014)
14. Lorenzini, M., Kim, W., De Momi, E., Ajoudani, A.: A real-time graphic interface for the monitoring of the human joint overloadings with application to assistive exoskeletons. In: International Symposium on Wearable Robotics, pp. 281–285. Springer (2018)
15. Marin, A.G., Shourijeh, M.S., Galibarov, P.E., Damsgaard, M., Fritzsch, L., Stulp, F.: Optimizing contextual ergonomics models in human-robot interaction. In: 2018 IEEE/RSJ International Conference on Intelligent Robots and Systems (IROS), pp. 1–9. IEEE (2018)
16. Maurice, P., Padois, V., Measson, Y., Bidaud, P.: Experimental assessment of the quality of ergonomic indicators for dynamic systems computed using a digital human model. International Journal of Human Factors Modelling and Simulation **5**(3), 190–209 (2016)
17. Peternel, L., Fang, C., Laghi, M., Bicchi, A., Tsagarakis, N., Ajoudani, A.: Human arm posture optimisation in bilateral teleoperation through interface reconfiguration. In: 2020 8th IEEE RAS/EMBS International Conference for Biomedical Robotics and Biomechatronics (BioRob), pp. 1102–1108. IEEE (2020)
18. Peternel, L., Fang, C., Tsagarakis, N., Ajoudani, A.: A selective muscle fatigue management approach to ergonomic human-robot co-manipulation. Robotics and Computer-Integrated Manufacturing **58**, 69–79 (2019)
19. Peternel, L., Kim, W., Babič, J., Ajoudani, A.: Towards ergonomic control of human-robot co-manipulation and handover. In: 2017 IEEE-RAS 17th International Conference on Humanoid Robotics (Humanoids), pp. 55–60. IEEE (2017)
20. Peternel, L., Schøn, D.T., Fang, C.: Binary and hybrid work-condition maps for interactive exploration of ergonomic human arm postures. Frontiers in Neurorobotics p. 114 (2021)
21. Portnoy, S., Halaby, O., Dekel-Chen, D., Dierick, F.: Effect of an auditory feedback substitution, tactilo-kinesthetic, or visual feedback on kinematics of pouring water from kettle into cup. Applied ergonomics **51**, 44–49 (2015)
22. Rahal, R., Matarese, G., Gabiccini, M., Artoni, A., Prattichizzo, D., Giordano, P.R., Pacchierotti, C.: Caring about the human operator: haptic shared control for enhanced user comfort in robotic telemanipulation. IEEE transactions on haptics **13**(1), 197–203 (2020)
23. Schneider, E., Irastorza, X.: OSH in figures: Work-related musculoskeletal disorders in the EU - Facts and Figures. Publications Office of the European Union (2010). https://doi.org/10.2802/10952
24. Shafti, A., Ataka, A., Lazpita, B.U., Shiva, A., Wurdemann, H.A., Althoefer, K.: Real-time robot-assisted ergonomics. In: 2019 International Conference on Robotics and Automation (ICRA), pp. 1975–1981. IEEE (2019)

25. Sigrist, R., Rauter, G., Riener, R., Wolf, P.: Augmented visual, auditory, haptic, and multimodal feedback in motor learning: a review. Psychonomic bulletin & review **20**(1), 21–53 (2013)
26. van der Spaa, L., Gienger, M., Bates, T., Kober, J.: Predicting and optimizing ergonomics in physical human-robot cooperation tasks. In: 2020 IEEE International Conference on Robotics and Automation (ICRA), pp. 1799–1805. IEEE (2020)
27. Vianello, L., Gomes, W., Stulp, F., Aubry, A., Maurice, P., Ivaldi, S.: Latent ergonomics maps: Real-time visualization of estimated ergonomics of human movements. Sensors **22**(11), 3981 (2022)

An Optimal Human-Based Control Approach for Mobile Human-Robot Collaboration

Andrea Pupa(✉), Francesco Breveglieri, and Cristian Secchi

Department of Sciences and Methods of Engineering, University of Modena and Reggio Emilia, Reggio Emilia, Italy
andrea.pupa@unimore.it

Abstract. In collaborative robotic applications, human and robot have to cooperate in executing a common job. During the collaboration, the operator's experience in the job is of paramount importance. Furthermore, it is essential to ensure that the two agents are always close enough to perform the assigned task, without jeopardizing the safety of the operator. In this paper, we propose an integrated architecture that allows the human operator to drive the collaboration based on its position. The robot is forced to stay close to the human during the execution of the task, but without colliding. Moreover the architecture is capable of changing online the size of the collaborative area, making it suitable for most of the collaborative jobs. The proposed approach is validated on a UR10e mobile manipulator which has been mounted on a MIR100 collaborative platform.

Keywords: Human–robot collaboration · Robot safety · Optimization and optimal control

1 Introduction

Collaborative robotics is becoming more and more important within industrial environments, changing the paradigm. The new industrial scenarios, in fact, are characterized by shared workspaces in which humans and robots collaborate to achieve a common goal. This possibility of interaction has made it possible to solve many manufacturing challenges and to remove physical barriers inside the work cells, increasing the flexibility.

The lack of physical barriers require to pay special attention in how guaranteeing safety. In fact, the safety standards have been updated to face the new collaborative scenario and must be respected. Several approaches have been presented in the literature to address human safety. In [1] the authors present an optimization problem where safety is treated as an hard constraint to be satisfied. The proposed strategy allows to proportional reduce the velocity while ensuring safety, with a consequent increment of the productivity. In [2] a safety framework for collaborative tasks is presented. Multiple robots share the workspace with

P. Borja et al. (Eds.): HFR 2022, 26, pp. 30–44, 2023.
https://doi.org/10.1007/978-3-031-22731-8_3

human operators and the velocity is scaled preventing that a safety index falls below a certain value. If the scaling procedure is not enough, an emergency stop is applied. In [3] the authors propose a two level framework for safe dynamic planning and replanning. The first layer plans and adapts online the trajectory considering only the robot limits. The second layer explicitly consider the safety standards adapting online the velocity along the path.

Humans and robots are different agents with different skills, for this reason it is important to have a strategy that optimal assigns the tasks, making the most of the human–robot collaboration (HRC). A lot of research been done in this direction. In [4] the authors model the task allocation problem as a Multimodal Multiprocessor Task Scheduling Problem in which the resources are the two agents, namely human and robot. In [5] the authors introduce a two-level framework for task assignment that is capable of dynamically handling task errors in an HRC scenario. These solution, however, assumes that the human requires a constant execution time for each task, which is a very strong assumption. The HRC scenarios, indeed, are highly dynamic, e.g. the human may accomplish the wrong task or take longer to perform the tasks. For this reason, the research is focusing on handling dynamic rescheduling strategy [6–8].

Assigning tasks between humans and robots in advance is not always the best choice. Sometimes it may be better to let the operator drive the collaboration, leveraging his experience and knowledge of the job. In [9] the authors present a two-level feedforward optimization strategy for offline subtask allocation for HRC. Their strategy is integrated with a mutual trust feedback procedure that is exploited to re-allocate online the subtasks. In [10] a weighted probabilistic state machines based algorithm that recognize the human intention is presented. The tasks are divided into two different categories, namely the communication of explicit and implicit intention, depending on whether the operator engages the robot or not. Depending on the type of estimated intention, the robot acts accordingly. In [11] the authors implement a Neural Network to estimate the desired trajectory of the human operator. This estimation is integrated into impedance control such that the robot arm follows the human operator. However, these solutions do not consider the probability of collision between humans and robots.

To increase the robot workspace, giving it more flexibility, HRC research has focused on the study of mobile manipulators [12]. Several works have been done in the field of trajectory planning and control of these collaborative platforms [13–16]. In [17] the authors present a two-layers framework for task assignment and motion planning for a bi-manual mobile manipulator. The task assignment exploits search trees in temporal windows to determine feasible task assignments of agents using task and spatial constraint-based heuristics. The tree branches are then used as input to plan the motion for each agent. In [18] a strategy that exploits concurrent and sequential task representation for task allocation in multi human–robot collaboration in industrial environments is presented.

Human intention-based strategies have been widely applied also in the field of collaborative mobile manipulators. In [19] the authors present a framework for human–robot collaboration with mobile manipulators. Their strategy allows to enhance the physical human–robot interaction while satisfying the imposed constrained. In [20] a hierarchical architecture that allows the user to generate commands via a P300-based brain computer interface is proposed. In [21] the authors present a gesture polysemy based control strategy. Thanks to the proposed method the operator is capable of controlling both the motion of mobile robot and the posture of robot using one hand. In [22] a hierarchical finite state machine that exploits voice command is proposed. The framework analyzes user's input to obtain the human intention which is then transferred to the mission plan for executing the tasks. However, these solutions are difficult to extend and do not allow the control of the size of the collaboration area. In the case of mobile platforms, in fact, it could be useful to restrict the work area, in order to have a more efficient collaboration.

In this paper we propose a novel control architecture for collaborative mobile manipulators that allows the human operator to actively lead the collaboration by dynamically determining the collaborative area. Using Control Barrier Functions (CBF) [23,24] the robotic system is constrained to move inside a desired collaborative volume that depends on the pose of the operator. The size of the collaborative volume can be dynamically determined, depending on external factors as, e.g., the job to execute. Moreover, the framework is endowed of a collision avoidance strategy to ensure safety during the collaboration. The proposed architecture provides several advantages to the control of mobile collaborative robots. First, it allows a control of the whole body of the robot, allowing to exploit all the DOFs of the system for achieving the desired performance unlike the approaches that exploit separately the mobile base and the manipulator. Second, thanks to the modularity of CBFs, it is possible to dynamically change the number of constraints that the mobile manipulator has to satisfy making it reactive to changes in the environment and/or in the task it has to satisfy. Finally, exploiting time-varying CBFs, it is possible to adapt online the collaborative volume in order to improve acceptability and performance of the human–robot collaborating team.

The main contributions of this paper are:

- A control architecture that allows the human operator to lead the collaboration, forcing the robot to stay close, while ensuring safety.
- A time-varying definition of the collaborative area that gives flexibility to the overall framework.
- A control architecture that is modular and can be easily extended.

The paper is organized as follows: in Sect. 2 the problem is detailed while in Sect. 3 the general CBF-based architecture is derived. In Sect. 4 the proposed control framework is defined, with the specific with the specific discussion of the necessary constraints. Finally in Sect. 5 the experimental validation of the is presented while in Sect. 6 some conclusions and future works are addressed.

2 Problem Statement

Consider an industrial workspace where a human operator and a mobile robot manipulator have to collaborate accomplishing a job. Defining the space coordinates of the robot as:

$$\mathbf{q} = \begin{bmatrix} \mathbf{q}_b \\ \mathbf{q}_a \end{bmatrix} \tag{1}$$

where $\mathbf{q}_b \in \mathbb{R}^{n_b}$ is a reduced representation of the configuration space of the differential drive mobile base, while $\mathbf{q}_a \in \mathbb{R}^{n_a}$ represents the space coordinates of the robot arm. The overall model of the robot can be described by the following equation:

$$\dot{\mathbf{x}} = J(\mathbf{q})\dot{\mathbf{q}} = \begin{bmatrix} J_b(\mathbf{q}_b) \; J_a(\mathbf{q}_a) \end{bmatrix} \begin{bmatrix} \dot{\mathbf{q}}_b \\ \dot{\mathbf{q}}_a \end{bmatrix} \tag{2}$$

where $\mathbf{x} \in \mathbb{R}^m$ and $\dot{\mathbf{x}} \in \mathbb{R}^m$ represent the end-effector pose and velocity, respectively, while $J(\mathbf{q}) \in \mathbb{R}^{m \times (n_b + n_a)}$ is the augmented Jacobian [25].

To ensure safety and to allow the human operator to lead the collaborative job, the workspace is enriched with a monitoring unit that allows to track the human movements. Many strategies are already available in literature that deal with the human tracking, e.g. skeleton tracking with multiple cameras [26], markers on the human body [27], machine learning techniques [28]. In particular, the robot position is defined safe if:

$$d(\sigma_{ri}(\bar{\mathbf{q}}), \sigma_{Hj}) \geq d_{safe} \quad \begin{aligned} &\forall i \in \{1, \ldots, n_b + n_a\}, \\ &\forall j \in \{1, \ldots, L\} \end{aligned} \tag{3}$$

where $\sigma_{ri}(\bar{\mathbf{q}})$ is the line segment representing the i-th link when the robot is in configuration $\bar{\mathbf{q}}$. σ_{Hj} it the line segment of the j-th human body limb, e.g. the human arm, and L is the number of the human body limbs. $d(\sigma_{ri}(\bar{\mathbf{q}}), \sigma_{Hj})$ is a function that computes the distance between the two line segments and d_{safe} represent the safety distance.

In this work, we aim at designing the overall control architecture shown in Fig. 1 that starting from a desired robot velocity:

- Generates the **optimal** control input $\dot{q}$ such that the robot stay close to the human operator while performing the job, maximizing the performances.
- Ensures safety for the human operator during the collaboration, implementing a collision avoidance behavior.
- Allows to adapt at runtime the size of the collaborative area, making it flexible and suitable fo many HRC scenarios.

3 Control Barrier Function Based Architecture

The proposed control architecture exploits the use of Control Barrier Functions (CBFs) to formulate a convex optimization problem. This is possible under the assumption that the robot limits, the task that the robot has to accomplish and

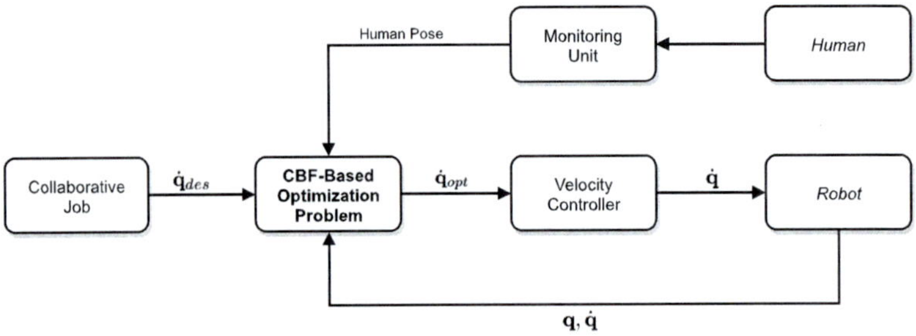

Fig. 1. The proposed overall architecture.

the safety can be represented as a time-varying constraints on the control input [29,30]. Solving online this optimization problem, it is possible to compute the best input that satisfies all the constraints, i.e. the robot velocities that allow to execute the task in an optimal way.

Starting from the robot model defined in (2) it is possible to define a tasks whose execution is expressed as the minimization of a non negative, possibly time-varying, continuously-differentiable cost function $C : \mathbb{R}^n \times \mathbb{R} \rightarrow \mathbb{R}$ with the following optimization problem:

$$\begin{aligned} \underset{\dot{\mathbf{q}}}{\text{minimize}} \quad & C(\boldsymbol{\sigma}, t) \\ \text{subject to} \quad & \dot{\mathbf{x}} = J(\mathbf{q})\dot{\mathbf{q}} \\ & \boldsymbol{\sigma} = k(\mathbf{q}, t) \end{aligned} \tag{4}$$

where $\boldsymbol{\sigma} \in \mathbb{R}^n$ is the output variable and represents the time-varying task variable.

Exploiting CBFs, the optimization problem (4) can be adapted to be computationally cheap, i.e. a convex optimization problem that can be solved online. Defining $\mathcal{C} \in \mathbb{R}^n$ as the subset where the task is considered executed, i.e. where $C(\boldsymbol{\sigma}, t) = 0$, it is possible to formulate a CBF $h : \mathbb{R}^n \times \mathbb{R} \rightarrow \mathbb{R}$ defined as $h(\boldsymbol{\sigma}, t) = -C(\boldsymbol{\sigma}, t)$. Since by definition h is non negative only in the region of execution of the task, i.e. when $C(\boldsymbol{\sigma}, t) = 0$, enforcing the non negativity of h is translated in the execution of the task $\boldsymbol{\sigma}$. This can be embedded in the following convex optimization problem:

$$\begin{aligned} \underset{\dot{\mathbf{q}}}{\text{minimize}} \quad & \frac{1}{2}||\dot{\mathbf{q}}_{des} - \dot{\mathbf{q}}||^2 \\ \text{subject to} \quad & \frac{\partial h}{\partial t} + \frac{\partial h}{\partial \boldsymbol{\sigma}}\frac{\partial \boldsymbol{\sigma}}{\partial x}J(\mathbf{q})\dot{\mathbf{q}} \geq -\alpha(h(\boldsymbol{\sigma}, t)) \end{aligned} \tag{5}$$

where $\dot{\mathbf{q}}_{des} \in \mathbb{R}^{n_b+n_m}$ are the desired robot velocities $\alpha(\cdot)$ is an extended class $\mathcal{K}$ function, i.e. a function $\phi : \mathbb{R} \rightarrow \mathbb{R}$ such that ϕ is strictly increasing and $\phi(0) = 0$.

The formulation in (5) can be easily extended to the execution of multiple tasks. Considering a set of M different tasks $\{T_1, \ldots, T_M\}$ that the robot has to accomplish, each of which represented by a the respective cost functions $C_1, \ldots, C_M$, the execution of these tasks can be encoded in the following optimization problem:

$$\begin{aligned} \underset{\dot{\mathbf{q}},\boldsymbol{\delta}}{\text{minimize}} \quad & \|\dot{\mathbf{q}}\|^2 + \beta\|\boldsymbol{\delta}\|^2 \\ \text{subject to} \quad & \frac{\partial h_i}{\partial t} + \frac{\partial h_i}{\partial \boldsymbol{\sigma}}\frac{\partial \boldsymbol{\sigma}}{\partial x} J(\mathbf{q})\dot{\mathbf{q}} \geq \\ & -\alpha(h_i(\boldsymbol{\sigma}, t)) - \delta_i \quad \forall i \in \{1, \ldots, M\} \end{aligned} \tag{6}$$

where $h_i(\boldsymbol{\sigma}, t) = -C_i(\boldsymbol{\sigma}, t)$ and $\boldsymbol{\delta} = [\delta_1, \ldots, \delta_M]$ is the vector of slack variables corresponding to each constraint, while $\beta \geq 0$ is a scaling factor. Each δ_i indicates how much the constraint corresponding to the i-th task can be relaxed. This is needed to to guarantee the feasibility of the problem when conflicting constraints are active at the same time.

4 Proposed CBF Framework

In this the proposed control framework is detailed. Starting from the general formulation of the CBFs-based architecture, it is possible to define the tasks that the robot has to accomplish during the collaboration. In this work three different tasks have been considered:

- **Human Follow**. The robot has to follow the human operator inside the workspace. The size of the collaborative area can change at runtime based on the collaborative job that the two agents has to perform.
- **Safety**. The robot has to avoid the human operator during the execution of the collaborative job.
- **Mobile Base Orientation**. The orientation of the robot has to be adjusted so that the mobile base always faces the human operator.
- **Manipulability**. The configuration of the robot has to be controlled in order to always maintain a certain level of manipulability, avoiding singularities.

These tasks are further detailed in the following sections.

4.1 Task 1: Human Follow

To ensure that the operator can work within the collaboration, the first task that must be carried out by the robot is to stay close to the man, following its movements within the workspace. This can be mathematically expressed as constraining the end-effector of the manipulator to be inside a sphere centered in a specific point of the human operator, e.g. the wrist. The situation is illustrated in Fig. 2a.

This constraint can be expressed with the following CBF:

$$h_{Hf}(\mathbf{q}, t) = R(t) - \mathcal{D}(\mathbf{x}(\mathbf{q}), \mathbf{x}_H(t)) \tag{7}$$

(a) (b) (c)

Fig. 2. Graphical representation of the CBF-based constraints. **a** Sphere representing the area in which the manipulator end-effector has to stay. **b** Sphere representing the area forbidden to the robot for safety reasons. **c** Circumference representing the area in which the controlled point of the I-O SFL has to stay.

where $R(t) \in \mathbb{R}$ is the radius of the sphere, i.e. the size of the collaborative area, and can be changed at runtime depending on the task that the human–robot team has to accomplish. $\mathbf{x} \in \mathbb{R}^3$ and $\mathbf{x}_H \in \mathbb{R}^3$ are the cartesian positions of the manipulator end-effector and of the analyzed human point, respectively. Lastly, $\mathcal{D}(\mathbf{x}(\mathbf{q}), \mathbf{x}_H(t))$ is a function that computes the euclidean distance between the two positions.

In order to impose the CBF constraint it is necessary to differentiate (7), obtaining:

$$\dot{h}_{Hf}(\mathbf{q}, t) = \dot{R} - \frac{(\mathbf{x}(\mathbf{q}) - \mathbf{x}_H)^T}{\|\mathbf{x}(\mathbf{q}) - \mathbf{x}_H\|}(J(\mathbf{q})\dot{\mathbf{q}} - \dot{\mathbf{x}}_H) \tag{8}$$

4.2 Task 2: Safety

During execution, it is crucial that the robot does not collide with the human operator, implementing collision avoidance behavior. As defined in (3), the robot configuration is considered safe if each link of the robot, both the manipulator and the mobile base, is at a safe distance from the human operator. Instead of considering each robot link, this relationship can be approximated by choosing N keypoints on the robot, i.e. the center of each link and the center of the mobile base, and imposing that the distance between these points and the operator is greater than the safety distance. Thus, it is possible to define N CBFs as:

$$h_{si}(\mathbf{q}, t) = \mathcal{D}(\mathbf{x}_i(\mathbf{q}), \mathbf{x}_H(t)) - d_{safe} \qquad \forall i \in \{1, \ldots, N\} \tag{9}$$

where $\mathbf{x}_i(\mathbf{q}) \in \mathbb{R}^3$ represents the cartesian position of the i-th keypoint and can be computed using partial forward kinematics [31]. A graphical representation of this constraint is shown in Fig. 2b. It is worth noting that the graphical representation is the same for all N CBFs. This is because, unlike (7), the safe set of these CBFs is the entire workspace but the sphere representing the human operator.

Similarly to Sect. 4.1, it is possible to obtain the constraint computing the derivatives as:

$$\dot{h}_{si}(\mathbf{q}, t) = \frac{(\mathbf{x}_i(\mathbf{q}) - \mathbf{x}_H)^T}{\|\mathbf{x}_i(\mathbf{q}) - \mathbf{x}_H\|}(J_i(\mathbf{q})\dot{\mathbf{q}} - \dot{\mathbf{x}}_H) \qquad \forall i \in \{1, \dots, N\} \tag{10}$$

4.3 Task 3: Mobile Base Orientation

Ensuring that the manipulator end-effector stays close to the human operator is not enough to guarantee that the robot adopts optimal behavior. Since the mobile base is a differential drive robot it may happen that the robot is not capable to follow the human movements. A differential drive robot, indeed, is subject to non-holonomic constraints, i.e. constraints that forbid the lateral motion of the robot. For this reason it has been implemented a third CBF that allows to control the orientation of the mobile base, increasing the performances of the algorithm.

Considering only the mobile base, the differential-drive robot model can be expressed as:

$$\dot{\mathbf{x}}_B = \begin{bmatrix} \cos\theta & 0 \\ \sin\theta & 0 \\ 0 & 1 \end{bmatrix} \begin{bmatrix} v \\ \omega \end{bmatrix} \tag{11}$$

where $\dot{\mathbf{x}}_B = \begin{bmatrix} \dot{x}_B & \dot{y}_B & \dot{\theta} \end{bmatrix}$ is the vector of the cartesian velocity ($\dot{x}_B \in \mathbb{R}$, $\dot{y}_B \in \mathbb{R}$) and the angular velocity ($\dot{\theta} \in \mathbb{R}$) of the midpoint of the wheel axis B. $v \in \mathbb{R}$ and $\omega \in \mathbb{R}$ are the linear and angular velocity of the mobile robot and represents the control inputs, namely $\dot{\mathbf{q}}_b$ in (2). The robot model is shown in Fig. 3a.

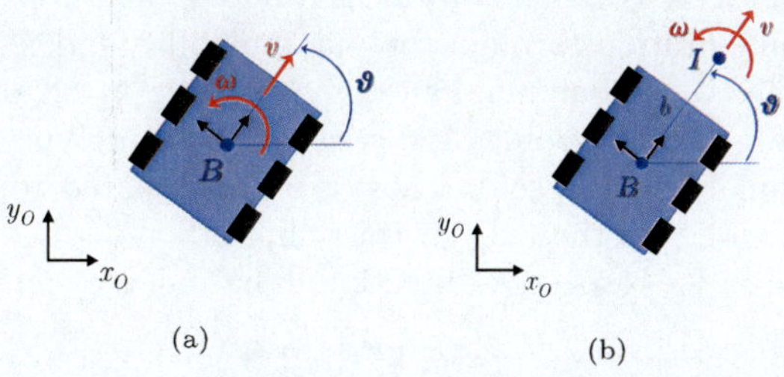

Fig. 3. Differential drive robot model. **a** Control at the midpoint of the wheel axis B. **b** I-O SFL Control.

Instead of controlling the midpoint of the wheel axis, which must respect the non-holonomic constraints, it is possible to command a new reference point positioned on the normal axis of the passing wheels for B, at a distance b from it, see Fig. 3b. This control strategy is called *Input–Output State Feedback Linearization* (I-O SFL) [32]:

$$\dot{\mathbf{x}}_I = \begin{bmatrix} \dot{x}_I \\ \dot{y}_I \\ \dot{\theta} \end{bmatrix} = T_B^I \begin{bmatrix} v \\ \omega \end{bmatrix} = \begin{bmatrix} \cos\theta & -b\sin\theta \\ \sin\theta & b\cos\theta \\ 0 & 1 \end{bmatrix} \begin{bmatrix} v \\ \omega \end{bmatrix} \tag{12}$$

Once the I-O SFL has been applied, it is possible to constrain the new control point I to remain within the projection on the plane of the sphere defined in Sect. 4.1. This solution allows to guarantee that the mobile base is always oriented towards the human operator, ensuring greater freedom of movement. The situation is illustrated in Fig. 2c.

The CBF that represents this constraint is defined as:

$$h_o(\mathbf{q}_b, t) = R(t) - \mathbf{D}(\mathbf{x}_I(\mathbf{q}_b), \mathbf{x}_{HP}(t)) \tag{13}$$

where $\mathbf{x}_{HP} \in \mathbb{R}^2$ are the coordinates of the human point projected on the plane while $\mathbf{D}(\mathbf{x}_I, \mathbf{x}_{HP}(t))$ is the euclidean distance on the plane.

Differentiating (13) it is possible to obtain:

$$\dot{h}_o(\mathbf{q}_b, t) = \dot{R} - \frac{(\mathbf{x}_I(\mathbf{q}_b) - \mathbf{x}_{HP})^T}{\|\mathbf{x}_I(\mathbf{q}_b) - \mathbf{x}_{HP}\|}(J_b(\mathbf{q}_b)T_B^I\dot{\mathbf{x}}_I - \dot{\mathbf{x}}_{HP}) \tag{14}$$

4.4 Task 4: Manipulability

To further improve the behavior of the robot it would be better to avoid singularity configurations, which could prevent the execution of the task. In general, for a robotic manipulator, it is possible to measure its ability to perform a specific task in a given configuration through the manipulability index [33]. This index indicates how easily the robot manipulator can move in all directions of the work area. From a geometric point of view it can be represented with an ellipsoid, called manipulability ellipsoid, whose axes are related to the eigenvelues of the jacobian. The larger the axes of this ellipsoid, the greater the ability of the robot to move in its respective direction. The index of manipulability can be defined as:

$$\mu = \sqrt{\det(J_a J_a^T)} \tag{15}$$

which is equal to 0 when the robot is in a singularity configuration, $\det(J_a) = 0$. To avoid singularities it is sufficient to guarantee that:

$$\mu \geq \mu_{\min} \tag{16}$$

where $\mu_{\min}$ is the minimum admissible value of the manipulability index and must be designed according to the task.

Computing and constraining this index online can be computationally demanding, since it is a very complex non linear function. Thus, in [34] the authors propose an approximated index with the same local minimum as (15):

$$\mu_{\sim} = \frac{1}{2}\sum_{i=2}^{n_a-1} \sin^2(\mathbf{q}_{a,i}) \tag{17}$$

where it is important to underline that the first and the last joint are not used in computing the index. This is because these two joints do not affect the singularity of the manipulator.

Imposing $h_m(\mathbf{q}_a, t) = \mu_{\sim}$, it is possible to obtain:

$$\dot{h}_m(\mathbf{q}, t) = \sum_{i=2}^{n_a-1} \sin(\mathbf{q}_{a,i})\cos(\mathbf{q}_{a,i})\dot{\mathbf{q}}_{a,i} \tag{18}$$

4.5 Optimization Problem

Starting from the definition in (6), it is possible to use all the constraints in order to build the following optimization problem:

$$\begin{aligned}
\underset{\dot{\mathbf{q}},\boldsymbol{\delta}}{\text{minimize}} \quad & \frac{1}{2}\left[\mathbf{w}_b^T \ \mathbf{w}_m^T\right] \|\dot{\mathbf{q}}_{des} - \dot{\boldsymbol{q}}\|^2 + \boldsymbol{\beta}^T\|\boldsymbol{\delta}\|^2 \\
\text{subject to} \quad & \dot{h}_{Hf}(\mathbf{q}, t) \geq -\alpha_{Hf}(h_{Hf}(\mathbf{q}, t)) + \delta_{Hf} \\
& \dot{h}_{si}(\mathbf{q}, t) \geq -\alpha_{si}(h_{si}(\mathbf{q}, t)) + \delta_{si} \qquad \forall i \in \{1, \ldots, N\} \\
& \dot{h}_o(\mathbf{q}_b, t) \geq -\alpha_o(h_o(\mathbf{q}_b, t)) + \delta_o \\
& \dot{h}_m(\mathbf{q}_a, t) \geq -\alpha_m(h_m(\mathbf{q}_a, t)) + \delta_m \\
& \mathbf{q}_{\min} \leq \mathbf{q}_{act} + \dot{\mathbf{q}}T \leq \mathbf{q}_{\max} \\
& \dot{\mathbf{q}}_{\min} \leq \dot{\mathbf{q}} \leq \dot{\mathbf{q}}_{\max} \\
& \ddot{\mathbf{q}}_{\min} \leq \frac{\dot{\mathbf{q}} - \dot{\mathbf{q}}_{act}}{T} \leq \ddot{\mathbf{q}}_{\max}
\end{aligned} \tag{19}$$

where $\mathbf{w}_b \in \mathbb{R}^{n_b}$ and $\mathbf{w}_m \in \mathbb{R}^{n_a}$ are the weights associated to the movement of the mobile base and the robot manipulator, respectively. These weights are exploited to give different priority of movement to the different robot DOF-s. $\boldsymbol{\beta} \in \mathbb{R}^{N+3}$, instead, represents the weights associated to the slack variables, necessary to prioritize the tasks. $\dot{\mathbf{q}}_{des} \in \mathbb{R}^{n_b+n_a}$ represents the desired robot velocities. $\mathbf{q}_{\min} \in \mathbb{R}^{n_b+n_a}$ and $\mathbf{q}_{\max} \in \mathbb{R}^{n_b+n_a}$ are the position limits and T is the cycle time. $\dot{\mathbf{q}}_{\min} \in \mathbb{R}^{n_b+n_a}$ and $\dot{\mathbf{q}}_{\max} \in \mathbb{R}^{n_b+n_a}$ are the joint velocity lower bounds and the joint velocity upper bounds, respectively. While $\ddot{\mathbf{q}}_{\min} \in \mathbb{R}^{n_b+n_a}$ and $\ddot{\mathbf{q}}_{\max} \in \mathbb{R}^{n_b+n_a}$ are the acceleration limits. Lastly, $\mathbf{q}_{act} \in \mathbb{R}^{n_b+n_a}$ and $\dot{\mathbf{q}}_{act} \in \mathbb{R}^{n_b+n_a}$ are the actual robot positions and velocity, respectively.

The first four constraints are the four CBF tasks, while the last three constraints represent the robot kinematic limits, i.e. minimum and maximum velocities and accelerations.

The optimization problem (19) is a convex problem and, therefore, computationally cheap. Thanks to its convexity, the solution obtained by the solver is always the global minimum of the cost function, i.e. the robot velocities that allow you to perform the job in an optimal way, maximizing the performance of the HRC. Moreover, thanks to the use of slack variables $\boldsymbol{\delta}$, the optimization problem has always a feasible solution. It is worth noting that, since it relies on the use of CBFs, the optimization problem can be easily extended to perform other tasks, making the architecture very flexible.

5 Experiments

The proposed architecture has been experimentally validated in a HRC scenario. During the experiments, the human operator cooperates with a UR10e, a 6-DoF collaborative robot, which has been mounted on a MIR100, a differential-drive collaborative mobile base. To track the human position eight OptiTrack Prime^{x} cameras with the OptiTrack Motive software[1] have been used, while the size of the collaborative area is changed at runtime exploiting an Alexa virtual assistant [35]. The complete setup for the experiments is shown in Fig. 4.

Fig. 4. Setup of the experiments.

All the software components were developed using ROS Melodic Morenia. The CBF-based optimization problem has been implemented exploiting the C code generated by Matlab Coder[2]. Concerning the frequencies, the communication with the robot works at 500 Hz while the optimization problem converges in 1 ms. The OptiTrack, instead, works at a frequency of 240 Hz.

In the experiment the human and the robot have to perform a collaborative transportation, whose snapshots are showed in Fig. 5. During this experiment,

[1] https://www.optitrack.com/.
[2] https://mathworks.com/products/matlab-coder.html.

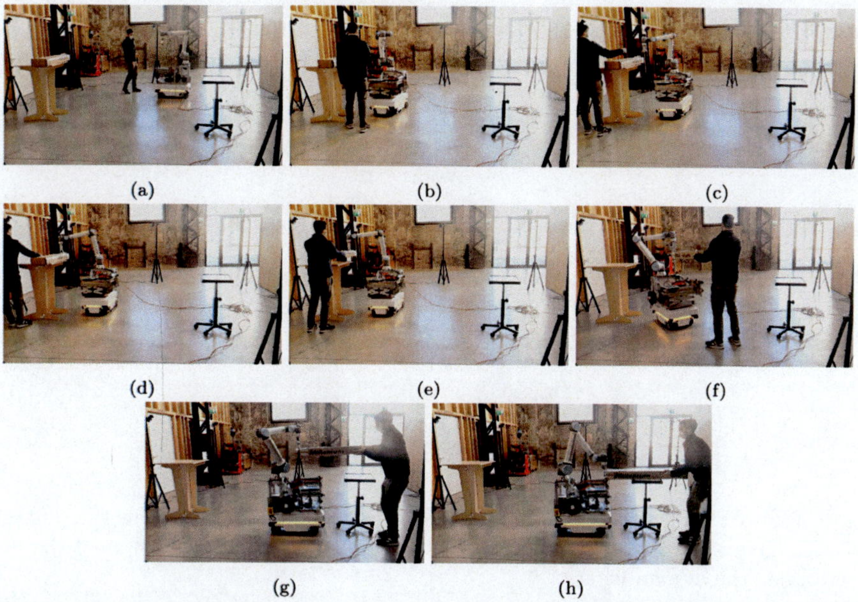

Fig. 5. Snapshots of the experiment. **a** Approaching phase. **b** Refinement of the manipulator position. **c** Picking phase. **d** Pick of the load. **e** Lifting of the load. **f** Collaborative transportation. **g** Placing phase. **h** Place of the load.

the optimization problem in (19) has been extended with an orientation constraint for the manipulator. This was necessary for the nature of the experiment, but could be ideally avoided by using a proper control strategy that generates a proper $\dot{\mathbf{q}}_{des}$. Initially, the two actors are very far from the load to be picked. For this reason, the human operator starts approaching the object, see Fig. 5a. In this approaching phase, it is preferable to move only the mobile base, making the most out of the overall platform. This is achieved by setting $\mathbf{w}_m \gg \mathbf{w}_b$. Subsequently, the human operator guides the robot to pick the object, see Fig. 5b. It is worth noting that in this phase, only the mobile manipulator is moved and to achieve a better control the CBF defined in (7) is initially centered in the robot end-effector and follows the human operator movements. After the robot has picked up the object, the human operator begins to lift the load and, using the same control used for the picking phase, the operator gets help from the robot. This is illustrated in Fig. 5c, d. Once the load has been completely lifted, the human operator starts approaching the placing location and the mobile base follows consequently. See Fig. 5e After reaching the desired position, the human operator starts to place the object while the robot helps. This is shown in Fig. 5g, h. It is worth noting that the weights associated to the slack variables, $\boldsymbol{\beta}$, are equal to 1.

A graph demonstrating that the robot end-effector stays inside the safe set is illustrated in Fig. 6. It is worth noting that there are two moments in which the distance is outside the boundaries, in the first phase and around $t = 85\ s$. This is due to the fact that the control was momentarily deactivated, in the initial phase and while the human operator was picking the load.

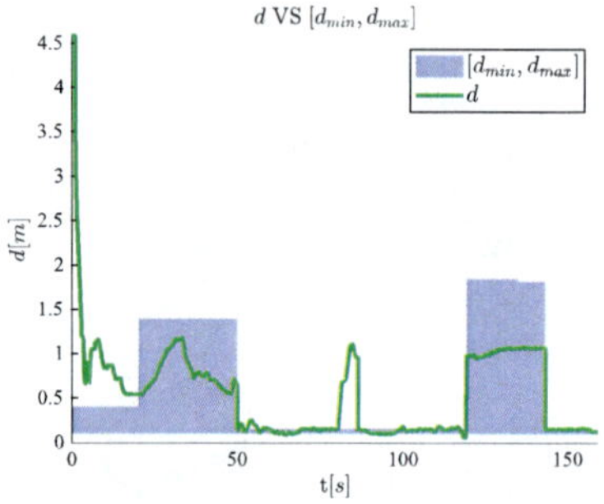

Fig. 6. Plot of the CBF at the robot end-effector. The blu area represent the safe set, while the green line represent the distance between the ee and the center of the two spheres, i.e., the human wrist.

6 Conclusions

In this paper a control architecture for HRC scenarios has been built. The proposed framework exploits the use of time varying CBFs to allow the human operator to actively lead the collaboration, without compromising the safety. Indeed, the collaborative area can be modified online based on the actual needs of the human operator. Furtherore, thanks to the modularity of CBFs, the framework can be easily extended making it suitable for many industrial application. The experimental validation has been performed on a collaborative transportation job, proving the effectiveness of the control strategy.

Future works aim at automatically computes the weights of the optimization problem (19) in order to always have the best robot behaviour. Furthermore, the representation of the collaborative area through a sphere could be problematic, since it does not give control on the manipulator orientation. This could be solved by the use of ellipsoids. Lastly, the overall architecture could be tested in a real industrial scenario.

Acknowledgement. This project has received funding from the European Union's Horizon 2020 research and innovation programme under grant agreement No. 818087 (ROSSINI).

References

1. Zanchettin, A.M., Ceriani, N.M., Rocco, P., Ding, H., Matthias, B.: Safety in human–robot collaborative manufacturing environments: metrics and control. IEEE Trans. Autom. Sci. Eng. **13**(2), 882–893 (2015)
2. Lippi, M., Marino, A.: Human multi-robot safe interaction: A trajectory scaling approach based on safety assessment. IEEE Transactions on Control Systems Technology (2020)
3. Pupa, A., Arrfou, M., Andreoni, G., Secchi, C.: A safety-aware kinodynamic architecture for human–robot collaboration. IEEE Robot. Autom. Lett. **6**(3), 4465–4471 (2021)
4. Ferreira, C., Figueira, G., Amorim, P.: Scheduling human-robot teams in collaborative working cells. Int. J. Prod. Econ. **235**, 108094 (2021), https://www.sciencedirect.com/science/article/pii/S0925527321000700
5. Johannsmeier, L., Haddadin, S.: A hierarchical human–robot interaction-planning framework for task allocation in collaborative industrial assembly processes. IEEE Robot. Autom. Lett. **2**(1), 41–48 (2016)
6. Casalino, A., Zanchettin, A.M., Piroddi, L., Rocco, P.: Optimal scheduling of human-robot collaborative assembly operations with time petri nets. IEEE Trans. Autom. Sci. Eng. (2019)
7. Pupa, A., Talignani Landi, C., Bertolani, M., Secchi, C.: A dynamic architecture for task assignment and scheduling for collaborative robotic cells. In: Proceedings of the 13th International Workshop on Human-Friendly Robotics (2020)
8. Pupa, A., Van Dijk, W., Secchi, C.: A human-centered dynamic scheduling architecture for collaborative application. IEEE Robot. Autom. Lett. **6**(3), 4736–4743 (2021)
9. Rahman, S.M., Wang, Y.: Mutual trust-based subtask allocation for human-robot collaboration in flexible lightweight assembly in manufacturing. Mechatronics **54**, 94–109 (2018)
10. Awais, M., Henrich, D.: Human-robot collaboration by intention recognition using probabilistic state machines. In: 19th International Workshop on Robotics in Alpe-Adria-Danube Region (RAAD 2010). pp. 75–80 (2010)
11. Li, Y., Ge, S.S.: Human–robot collaboration based on motion intention estimation. IEEE/ASME Trans. Mechatron. **19**(3), 1007–1014 (2013)
12. Yamamoto, Y., Yun, X.: Coordinating locomotion and manipulation of a mobile manipulator. In: [1992] Proceedings of the 31st IEEE Conference on Decision and Control. pp. 2643–2648. IEEE (1992)
13. Hu, C., Chen, W., Wang, J., Wang, H.: Optimal path planning for mobile manipulator based on manipulability and localizability. In: 2016 IEEE International Conference on Real-time Computing and Robotics (RCAR). pp. 638–643. IEEE (2016)
14. Varela-Aldás, J., Andaluz, V.H., Chicaiza, F.A.: Modelling and control of a mobile manipulator for trajectory tracking. In: 2018 International Conference on Information Systems and Computer Science (INCISCOS). pp. 69–74. IEEE (2018)
15. Avanzini, G.B., Zanchettin, A.M., Rocco, P.: Constrained model predictive control for mobile robotic manipulators. Robotica **36**(1), 19–38 (2018)
16. Khan, A.H., Li, S., Chen, D., Liao, L.: Tracking control of redundant mobile manipulator: an rnn based metaheuristic approach. Neurocomputing **400**, 272–284 (2020)
17. Thakar, S., Kabir, A., Bhatt, P.M., Malhan, R.K., Rajendran, P., Shah, B.C., Gupta, S.K.: Task assignment and motion planning for bi-manual mobile manipulation. In: 2019 IEEE 15th International Conference on Automation Science and Engineering (CASE). pp. 910–915. IEEE (2019)

18. Karami, H., Darvish, K., Mastrogiovanni, F.: A task allocation approach for human-robot collaboration in product defects inspection scenarios. In: 2020 29th IEEE International Conference on Robot and Human Interactive Communication (RO-MAN). pp. 1127–1134. IEEE (2020)
19. Navarro, B., Cherubini, A., Fonte, A., Poisson, G., Fraisse, P.: A framework for intuitive collaboration with a mobile manipulator. In: 2017 IEEE/RSJ International Conference on Intelligent Robots and Systems (IROS). pp. 6293–6298. IEEE (2017)
20. Gillini, G., Di Lillo, P., Arrichiello, F., Di Vito, D., Marino, A., Antonelli, G., Chiaverini, S.: A dual-arm mobile robot system performing assistive tasks operated via p300-based brain computer interface. Indus. Robot: Int. J. Robot. Res. Appl. (2020)
21. Chen, M., Liu, C., Du, G.: A human–robot interface for mobile manipulator. Intell. Serv. Robot. **11**(3), 269–278 (2018)
22. Zhou, H., Min, H., Lin, Y., Zhang, S.: A robot architecture of hierarchical finite state machine for autonomous mobile manipulator. In: International Conference on Intelligent Robotics and Applications. pp. 425–436. Springer (2017)
23. Benzi, F., Secchi, C.: An optimization approach for a robust and flexible control in collaborative applications. arXiv preprint arXiv:2103.03082 (2021)
24. Ferraguti, F., Talignani Landi, C., Singletary, A., Lin, H.C., Ames, A., Secchi, C., Bonfe, M.: Safety and efficiency in robotics: the control barrier functions approach. IEEE Robot. Autom. Mag., pp. 2–14 (2022)
25. Sorour, M., Cherubini, A., Fraisse, P.: Motion control for steerable wheeled mobile manipulation. In: 2019 European Conference on Mobile Robots (ECMR). pp. 1–7. IEEE (2019)
26. Moon, S., Park, Y., Ko, D.W., Suh, I.H.: Multiple kinect sensor fusion for human skeleton tracking using kalman filtering. Int. J. Adv. Robot. Syst. **13**(2), 65 (2016)
27. Kofman, J., Wu, X., Luu, T.J., Verma, S.: Teleoperation of a robot manipulator using a vision-based human-robot interface. IEEE Trans. Indus. Electron. **52**(5), 1206–1219 (2005)
28. Fan, J., Xu, W., Wu, Y., Gong, Y.: Human tracking using convolutional neural networks. IEEE Trans. Neural Netw. **21**(10), 1610–1623 (2010)
29. Notomista, G., Egerstedt, M.: Constraint-driven coordinated control of multi-robot systems. In: 2019 American Control Conference (ACC). pp. 1990–1996. IEEE (2019)
30. Notomista, G., Mayya, S., Hutchinson, S., Egerstedt, M.: An optimal task allocation strategy for heterogeneous multi-robot systems. In: 2019 18th European Control Conference (ECC). pp. 2071–2076. IEEE (2019)
31. Ferraguti, F., Landi, C.T., Costi, S., Bonfè, M., Farsoni, S., Secchi, C., Fantuzzi, C.: Safety barrier functions and multi-camera tracking for human–robot shared environment. Robot. Auton. Syst. **124**, 103388 (2020)
32. d'Andréa Novel, B., Campion, G., Bastin, G.: Control of nonholonomic wheeled mobile robots by state feedback linearization. Int. J. Robot. Res. **14**(6), 543–559 (1995)
33. Yoshikawa, T.: Manipulability of robotic mechanisms. Int. J. Robot. Res. **4**(2), 3–9 (1985)
34. Caccavale, F., Lippiello, V., Muscio, G., Pierri, F., Ruggiero, F., Villani, L.: Grasp planning and parallel control of a redundant dual-arm/hand manipulation system. Robotica **31**(7), 1169–1194 (2013)
35. Ferrari, D., Benzi, F., Secchi, C.: Bidirectional communication control for human–robot collaboration (2022), https://arxiv.org/abs/2206.05202

Approximating Robot Reachable Space Using Convex Polytopes

Antun Skuric(✉), Vincent Padois, and David Daney

INRIA Sud-Ouest, Bordeaux, France
antun.skuric@inria.fr, vincent.padois@inria.fr, david.daney@inria.fr
http://team.inria.fr/auctus/

Abstract. This paper presents an approach for approximating the reachable space of robotic manipulators based on convex polytopes. The proposed approach predicts the reachable space over a given time horizon based on the robot's actuation limits and kinematic constraints. The approach is furthermore extended to integrate the robot's environment, assuming it can be expressed in a form of linear constraints, and to account for the robot's link geometry. The accuracy of the proposed method is evaluated using simulations of robot's nonlinear dynamics and it is compared against the cartesian space limits, usually provided by manufacturers in standard datasheets. The accuracy analysis results show that the proposed method has good performance for the time horizons up to $t_h \leq 250$ ms, encapsulating most of the simulated robot's reachable space while maintaining comparable volume. For a 7 dof robot, the method has an average execution time of 50ms, independent of the horizon time, potentially enabling real-time applications.

Keywords: Reachability analysis · Convex polytopes · Collaborative robotics

1 Introduction

Collaborative robots, designed for safe physical interaction with humans, have a great potential to find their way in many spheres of industry, research and potentially become a part of our everyday lives. Apart from replacing rigid and dangerous manipulators in the industrial applications, their main potential is creating new robotic assistance scenarios leveraging high degree of physical human-robot interaction. Such scenarios demand not only many safety guarantees, but a high degree of operator's situation awareness both in terms of robot's behaviour and its physical abilities in real-time.

A metric unifying the robot's physical abilities and its movement capacity, important for safety and performance analysis, is called the robot's *reachable space* [1]. It represents the space of reachable robot's positions (e.x. end-effector

This work was funded by the BPI France Lichie project.

P. Borja et al. (Eds.): HFR 2022, 26, pp. 45–60, 2023.
https://doi.org/10.1007/978-3-031-22731-8_4

cartesian positions) for a certain time horizon, given different assumptions on its physical abilities. As robotic systems are highly non-linear and featuring complex dynamics, the true reachable space of robots is very complex to characterise and calculate. Therefore, approximation techniques are necessary in order to yield practical solutions. Pereira and Athhoff [6] have developed an interval analysis based approach, later improved by Shrepp et al. [7], to approximate the human arm reachable space, modeled as a serial robotic manipulator, using a set of spheres and cylinders. However, in many cases the approximation of the reachable space using these shapes is impractical due to their non-linear nature.

In this paper, an approximation approach of the robot's reachable space using convex polytopes is proposed, leveraging several of their key characteristics. Convex polytopes can be represented as a set of linear inequalities $\boldsymbol{A}\boldsymbol{x} \leq \boldsymbol{b}$, which may be directly used with different optimisation techniques. Additionally, this set of linear inequalities can be intuitively extended with different environmental and user defined constraints. Since the reachable space is often low dimensional ($\leq$3D) these polytopes can be easily visualised, having the structure of a triangulated mesh, and potentially provide a human operator with valuable information about the robot's capacity. Furthermore, operations over polytopes such as Minkowski sum and intersection are well defined and efficient to calculate, enabling for the intuitive extension of the proposed method to account for robot's link geometry. Finally, the efficiency of polytope enumeration techniques for low dimensional polytopes has a potential to make this metric real-time capable.

Even though convex polytopes are widely used to characterise different metrics of robot's physical abilities [9]: force, acceleration and velocity, to the best of our knowledge they have not yet been used to characterise the reachable space of a robotic manipulator. The proposed approach is partially inspired by the works of Long et al. [5] on the constrained manipulability (velocity) polytopes.

Depending on whether the application requires an under or over approximation, or perhaps time-efficiency, the evaluation metrics of the accuracy of the reachable space approximation will differ significantly. In this paper, in addition to the execution time analysis, three accuracy measures are chosen to enable the evaluation of the proposed approach's performance and study its limitations.

The paper has the following structure. In the Sect. 2 the formal definition of the approach is introduced, followed by the Sect. 3, describing the procedure of benchmarking and the accuracy analysis. In the Sect. 4, the results of the accuracy and execution time analysis of the approach are given and the Sect. 5 discusses the main limitation as well as potential applications.

2 Convex Polytope of Reachable Space

Robot dynamics in joint space can be expressed as

$$M(\boldsymbol{q})\ddot{\boldsymbol{q}} + \underbrace{C(\boldsymbol{q}, \dot{\boldsymbol{q}})\dot{\boldsymbol{q}} + \boldsymbol{\tau}_g(\boldsymbol{q})}_{\boldsymbol{\tau}_d} = \boldsymbol{\tau} \tag{1}$$

where M is the mass matrix, C is coriolis matrix and $\boldsymbol{\tau}_g$ is the gravity vector. $\boldsymbol{\tau}$ is the applied joint torque vector and $\boldsymbol{\tau}_d$ is the equivalent joint torque vector due to the coriolis and gravity effects.

For a robot with n actuated degrees of freedom (DOF), the robot's joint torque $\boldsymbol{\tau}$, velocity $\dot{\boldsymbol{q}}$ and joint angles $\boldsymbol{q}$ are n-dimensional vectors, limited by the robot's hardware[1]

$$\boldsymbol{\tau} \in [\boldsymbol{\tau}_{\min}, \boldsymbol{\tau}_{\max}], \quad \dot{\boldsymbol{q}} \in [\dot{\boldsymbol{q}}_{\min}, \dot{\boldsymbol{q}}_{\max}], \quad \boldsymbol{q} \in [\boldsymbol{q}_{\min}, \boldsymbol{q}_{\max}] \tag{2}$$

For a given moment in time k and for given robot configuration $\boldsymbol{q}_k$, the affine relationship between joint torques $\boldsymbol{\tau}$ and accelerations $\ddot{\boldsymbol{q}}_k$ can be expressed as

$$\ddot{\boldsymbol{q}}_k = M^{-1}(\boldsymbol{q}_k)(\boldsymbol{\tau} - \boldsymbol{\tau}_d) = M_k^{-1}(\boldsymbol{\tau} - \boldsymbol{\tau}_d) \tag{3}$$

Considering fixed joint acceleration $\ddot{\boldsymbol{q}}_k$ during given horizon length t_h, an approximation of the robot's joint velocity $\dot{\boldsymbol{q}}_{k+1}$ and position $\boldsymbol{q}_{k+1}$, at the end of horizon, can be calculated using numerical integration (forward Euler method)

$$\dot{\boldsymbol{q}}_{k+1} = M_k^{-1} t_h(\boldsymbol{\tau} - \boldsymbol{\tau}_d) + \dot{\boldsymbol{q}}_k, \qquad \boldsymbol{q}_{k+1} = M_k^{-1}\frac{t_h^2}{2}(\boldsymbol{\tau} - \boldsymbol{\tau}_d) + \dot{\boldsymbol{q}}_k t_h + \boldsymbol{q}_k \tag{4}$$

This linear numerical integration (4) considers the robot's dynamics (1), and the applied joint torque $\boldsymbol{\tau}$, fixed during the horizon time t_h. Such assumption is reasonable only for short horizon times t_h which in term represents a limitation of the proposed method.

The relationship between the m-dimensional task space velocity and acceleration of certain frame on the robot (for example end-effector frame) and the n-dimensional joint space equivalents is defined through the corresponding jacobian matrix $J(\boldsymbol{q})$ and its time derivative $\dot{J}(\boldsymbol{q})$.

$$\dot{\boldsymbol{x}}_k = J(\boldsymbol{q}_k)\dot{\boldsymbol{q}}_k = J_k\dot{\boldsymbol{q}}_k, \qquad \ddot{\boldsymbol{x}}_k = J_k\ddot{\boldsymbol{q}}_k + \dot{J}_k\dot{\boldsymbol{q}}_k \tag{5}$$

Finally, given the horizon of interest t_h, the predicted cartesian position $\boldsymbol{x}_{k+1}$ can then be expressed as

$$\begin{aligned}\boldsymbol{x}_{k+1} &= \ddot{\boldsymbol{x}}_k\frac{t_h^2}{2} + \dot{\boldsymbol{x}}_k t_h + \boldsymbol{x}_k \\ &= J_k M_k^{-1}\frac{t_h^2}{2}\boldsymbol{\tau} \underbrace{\underbrace{- J_k M_k^{-1}\frac{t_h^2}{2}\boldsymbol{\tau}_d}_{\Delta\boldsymbol{x}_{k,dyn}} \underbrace{+ \dot{J}_k\dot{\boldsymbol{q}}_k\frac{t_h^2}{2} + \dot{\boldsymbol{x}}_k t_h}_{\Delta\boldsymbol{x}_{k,vel}} + \boldsymbol{x}_k}_{\boldsymbol{x}^*_{k+1}}\end{aligned} \tag{6}$$

where the $\boldsymbol{x}^*_{k+1} = \Delta\boldsymbol{x}_{k,dyn} + \Delta\boldsymbol{x}_{k,vel} + \boldsymbol{x}_k$ is a predicted position vector based on the joint configuration $\boldsymbol{q}_k$, joint velocity $\dot{\boldsymbol{q}}_k$ (or cartesian velocity $\dot{\boldsymbol{x}}_k = J\dot{\boldsymbol{q}}_k$) and

[1] There is a coupling between the velocity and torque limits of an actuator related to its power. Yet, it is common practice to consider a subset of these limits such as torque and velocity can be chosen independantly one from the other.

the joint torque $\boldsymbol{\tau}_d$ in the current time-step k, while the first term describes the influence of the applied joint torques $\boldsymbol{\tau}$.

The convex polytope of reachable cartesian space $\mathcal{P}_x$ can then be defined as a set of all the possible cartesian positions $\boldsymbol{x}_{k+1}$ at the end of the horizon t_h, achieved by any combination of joint torques $\boldsymbol{\tau}$ within robot's physical limitations (2), given the robot's current state $\dot{\boldsymbol{q}}_k$, $\boldsymbol{q}_k$ and $\boldsymbol{\tau}_d$.

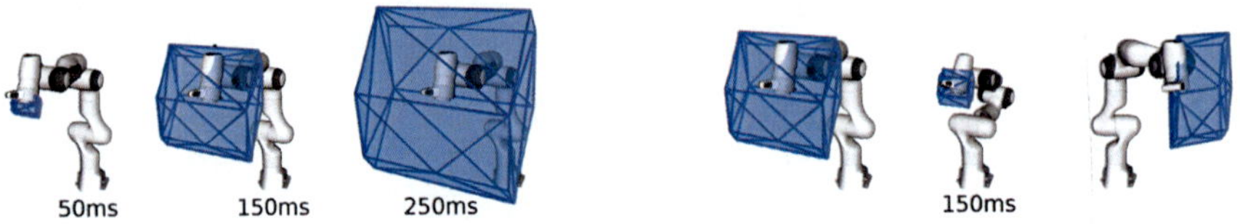

Fig. 1. Three images on the left show the comparison of the size of the reachable space polytope $\mathcal{P}_x$ of a *Franka Emika Panda*(More information about the Panda robot at https://frankaemika.github.io/docs/) robot's end-effector for 3 horizons (left to right) t_h = 0.05, 0.15 and 0.25s, for the same configuration. Whereas the three figures on the right show the constraining effect of the joint position limits on the $\mathcal{P}_x$ for a fixed horizon time (t_h=0.15 s). Third image from the right shows the robot in its initial position $\boldsymbol{q} = [0, 0, 0, -\pi/2, 0, 3\pi/5, 0]$, close to the center of the joint ranges. In the second image form the right, the robot's joints q_1 and q_3 at their limits, $\boldsymbol{q} = [0, -1.59, 0, -2.9, 0, 3\pi/5, 0, 0]$. On the last image, robot's joints q_0 and q_2 are at their limits $\boldsymbol{q} = [-2.72, 0, -2.72, -\pi/2, 0, 13\pi/5, 0, 0]$, preventing the robot to rotate around the z axis in one direction.

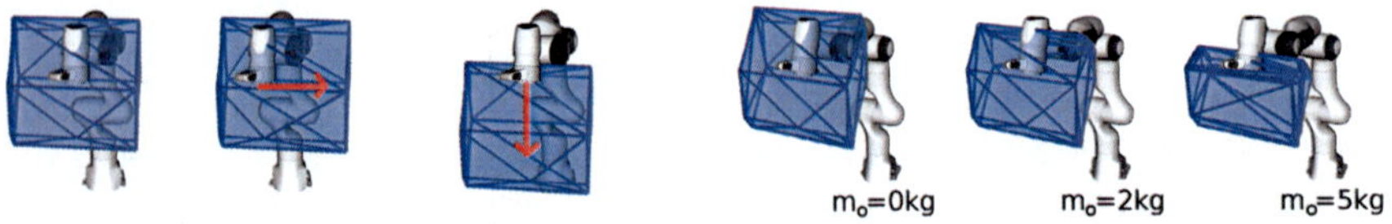

Fig. 2. Three images on the left show the shifting effect on the reachable space polytope $\mathcal{P}_x$ produced by certain cartesian velocity at the beginning of the horizon, from left to right: $\dot{\boldsymbol{x}}_k = [0, 0, 0]$, $[0, 1, 0]$ and $[0, 0, -1]m/s$. Three images on the right show the reducing effect on the polytope $\mathcal{P}_x$ by three different carried object masses (left to right) m_o = 0, 2 and 5kg. Horizon used is t_h=0.15s and the robot is in initial configuration.

$$\begin{aligned}
\mathcal{P}_x = \{\boldsymbol{x}_{k+1} \in \mathbf{R}^m \quad | \quad \boldsymbol{x}_{k+1} &= J_k M_k^{-1} \frac{t_h^2}{2} \boldsymbol{\tau} + \boldsymbol{x}_{k+1}^*, \\
\boldsymbol{\tau} &\in [\boldsymbol{\tau}_{\min}, \boldsymbol{\tau}_{\max}], \\
M_k^{-1} t_h (\boldsymbol{\tau} - \boldsymbol{\tau}_d) + \dot{\boldsymbol{q}}_k &\in [\dot{\boldsymbol{q}}_{\min}, \dot{\boldsymbol{q}}_{\max}], \\
M_k^{-1} \frac{t_h^2}{2} (\boldsymbol{\tau} - \boldsymbol{\tau}_d) + \dot{\boldsymbol{q}}_k t_h + \boldsymbol{q}_k &\in [\boldsymbol{q}_{\min}, \boldsymbol{q}_{\max}]\}
\end{aligned} \tag{7}$$

Figures 1 and 2 showcase the influences of different horizon lengths t_h, robot constraints and robot movement $(\dot{\boldsymbol{q}}_k, \boldsymbol{\tau}_d)$ on the shape of $\mathcal{P}_x$. Section 2.4 describes an approach to enumerate the reachable space polytope.

2.1 Influence of the Carried Object

If the robot is carrying a payload, an object with a mass m_o and inertia i_o attached to its end-effector, this object will have an influence on the robot's dynamics, modifying the mass matrix M, coriolis matrix C and the gravity vector $\boldsymbol{\tau}_g$ [4]. The augmented dynamical model of the robot will have reduced acceleration capabilities due to the added effort necessary for the object's movements.

Figure 2 shows the influence of different object masses m_o, on the resulting reachable space polytope $\mathcal{P}_x$ for the *Franka Emika Panda* robot.

2.2 Integration of the Environment

If the robot's environment can be defined as a set of convex inequalities

$$A_e\boldsymbol{x} \leq b_e \tag{8}$$

these constraints can be directly transformed to the constraints of the joint torque $\boldsymbol{\tau}$ using the Eq. (6)

$$A_e\boldsymbol{x}_{k+1} = A_e J_k M_k^{-1}\frac{t_h^2}{2}(\boldsymbol{\tau} - \boldsymbol{\tau}_d) + A_e\boldsymbol{x}_{k+1}^* \leq b_e \tag{9}$$

and then included in the $\mathcal{P}_x$ calculation (7)

$$\begin{aligned}
\mathcal{P}_x = \{\boldsymbol{x}_{k+1} \in \mathbf{R}^m \quad | \quad \boldsymbol{x}_{k+1} = J_k M_k^{-1}\frac{t_h^2}{2}\boldsymbol{\tau} + \boldsymbol{x}_{k+1}^*,& \\
\boldsymbol{\tau} \in [\boldsymbol{\tau}_{\min}, \boldsymbol{\tau}_{\max}]\,,& \\
M_k^{-1}t_h(\boldsymbol{\tau} - \boldsymbol{\tau}_d) + \dot{\boldsymbol{q}}_k \in [\dot{\boldsymbol{q}}_{\min}, \dot{\boldsymbol{q}}_{\max}]\,,& \\
M_k^{-1}\frac{t_h^2}{2}(\boldsymbol{\tau} - \boldsymbol{\tau}_d) + \dot{\boldsymbol{q}}_k t_h + \boldsymbol{q}_k \in [\boldsymbol{q}_{\min}, \boldsymbol{q}_{\max}]\,,& \\
A_e J_k M_k^{-1}\frac{t_h^2}{2}(\boldsymbol{\tau} - \boldsymbol{\tau}_d) + A_e\boldsymbol{x}_{k+1}^* \leq b_e \ \}&
\end{aligned} \tag{10}$$

Figure 3 showcases the influence of the environmental constraints on the shape of the $\mathcal{P}_x$.

2.3 Integration of Robot's Link Geometry

If the robot link l_i is considered to be a straight line, to calculate the predicted reachable space of a robot's link l_i the polytope $\mathcal{P}_x$ can be calculated at the start $\mathcal{P}_{xs}$ and the end $\mathcal{P}_{xe}$ of the line l_i. Finally the reachable space of this idealised

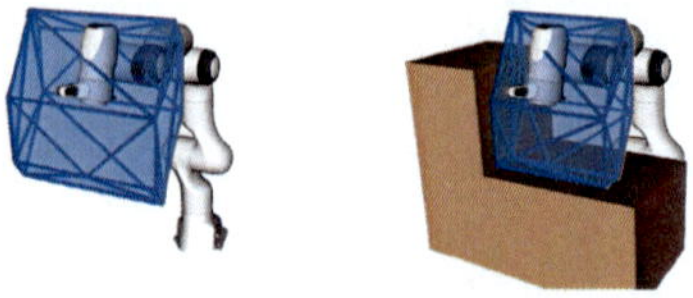

Fig. 3. Resulting reachable space polytope $\mathcal{P}_x$ for a *Franka Emika Panda* robot's end-effector when integrating the environmental constraints, horizon time used is t_h=0.15s. Environment is defined as $z \geq 0.5$m and $y \geq -0.2$m.

robot link l_i can then be calculated as the *convex-hull* (CH) of the polytopes $\mathcal{P}_{xs}$ and $\mathcal{P}_{xe}$

$$\mathcal{P}_{xli} = CH\left(\mathcal{P}_{xs},\ \mathcal{P}_{xe}\right) \tag{11}$$

If, instead of a straight line, a space captured by a robot's link l_i can be expressed as a convex set of constraints or a polytope

$$\mathcal{L}_i = \left\{\boldsymbol{x}_l \mid A_l\boldsymbol{x}_l \leq b_l\right\} \tag{12}$$

the reachable space polytope $\mathcal{P}_{xli}$ can be extended to account for this link geometry by first finding the vertices of the polytope $\mathcal{L}$, followed by computing the reachable space polytope $\mathcal{P}_x$ for each one of the k vertices, and finally computing their convex hull.

$$\mathcal{P}_{xli} = CH\left(\mathcal{P}_{x1},\ \dots\ ,\mathcal{P}_{xk}\right) \tag{13}$$

By evaluating the polytope $\mathcal{P}_{xli}$ for each robot's link one can obtain an efficient approximation of the envelope of the reachable space of the entire robot, as shown on Fig. 4.

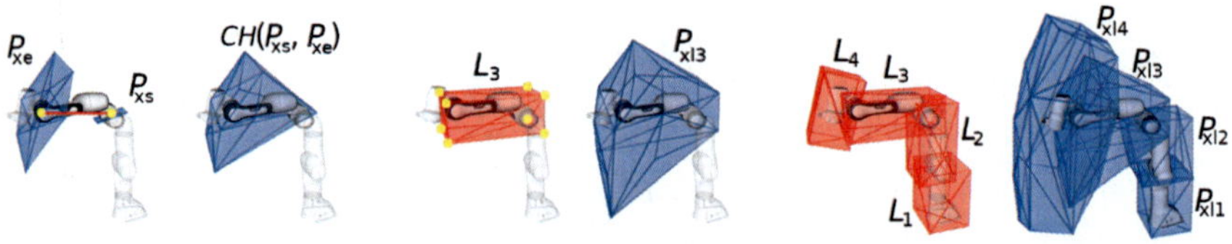

Fig. 4. First four images show the construction of the reachable space polytope of the panda robot's link 3, $\mathcal{P}_{xl3}$. From left to right they show, robot's link 3 as a line segment (in red) with polytopes $\mathcal{P}_{xs}$ and $\mathcal{P}_{xe}$, their convex-hull, an example of the enveloping space $\mathcal{L}_3$ of the link 3 (in red), and the final polytope $\mathcal{P}_{xl3}$ calculated as a convex hull of the reachable space polytopes of each one of the 8 vertices (shown in yellow) of $\mathcal{L}_3$. Last two images further show an example of 4 enveloping spaces $\mathcal{L}_i$ for each one of the robot's links, and their reachable space polytopes $\mathcal{P}_{xli}$. Robot is in its initial configuration, and the horizon time used is 150ms.

2.4 Enumerating Reachable Space Polytope

The reachable space polytope $\mathcal{P}_x$ belongs to the family of over-determined problems

$$\mathcal{P}_x = \{\boldsymbol{x} \in \mathbf{R}^m \quad | \quad \boldsymbol{x} = P\boldsymbol{\tau}, \quad A\boldsymbol{\tau} \leq \boldsymbol{b} \ \}, \quad \boldsymbol{\tau} \in R^n, \quad n \geq m \tag{14}$$

where the inequality constraint $A\boldsymbol{\tau} \leq \boldsymbol{b}$ encapsulates all the robot constraints (2) as well as the environmental constraints (9), and the matrix $J_k M_k^{-1} \frac{t_h^2}{2}$ becomes the projection matrix P. Vector $\boldsymbol{\tau}$ corresponds to the joint torque vector and the vector $\boldsymbol{x}$ corresponds to the difference $\boldsymbol{x}_{k+1} - \boldsymbol{x}_{k+1}^*$.

In order to fully define and simplify this formulation, the set of all the vertices ($\mathcal{V}$-representation) or all the half-spaces ($\mathcal{H}$-representation) of this polytope has to be found. Vertex representation of a polytope is a list of all of its vertices $\boldsymbol{v}_i$ and is commonly used for visualisation purposes $\mathcal{V} = \{\boldsymbol{v}_0, \boldsymbol{v}_1, \boldsymbol{v}_2, \dots\}$, while half-space representation is defined as a list of inequalities $H\boldsymbol{x} \leq \boldsymbol{d}$, each corresponding to one of the polytope facets. This matrix inequality equation is more suitable for applications involving different optimisation strategies.

The reachable space polytope (14) is a particular case of the projection of a high-dimensional polytope onto a lower dimensional space. Therefore, the set of constraints $A\boldsymbol{\tau} \leq \boldsymbol{b}$ form a $\mathcal{H}$-representation of a polytope $\mathcal{P}_\tau$ in the n-dimensional joint torque space.

$$\mathcal{P}_\tau = \{\boldsymbol{\tau} \in \mathbf{R}^n \quad | \quad A\boldsymbol{\tau} \leq \boldsymbol{b} \ \} \tag{15}$$

This polytope is then projected using the projection matrix P into the m-dimensional cartesian space, usually much lower dimensional, forming the reachable space polytope $\mathcal{P}_x$.

$$\mathcal{P}_x = \{\boldsymbol{x} \in \mathbf{R}^m \quad | \quad \boldsymbol{x} = P\boldsymbol{\tau}, \quad \boldsymbol{\tau} \in \mathcal{P}_\tau \ \} \tag{16}$$

The most straight forward way of approaching the enumeration of this polytope is to first enumerate the vertices of the polytope $\mathcal{P}_\tau$, project them to the cartesian space using the projection matrix P, followed by a Convex-Hull algorithm to extract the real vertices $\boldsymbol{v}_i$ of the polytope $\mathcal{P}_x$ from the projected points and its half-space representation.

In order to avoid enumerating all the vertices of the high-dimensional joint torque polytope $\mathcal{P}_\tau$, as the most complex operation of this polytope enumeration, *Iterative convex hull method* [10] (ICHM) is used, allowing to directly enumerate low-dimensional cartesian polytope $\mathcal{P}_x$. This iterative method is based on successively applying *Linear programming* (LP) and *Convex-Hull* (CH) algorithms and is able to find both vertex and half-space representation of this polytope at the same time. ICHM algorithm is defined for a family of sets:

$$\{\ \boldsymbol{x} \in R^m \mid A\boldsymbol{x} = B\boldsymbol{y}, \quad C\boldsymbol{y} \leq \boldsymbol{d}\ \}, \qquad \boldsymbol{y} \in R^n, \quad m \leq n \tag{17}$$

Therefore the usage of the ICHM algorithm for enumerating the reachable space polytope is rather straight forward, by setting its matrix A to identity $A = I_{m \times m}$, matrix B becomes the projection matrix P, the inequality constraint $C\boldsymbol{y} \leq \boldsymbol{d}$ then becomes the set of the constraints $A\boldsymbol{\tau} \leq \boldsymbol{b}$.

3 Analysing the Approximation Accuracy

Defining metrics of interest for accuracy analysis is highly dependent of the applications these methods will be used on. If the application requires an under-approximation of the reachable set or an over-approximation, the metrics of interest will not be the same. In this paper, the algorithms accuracy is analysed using three different quantitative metrics, each one providing an insight in different characteristics and limitations of the method.

Additionally as this approximation method has a potential to be used for real-time applications, execution time analysis is performed as well.

In the extent of the proposed experiments, the considered reachable space is the reachable space of the robot's end effector.

3.1 Metrics Definition

To analyse the approximation accuracy of the reachable space polytope $\mathcal{P}_x$, a simple discrete non-linear robot dynamics simulation, subject to the constraints (2) is employed. Simulation is carried out by discretizing the Eq. (1) to determine the instantaneous joint acceleration $\ddot{\boldsymbol{q}}_k$ and followed by its numerical integration to determine joint velocity $\dot{\boldsymbol{q}}_{k+1}$ and position $\boldsymbol{q}_{k+1}$ in the step $k+1$.

$$\begin{aligned}
\ddot{\boldsymbol{q}}_k &= M^{-1}(\boldsymbol{q}_k)\left(\boldsymbol{\tau} - C(\boldsymbol{q}_k, \dot{\boldsymbol{q}}_k)\dot{\boldsymbol{q}}_k + \boldsymbol{\tau}_g(\boldsymbol{q}_k)\right) \\
\dot{\boldsymbol{q}}_{k+1} &= \ddot{\boldsymbol{q}}_k \Delta t + \dot{\boldsymbol{q}}_k \\
\boldsymbol{q}_{k+1} &= \ddot{\boldsymbol{q}}_k \frac{\Delta t^2}{2} + \dot{\boldsymbol{q}}_k \Delta t + \boldsymbol{q}_k
\end{aligned} \tag{18}$$

where Δt is a fixed simulation sampling time. The cartesian position in each step $\boldsymbol{x}_k$ is then calculated by evaluating the robot's forward kinematics

$$\boldsymbol{x}_k = f_{DK}(\boldsymbol{q}_k). \tag{19}$$

Each simulation roll-out considers a fixed joint torque vector $\boldsymbol{\tau}$ and simulates the robot dynamics through the horizon time t_h performing $N = t_h/\Delta t$ steps.

Each vertex $\boldsymbol{v}_i$ of reachable space polytope $\mathcal{P}_x$, calculated for a joint position $\boldsymbol{q}_k$ and a horizon t_h, is generated by a joint torque vector $\boldsymbol{\tau}_i$, belonging to a subset of vertices of the joint torque polytope $\mathcal{P}_\tau$.

To evaluate the accuracy of the proposed method, the difference in reached space produced by the constant liner model (polytope $\mathcal{P}_x$) and time varying non-linear model of the robot is evaluated, for the same set of applied joint torques. The applied joint torques chosen for the experiments are the joint torques $\boldsymbol{\tau}_i$ generating the vertices $\boldsymbol{v}_i$ of the reachable space polytope $\mathcal{P}_x$ calculated by the linear model.

As shown on Fig. 5, for each of the n_v vertices $\boldsymbol{v}_i$ of the polytope $\mathcal{P}_x$ the simulation (18) is performed and all the cartesian positions $\boldsymbol{x}_k$ of the robot in each of the $N = t_h/\Delta t$ sample times are retained for the further analysis.

$$\mathcal{X} = \{\underbrace{\boldsymbol{x}_{0,0} \ \dots \ \boldsymbol{x}_{0,N}}_{\boldsymbol{\tau}_0},\ \underbrace{\boldsymbol{x}_{1,0} \ \dots \ \boldsymbol{x}_{1,N}}_{\boldsymbol{\tau}_1},\ \dots,\ \underbrace{\boldsymbol{x}_{n_v,0} \ \dots \ \boldsymbol{x}_{n_v,N}}_{\boldsymbol{\tau}_{n_v}}\}$$

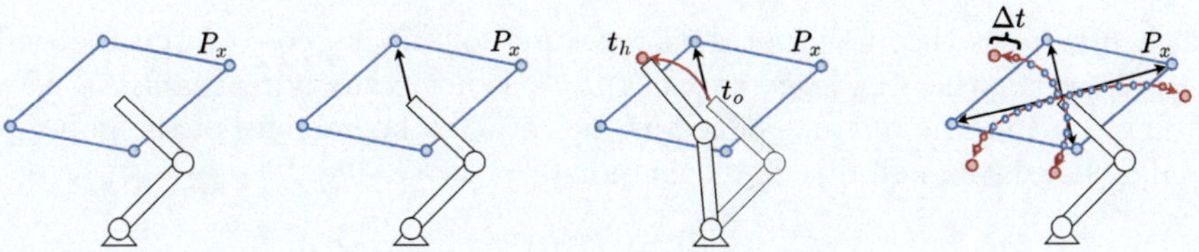

Fig. 5. Images show the steps (left to right) of the nonlinear simulation procedure on a simplified 2 link planar robot. First the polytope $\mathcal{P}_x$ of the end-effector is determined. For each vertex of the $\mathcal{P}_x$, joint torque τ generating the vertex is applied to the robot simulation (Figs. 2 and 3). All the robot's end-effector positions in all the simulation time steps Δt are saved for further analysis.

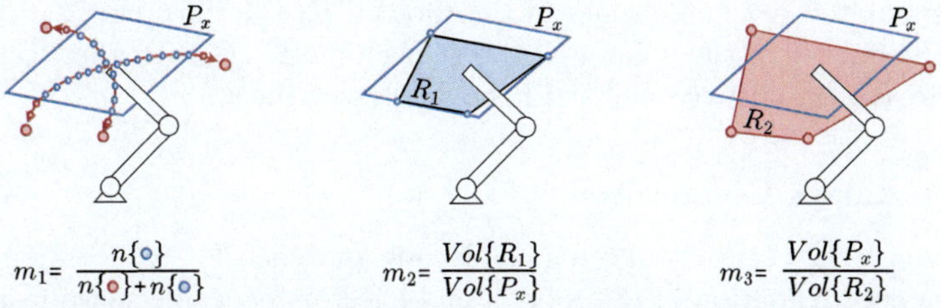

Fig. 6. Depiction of the defined quantitative metrics for a simple 2 link planar robot.

First metric m_1, as shown on Fig. 6, is defined as a ratio of number of the simulated cartesian robot positions $\boldsymbol{x}_k \in \mathcal{X}$ inside the polytope $\mathcal{P}_x$ with respect to the overall number of simulated positions.

$$m_1 = \frac{|\mathcal{X} \cap \mathcal{P}_x|}{|\mathcal{X}|} = \frac{|\mathcal{X} \cap \mathcal{P}_x|}{N \cdot n_v} \tag{20}$$

The metric m_1 gives an insight on how well the polytope $\mathcal{P}_x$ encapsulates the real reachable space determined with simulations. This metric might be used to asses the confidence in the polytope $\mathcal{P}_x$ for safety applications, the closer this metric is to 1, the less chance that the robot can exit the space bounded by $\mathcal{P}_x$.

The second metric, m_2, is defined as the ratio of the volumes of the polytope $\mathcal{P}_x$ and the convex hull CH of the points $\boldsymbol{x}_k \in \mathcal{X} \cap \mathcal{P}_x$,

$$R_1 = CH(\mathcal{X} \cap \mathcal{P}_x)$$

representing a subset of the points reached by the robot in the simulations found inside of the polytope $\mathcal{P}_x$.

$$m_2 = \frac{Vol\{R_1\}}{Vol\{P_x\}} \tag{21}$$

The metric m_2 gives an insight on how much volume of the polytope $\mathcal{P}_x$ is actually attainable by the robot, with respect to the simulations. This metric is designed to asses the quality of the approximation, if the score on this metric is low, this means that the large part of the $\mathcal{P}_x$ is not actually reachable.

The third metric, m_3, is defined as the ratio of the volumes of the polytope $\mathcal{P}_x$ and the convex hull CH of all the points $\boldsymbol{x}_k \in \mathcal{X}$,

$$R_2 = CH(\mathcal{X})$$

reached by the robot in the simulations.

$$m_3 = \frac{Vol\{P_x\}}{Vol\{R_2\}} \tag{22}$$

The metric m_3 gives insight about the ratio of the volumes of the polytope $\mathcal{P}_x$ and the simulated reachable space of the robot $CH(\mathcal{X})$. This metric is designed to asses the quality of the volume of the $\mathcal{P}_x$ as a metric. The further this metric from 1 the less confidence one has in the volume of the $\mathcal{P}_x$.

3.2 Benchmark Comparison

To benchmark the accuracy of convex polytope approach $\mathcal{P}_x$, it is compared with the coarse approximation of the robot's end-effector reachable space using fixed cartesian velocity $\dot{\boldsymbol{x}}$ and acceleration $\ddot{\boldsymbol{x}}$ limits specified by robot manufacturer.

$$\ddot{\boldsymbol{x}} \in [\, \ddot{\boldsymbol{x}}_{\min}, \ddot{\boldsymbol{x}}_{\max}], \qquad \dot{\boldsymbol{x}} \in [\, \dot{\boldsymbol{x}}_{\min}, \dot{\boldsymbol{x}}_{\max}] \tag{23}$$

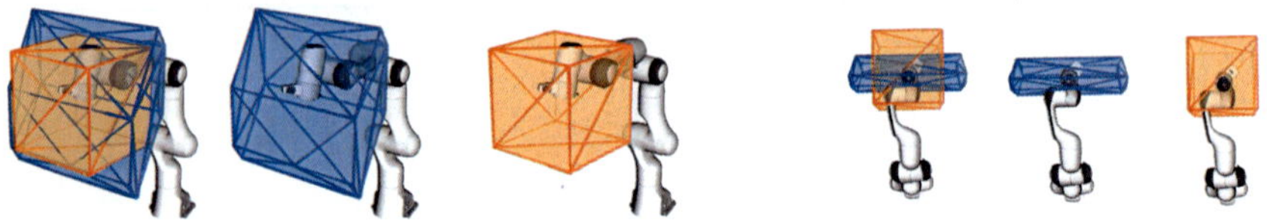

Fig. 7. Comparison of the end-effector reachable space approximation using the convex polytope $\mathcal{P}_x$ (blue) and the cube $\mathcal{C}_x$ (orange) for the *Franka Emika Panda* robot with the horizon time t_h=200ms. They are both evaluated in robot's initial position and the figures show two different views.

The cartesian space limitations (23) assume that the robot's acceleration $\ddot{\boldsymbol{x}}$ and velocity $\dot{\boldsymbol{x}}$ capacity is constant, which is a not true since the robot's capacity depends highly on robot's joint configuration $\boldsymbol{q}$ and velocity $\dot{\boldsymbol{q}}$ [2,9]. However, they are very convenient and widely used in many applications. A simple convex space $\mathcal{C}_x$ (cube in 3d) of reachable end-effector positions given the cartesian space limitations (23) can be calculated

$$\begin{aligned} \mathcal{C}_x = \{\, \boldsymbol{x}_{k+1} \in R^m \mid \boldsymbol{x}_{k+1} = \ddot{\boldsymbol{x}}_k \frac{t_h^2}{2} + \dot{\boldsymbol{x}}_k t_h + \boldsymbol{x}_k \\ \ddot{\boldsymbol{x}}_k \in [\, \ddot{\boldsymbol{x}}_{\min}, \ddot{\boldsymbol{x}}_{\max}] \\ \ddot{\boldsymbol{x}}_k t_h \in [\, \dot{\boldsymbol{x}}_{\min}, \dot{\boldsymbol{x}}_{\max}] \,\} \end{aligned} \tag{24}$$

Figure 7 shows the visual comparison of the two approaches: $\mathcal{P}_x$ and $\mathcal{C}_x$ calculated for one pose of the *Franka Emika Panda* robot.

4 Results

The proposed analysis of the reachable space polytope accuracy is illustrated on the simulation of the collaborative robotic manipulator *Franka Emika Panda* with 7 degrees of freedom. Robot modeling and kinematics as well as simulations of robot dynamics are built using the python implementation of the ROBOTICS-TOOLBOX [3] package.

The evaluation of the reachable space polytope $\mathcal{P}_x$ and $\mathcal{C}_x$ is carried out using the ICHM algorithm's efficient python implementation within the PYCAPACITY [8] package. The ICHM's approximation accuracy used for all the evaluations is δ=1mm.

Eight different horizon lengths t_h are chosen for the experiments

$$t_h = [0.05s,\ 0.15s,\ 0.25s,\ 0.5s,\ 0.75s,\ 1.0s,\ 1.5s,\ 2.0s]$$

and the simulation sample time has been set to $\Delta t = 5$ ms.

The proposed method implementation as well as the code used for the experiments is fully open-source and can be found in the gitlab repository[2]. The gitlab repository additionally contains a Robot operating system (ROS) implementation of the proposed method for more interactive evaluation.

All the simulations are run on a computer with 1.90GHz Intel i7-8650U processor and 32Gb of RAM memory.

4.1 Execution Time

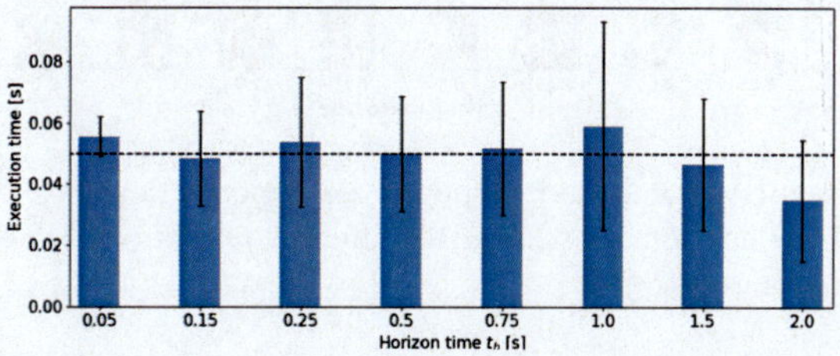

Fig. 8. Execution time of the polytope $\mathcal{P}_x$ enumeration using ICHM algorithm. Average and standard deviation calculated over 1000 executions for each of the 8 horizon lengths.

Figure 8 shows the polytope $\mathcal{P}_x$ execution time averaged over 1000 random robot configurations for each of the 8 horizon lengths. From the figure it can

[2] https://gitlab.inria.fr/auctus-team/people/antunskuric/reachable_space/.

be seen that the execution time is relatively consistent through all the horizon lengths, with the average value of around 50 ms. The constant execution time is expected because the horizon length t_h does not have a big influence on the geometrical complexity of the polytope (number of vertices and faces) but its scale, and the scale itself does have a big influence the execution time.

Figure 9 shows the ICHM execution time for the 8 horizon lengths with different numbers of environmental constraints $A_e\boldsymbol{x} \leq \boldsymbol{b}_e$. All the environmental constraints are generated randomly and all the results are averaged over 1000 random robot configurations for each horizon length t_h and all 4 different numbers of environmental constraints: 0 (no environment constraints), 10, 100, 500 and 1000. Results show that up to 10 added environmental constraints the execution time of the method does not change significantly, having the average execution time around 50ms. With 100 environmental constraints the average execution time increases around 20% to around 70 ms, for 500 the average execution is 100ms and with 1000 constraints the average execution time more than doubles, to 150 ms.

The execution time analysis of the ICHM algorithm has shown that the average polytope $\mathcal{P}_x$ enumeration takes between 50 and 150ms, which opens many doors for potential real-time applications.

The average execution time of the nonlinear simulation for $t_h = 50$ ms and $\Delta t = 5$ ms is around 60s, and it grows to about 240s for $t_h = 2$ s. The cube $\mathcal{C}_x$ average execution time stays constant for all horizon lengths at around 2ms.

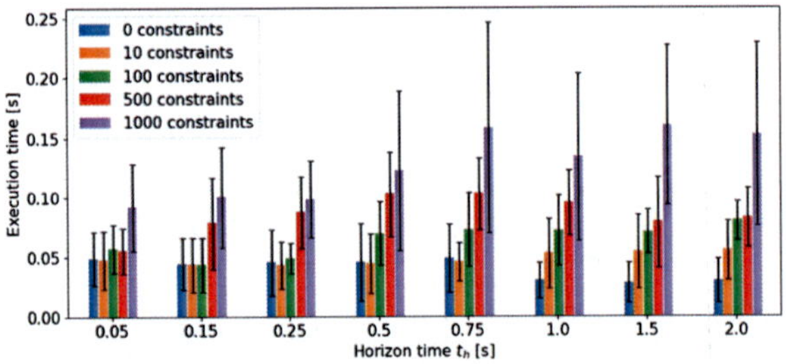

Fig. 9. Average execution time and standard deviation calculated over 1000 executions for each of the 8 horizon lengths, with different number of added environmental constraints, ranging from 0 to 1000.

4.2 Accuracy Analysis

Figure 10 shows the evolution of the three metrics m_1, m_2 and m_3 with respect to the horizon length t_h for the $\mathcal{P}_x$ and $\mathcal{C}_x$, averaged over 100 random robot configurations.

The first graph on the Fig. 10 shows the metric m_1. It can be seen that the ratio of simulated robot positions $\boldsymbol{x}_k \in \mathcal{X}$ inside the polytope $\mathcal{P}_x$ decreases with the increasing horizon length t_h, while it increases for the $\mathcal{C}_x$. For larger horizon lengths the robot is able to move far from the initial joint configuration $\boldsymbol{q}$ at the beginning of the horizon, making the assumption of linearity of the robot model, made for calculating $\mathcal{P}_x$, largely inaccurate. The approach $\mathcal{C}_x$, on the other hand, improves with the increase in horizon length, since the space bounded by the cube $\mathcal{C}_x$ increases linearly with the horizon length t_h, at some point it encapsulates the whole workspace of the robot. However, it can be seen that even though the ratio m_1 decreases, the polytope $\mathcal{P}_x$ still contains the most ($\geq$60%) of the simulated robot's reached positions $\boldsymbol{x}_k \in \mathcal{X}$. Finally, it can also be seen that the variance of the m_1 increases considerably, lowering the confidence of the polytope $\mathcal{P}_x$ for longer horizons.

Evolution of the metric m_2 is shown on the middle graph of the Fig. 10 . It can be seen that the volume of all the reached robot positions inside polytope $\mathcal{P}_x$ and $\mathcal{C}_x$ decreases considerably with the increase of the horizon time t_h. For the horizon times under $t_h \leq 0.25$s the majority ($\geq$50%) of the reachable space polytope $\mathcal{P}_x$ has been reached by the robot simulations. However for the longer horizon times $t_h \geq 0.5$s the reached volume ratio m_2 drops under 20% of the total volume of the $\mathcal{P}_x$. In the case of $\mathcal{C}_x$, the majority ($\geq$50%) of the volume of the $\mathcal{C}_x$ is never reached by the robot for all tested horizon lengths t_h.

The right graph of the Fig. 10 shows the evolution of the metric m_3. The graph shows that the volume of the polytope $\mathcal{P}_x$ and the cube $\mathcal{C}_x$ becomes several times higher than the volume of the R_2 with the increase in the horizon time t_h. For the horizon time of $t_h = 2$ s the average volume of $\mathcal{P}_x$ is 12 times larger than the volume of R_2 and the volume $\mathcal{C}_x$ is more than 50 times larger than R_2. At the same time the average reached volume R_1 inside the $\mathcal{P}_x$ and $\mathcal{C}_x$ consists of under 10% of their volume (metric m_2), which potentially makes both polytope $\mathcal{P}_x$ and $\mathcal{C}_x$ impractical for many of applications.

However, for the shorter horizon times $t_h \leq 0.25$ s the ratio m_3 is very close to 1 for $\mathcal{P}_x$, which means that the reachable space polytope's $\mathcal{P}_x$ volume corresponds well to the actual reached volume of the robot R_2.

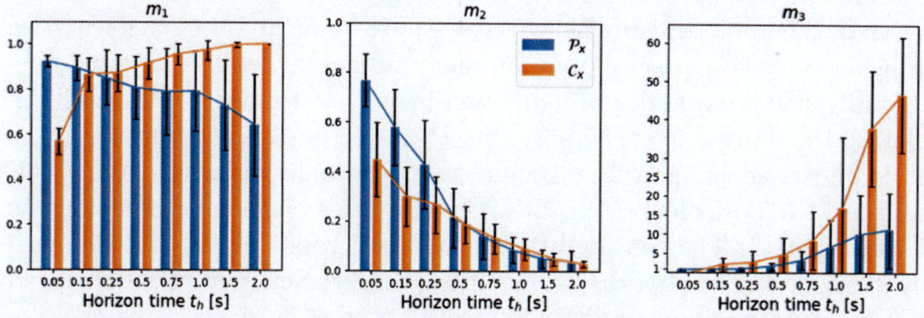

Fig. 10. Accuracy measures calculated for $\mathcal{P}_x$ and $\mathcal{C}_x$, averaged over 100 random poses of the *Franka Emika panda* robot for each of 8 horizon lengths.

Overall, this numerical analysis shows that the reachable space approximation based on the convex polytope $\mathcal{P}_x$ works reasonably well for low horizon times $t_h \leq 250$ ms, while for the longer horizon times $t_h > 250$ ms the assumption of the linearity of the system produces a large error in the reachable space estimation, producing the polytope $\mathcal{P}_x$ with volume several times higher than the real reachable space of the robot (shown by the metric m_3), most of which is not reachable by the robot at all (showed by the metric m_2). Finally, the results show that the polytope $\mathcal{P}_x$, even though a coarse approximation, has better performance across all the defined metrics then the $\mathcal{C}_x$ based on the constant cartesian limits, at the same time having smaller average volume for the all the horizon times. These results remind, once again, that robot's cartesian capacity is not constant and by modeling their evolution, even in a coarse manner, a substantial improvement in the estimation accuracy can be reached.

5 Discussion

Approximating the reachable space of the robot using convex polytopes $\mathcal{P}_x$ is an efficient and relatively fast way of gaining information about the potential robot's position for the horizon of interest. Even though this approximation approach is neither an under or an over approximation of the true reachable space of the robot, as shown in the numerical analysis, for shorter horizon times $t_h \leq 250$ ms this metric is reasonably precise. Due to its efficiency, easy and intuitive integration of different environmental constraints, possibility to consider the robot's link geometry and the fact that its output has the form of a convex set of inequality constraints (convex polytope $\mathcal{P}_x$), it has a potential to be a useful tool for robot performance and safety analysis. However there are several important limitations of this approach.

5.1 Main Limitations

First of all, the reachable space of a robot is highly non-convex and non-linear. As shown in the numerical analysis Sect. 4, approximating this space with convex polytopes is relatively precise only for small horizon times $t_h \leq 250$ ms.

Secondly, robot dynamics is highly non-linear and time variant as well. When calculating the reachable space polytope $\mathcal{P}_x$, as formulated in Sect. 2, the robot model is linearised in the robot's joint configuration $\boldsymbol{q}_k$ and velocity $\dot{\boldsymbol{q}}_k$ at the beginning of the horizon and this linearised model is considered constant during the horizon length. The consideration of constant robot model and its linearity, is only valid for small robot displacements around the current robot state, which is again a strong assumption and only valid for short horizon times.

Finally, the polytope formulation from the Sect. 2, considers only constant joint torques $\boldsymbol{\tau}$ applied to the robot during the whole horizon length t_h. The polytope $\mathcal{P}_x$ calculates the joint torque vectors $\boldsymbol{\tau}$ that will produce the largest possible displacements in cartesian space given the robot and environment constraints. This calculation is based on the robot's linearised model at the beginning of the horizon, which in term, once again, limits the precision of the approximation to the small displacements around the initial robot state, preventing longer horizon lengths t_h, as confirmed in the analysis Sect. 4.

5.2 Potential Applications

One promising field of applications for the polytope based reachable space metric is the human-robot interaction domain. The polytope is essentially a triangulated mesh and can be easily visualised, which could be an easy and intuitive way to give insight to the human operators about robot's real-time capabilities. It has a potential to be used for the haptic control applications, as human operators might not be always aware of how close the robot is to the limitation of its capabilities or what is the possible space the robot can be in while teleoperating. Furthermore, this metric could be potentially used to to increase human operators awareness about the robot's state during the co-manipulation by visualising the polytope $\mathcal{P}_x$ in real-time and even potentially increase the interaction safety.

The numerical analysis has shown that for shorter horizon times $t_h \leq 250$ ms the volume of the polytope $\mathcal{P}_x$ is comparable to the volume of the simulated reached space R_2, as shown on the Fig. 10. The volume of the $\mathcal{P}_x$ could potentially be used as a performance metric for robotic workspace analysis and for applications such as robot design.

Finally, as the polytope $\mathcal{P}_x$ can be represented as a set of convex inequalities $H\boldsymbol{x} \leq \boldsymbol{d}$, this metric could potentially be used as a set of constraints for robot control. For example in case of the Model Predictive Control (MPC) where, in many cases, robot's dynamics is already considered linear and constant during given horizon time, which corresponds well with the assumptions of the polytope $\mathcal{P}_x$ formulation.

6 Conclusion

In this paper a convex polytope based robot's reachable space approximation metric is introduced. The metric predicts the robot's reachable space for a given time horizon based on the robot's actuation capabilities and its kinematic constraints. The proposed metric can intuitively integrate the environmental constraints as well as the robot's link geometry. With the execution time of around 50ms, for standard 7dof collaborative robot, it has a potential to be used for real-time applications.

However, the metric relies on a linearised robot model, which limits its use to shorter horizon times. The analysis has shown that the accuracy of the approximation decreases significantly for horizon times longer than 250 ms.

References

1. Althoff, M., Krogh, B.H.: Reachability analysis of nonlinear differential-algebraic systems. IEEE Trans. Autom. Control **59**(2), 371–383 (2014). https://doi.org/10.1109/TAC.2013.2285751
2. Bowling, A., Khatib, O.: The dynamic capability equations: a new tool for analyzing robotic manipulator performance. IEEE Trans. Robot. **21**(1), 115–123 (2005). https://doi.org/10.1109/TRO.2004.837243
3. Corke, P., Haviland, J.: Not your grandmother's toolbox-the robotics toolbox reinvented for python. In: 2021 IEEE International Conference on Robotics and Automation (ICRA). pp. 11357–11363. IEEE (2021)
4. Hamad, M., Mansfeld, N., Abdolshah, S., Haddadin, S.: The role of robot payload in the safety map framework. In: 2019 IEEE/RSJ International Conference on Intelligent Robots and Systems (IROS). pp. 195–200 (2019). https://doi.org/10.1109/IROS40897.2019.8968022
5. Long, P., Padir, T.: Evaluating robot manipulability in constrained environments by velocity polytope reduction. In: 2018 IEEE-RAS 18th International Conference on Humanoid Robots (Humanoids). pp. 1–9 (2018). https://doi.org/10.1109/HUMANOIDS.2018.8624962
6. Pereira, A., Althoff, M.: Calculating human reachable occupancy for guaranteed collision-free planning. In: 2017 IEEE/RSJ International Conference on Intelligent Robots and Systems (IROS). pp. 4473–4480 (2017). https://doi.org/10.1109/IROS.2017.8206314
7. Schepp, S.R., Thumm, J., Liu, S.B., Althoff, M.: Sara: a tool for safe human-robot coexistence and collaboration through reachability analysis. In: 2022 International Conference on Robotics and Automation (ICRA). pp. 4312–4317 (2022). https://doi.org/10.1109/ICRA46639.2022.9811952
8. Skuric, A.: A real-time capable task-space capacity calculation python package, https://pypi.org/project/pycapacity
9. Skuric, A., Padois, V., Daney, D.: On-line force capability evaluation based on efficient polytope vertex search. In: 2021 IEEE International Conference on Robotics and Automation (ICRA). pp. 1700–1706 (2021). https://doi.org/10.1109/ICRA48506.2021.9562050
10. Skuric, A., Padois, V., Rezzoug, N., Daney, D.: On-line feasible wrench polytope evaluation based on human musculoskeletal models: an iterative convex hull method. IEEE Robot. Autom. Lett. **7**(2), 5206–5213 (2022). https://doi.org/10.1109/LRA.2022.3155374

Complexity and Similarity Metrics for Unsupervised Identification of Programming Methods for Robot Cooking Tasks

Stefan Ilic(✉) and Josie Hughes

CREATE Lab, EPFL, 1015 Lausanne, Switzerland
stefan.ilic@epfl.ch,
josie.hughes@epfl.ch
https://www.epfl.ch/labs/create

Abstract. Using robotics to automate complex manipulation tasks such as cooking could offer significant advantages. Robotics could offer improved throughput, provide assistance for those that lack the required mobility or allow for creative robotic chefs. Generating a robotic system that utilizes vision systems, planning and dexterous manipulation would offer a *generalized solution* to kitchen tasks. However, there can be many components of the overall task that do not require this generalization. In addition, there remain many tasks for which it is quicker and preferable for a human to complete some sub-components of the task. By computing a complexity and similarity metrics from multiple examples of a human teaching the required motion we propose using unsupervised clustering to determine the approach method of implementation. This includes hard coding, use of vision, play back or human completion. We apply this analysis to the task of cooking pancakes in a kitchen environment, analyzing the trajectories, determining the method of completing the task using unsupervised methods and implementing the proposed approach.

Keywords: Human–robot collaboration · Robot cooking · Human-in-the-loop · Robot programming

1 Introduction

Robotic automation of manipulation tasks has potential for significant economic and social benefits in task domains including laboratories [4], factories, agriculture [6], kitchen environments [11] and even the home [3]. Manipulation tasks can vary significantly in complexity; some may be highly repetitious, some may depend on the location of objects in the environment, and others may require highly complex trajectories. This variation can mean that whilst some tasks can be efficiently and easily performed by robots, others may be easy for a human

P. Borja et al. (Eds.): HFR 2022, 26, pp. 61–73, 2023.
https://doi.org/10.1007/978-3-031-22731-8_5

but extremely challenging for robots [12]. Equally, the optimal means of programming or automating tasks may vary depending on if the task is highly repetitious, or the task requires visual feedback from the environment. To identify the best method of implementing and sharing a task between robot and human we require methods to analyze the task and label them with the optimal method. This approach brings together concepts within the research domains on human–robot collaboration [1] and intuitive or easy human programming of robots [14,18].

In this paper we develop approaches to identify the appropriate means of sharing a task between robots and humans, and the appropriate programming method for these tasks. We focus on kitchen task which is performed in a semi-structured environment where tasks show a range of characteristics with varying levels of complexity for robots and humans. Specifically we explore how to share tasks between robots and humans for effective automation, and also how different aspects of the robot's task should be programmed to minimize human input time (including programming and teaching) whilst maintaining task flexibility [8]. The area of designing human–robot sharing has been approached in a number of ways. Previous work has considered using complexity based metrics [16] and applying this to industrial application scenarios where they consider task sharing between robots and humans. In this work they contrast cycle times, adaptability and safety to determine the 'automation potential' of specific tasks [15]. In addition, task sharing or balancing between humans and robots has been considered for heavy weight tasks [7], and also human–robot cells have been optimized using AI based methods [9]. Whilst these methods have shown promising results for these constrained industrial environments, they mostly consider optimizing the time of the robot and human to complete the task. However, they do not consider the cost or time taken for programming the robot, and identification of the optimal means of programming the robot.

We propose taking a number of recordings of a human moving a robot arm to perform a task to gain a blue-print of description of the task. This is simple for the human to perform, but contains significant information about the task. By analyzing the recorded trajectories, metrics which can describe features of the sub-tasks can be computed. The metrics we introduce focus on identifying the complexity of robot programming required for a sub-task, and also the similarity between recordings to provide an indication in the variability of the sub-tasks. From these metrics, we look to minimize the programming time and also the involvement of the human in the automation of the process by categorizing tasks to be solved with four different means. This includes: hard-coding, playback from human recording, using computer vision to determine object location and also the human performing the task. These two metrics have been chosen to serve as features that cluster different sub-tasks within the feature space. As illustrated in Fig. 1, we propose that the modality of a task can be determined by identifying the different clusters within this feature space. This clustering can be performed using unsupervised k-means to provide a generalizable method which seeks to optimally cluster groups of tasks with similar characteristics.

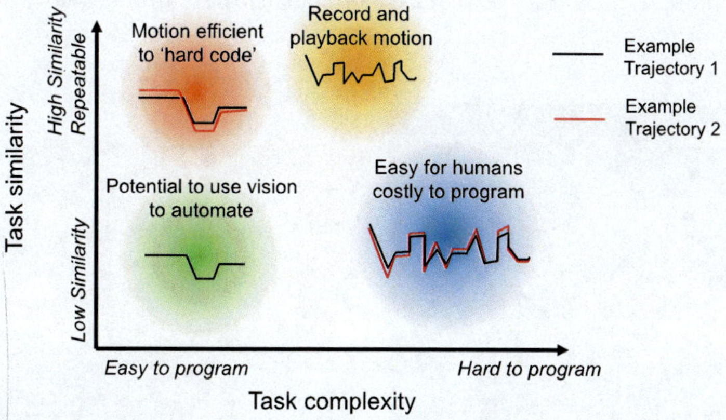

Fig. 1. The four different metrics we propose to enable different tasks to be assigned to different action types for performing the task.

To explore this approach we consider a kitchen task; that of making, cooking and adding toppings to pancakes. This task has some sub-tasks that are highly repetitious, and others that are dependent on the location of items in the workspace. In addition, some tasks are very simple to execute, for example picking up ingredients, where other tasks such as flipping the pancakes are complex and require complex trajectories. By analyzing the sub-tasks of the process, we identify the optimal way to perform the task using both the capabilities of the human and robot, and demonstrate the effectiveness of this approach. In the remainder of this paper we first introduce the methods before presenting the results of the implementation and analysis and concluding with a discussion of these results.

2 Methods

2.1 Problem Statement

To explore the role of task sharing between robot and humans we consider the case-study of the kitchen. This is a complex environment for robots to operate in, and also one where humans excel [17]. It is also an environment which humans are prepared to share with robots, as some tasks can be enjoyable and rewarding, yet there are also tasks that are frustrating, slow and repetitive [10]. The task we consider is that of cooking and making pancakes.

Using a 6 degree of freedom UR5 robot arm with a custom gripper, the motion path required to perform the task is recorded by a human moving the end-effector through the cooking task. This provides a quick way for the human to 'define' the task. Using these trajectories each sub-task can then be analyzed and the appropriate means for the task to be automated selected. This should minimize human time that is required for both programming and performing the task.

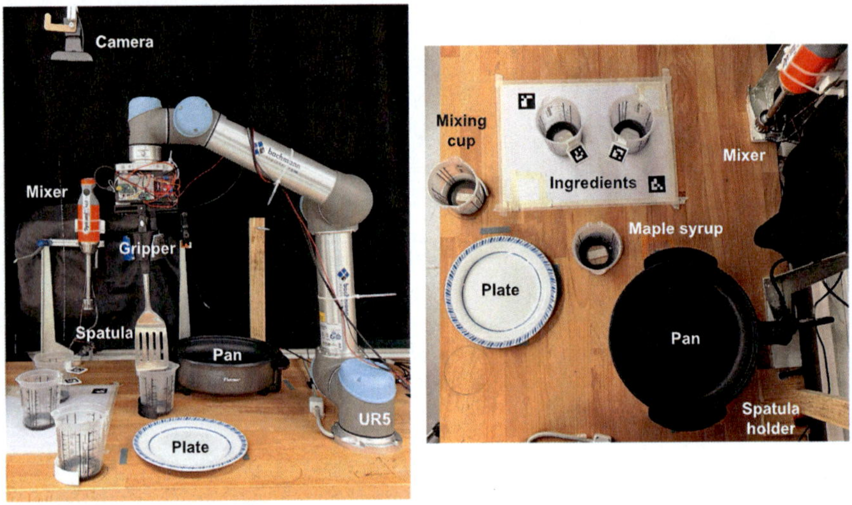

Fig. 2. Hardware setup used for the kitchen environment showing the robot with custom gripper, pan, spatula, mixer and ingredient cups.

2.2 Design of Task Completion and Sharing

From the human demonstration of the task, a method for enabling each subtask must be determined. These methods that we consider for the different subelements of the task include four different **action types**:

- **A1: Robot Motion Hard coded**. For motions which are highly repetitive and have only a few waypoints in their trajectory; it is quick, easy and repeatable method of robot programming but provides no flexibility or adaption to the environment.
- **A2: Robot Motion Recorded and played back**. For motions which are highly repetitious but challenging to program, the robot should be programmed by recording the motion and playing it back.
- **A3: Robot Motion with vision**. For tasks which have only a few waypoints and low complexity but have some dependency on objects in the environment (e.g., orientation or placement of the items), the robot should be programmed but the waypoints determined using visual feedback.
- **A4: Human Task Completion**. For tasks which are quick and easy for a human to complete, yet require significant robot programming and adaptation of the environment or location of objects (e.g., high variability between repetitions), it may be optimal for the human to perform these tasks.

These action types are denoted by $A = [A_1, ..., A_4]$. From the recorded motion of the robot we want to determine the appropriate action type to perform each sub-task. We propose two different metrics. First a *similarity metric* to identify sub-tasks that are repeated with high similarity regardless of changing location of objects in the environment. Second, a *complexity of programming* metric through which we try to identify number of waypoints required for robot programming and thus approximation of the time or 'cost' of programming. Whilst there have been many metrics developed for analyzing the similarity or complexity of robot task trajectories [5] or poses [13], we propose developing simple metrics which directly correspond to programming complexity and similarity.

We define a task T, for example cooking pancakes, as n sub-tasks where sub-tasks are manually determined and correspond to distinct phases, for example, mixing the batter or flipping the pancake, thus:

$$T_i = [T_{1,i}, T_{2,i}, T_{3,i}, ..., T_{n,i}] \tag{1}$$

where i denotes a specific recording of the task. Each T_i represents a vector storing the 6 degrees of freedom of motion in the cartesian and joint space for the time of the sub task t_{ni}. For each sub task n, the two metrics should be determined from which we can assign an action type A, that minimizes the cost of programming (C_p) and the cost of human input time (C_h).

2.3 Similarity Metrics

For each sub-task, we find the similarity between two recordings of the same task, i.e. between T_{in} and T_{jn} where i and j denote two repeats. To achieve this we first perform dynamic time-warping [2] to align the signals and account for differences in the recording speed when performed by the human. From this, we then calculate euclidean distance between the aligned time series to provide a measure of similarity. This is performed on the six channels of the joint space recordings. The average euclidean distance is then found across the m channels that experience a change of greater than a noise threshold which is 0.1 rad:

$$S_n = \frac{1}{m} \sum \|DTW(T_{in}) - T_{jn}\| \tag{2}$$

From this metric we can identify sub-tasks which are highly similar and thus well suited to hard-coding or playback, opposed to those that are dis-similar and thus are more likely to require automation with vision to be adaptive to changing environment, or, require a human to perform the sub-task.

2.4 Complexity Metric

The complexity metric is used to assess the complexity of manually programming of the sub-task. For this, we consider the total number of way-points or states in the trajectory of the motion. Each of these motions or way points must be

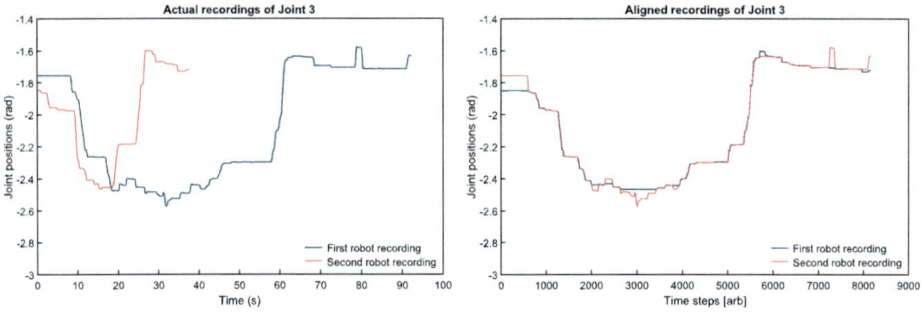

Fig. 3. Example of the similarity metric applied to the trajectory of a single joint.

considered when programming, thus the greater the number of these way points, the higher the complexity of robot programming. We approximate the number of way points by identifying significant changes in the cartesian position recordings, and counting the occurrence, removing any repeats within a given time window (e.g., within a 0.5 s period).

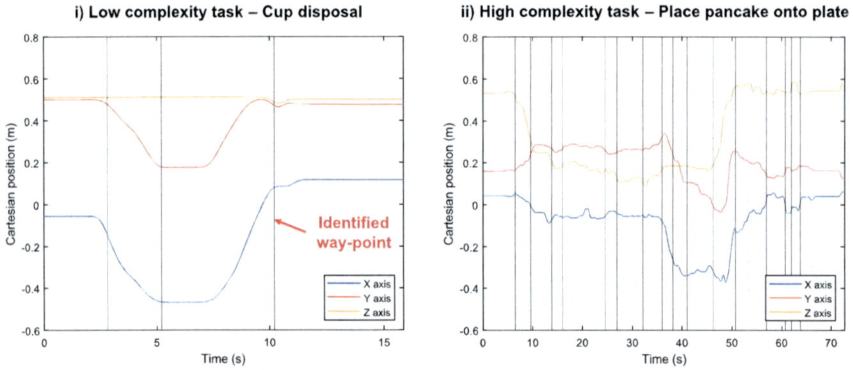

Fig. 4. Example of complexity metrics applied to different tasks which have different numbers of observable way-points.

This metric provides a estimation of the number of way-points that would be required when manually programming. An example of the determination of these thresholds is shown in Fig. 4 for two different motions with different numbers of way-points. This illustrates that for the higher complexity tasks a far greater number of way-points are identified.

2.5 Clustering in the Feature Space

The complexity and similarity metrics have been chosen to be orthogonal, and to allow tasks to be structured for clustering. By identifying the metrics for all sub-tasks these will then be clustered based on these two metrics. This will be performed using k-means clustering based upon the squared euclidean distance and where the number of clusters is set to 4, reflecting the four different action types. The metrics will first be normalized by the maximum recorded value to approximately balance the two metrics. The goal is that clusters will geometrically reflect the approximate distribution shown in Fig. 1 and thus can be labeled appropriately.

2.6 Experimental Setup and Hardware

The robot setup (as seen in Fig. 2) shows the 'kitchen' setup which includes a UR5 with a motorized custom gripper designed for interfacing with cups. By setting the robot arm to 'free-drive mode', it can be moved by the human and the positions recorded. The human also labels the start and end of each task using a keyboard during this process.

To control the robot arm we have the three action types A. For A_1, play back mode, the joints are recorded, smoothed and can then be played back. For the second action type, A_2, hard-coding, the key waypoints are identified from the recorded trajectories and these key joint locations are manually programmed. Finally, for action type A_3 there is the ability to use vision markers. In these tasks, April tags are placed on the objects used in the tasks. Using a camera placed above the work space of the robot, the x-y location of these markers in the work space can be identified and transformed to robot coordinates by calibrating the robot with respect to pixels in camera space. The robot waypoints can then be set, with the variable waypoints set to be a function of the location of the visual markers. Finally, for tasks determined to be performed by humans the robot is simply paused.

3 Results

In this section we demonstrate the analysis performed on recordings, the decisions made using these metrics and also the results from the automation of the process. For the cooking process we include making pancake batter, mixing, pouring, flipping and finally adding toppings. We record two examples of the process where human initially places the ingredients and plates in the workspace, as typical for a kitchen environment. From these we determine the optimal way to complete the task with robot and humans.

3.1 Analysis of Task Recordings

An example recording is shown in Fig. 5 for all the sub-tasks. Visually clear characteristics for the different sections can be observed and thus explained using similarity and complexity metrics, as shown on Fig. 6.

Highest level of complexity is identified in sub-tasks such as flipping of pancakes which have complex trajectories with 15+ different way points and several

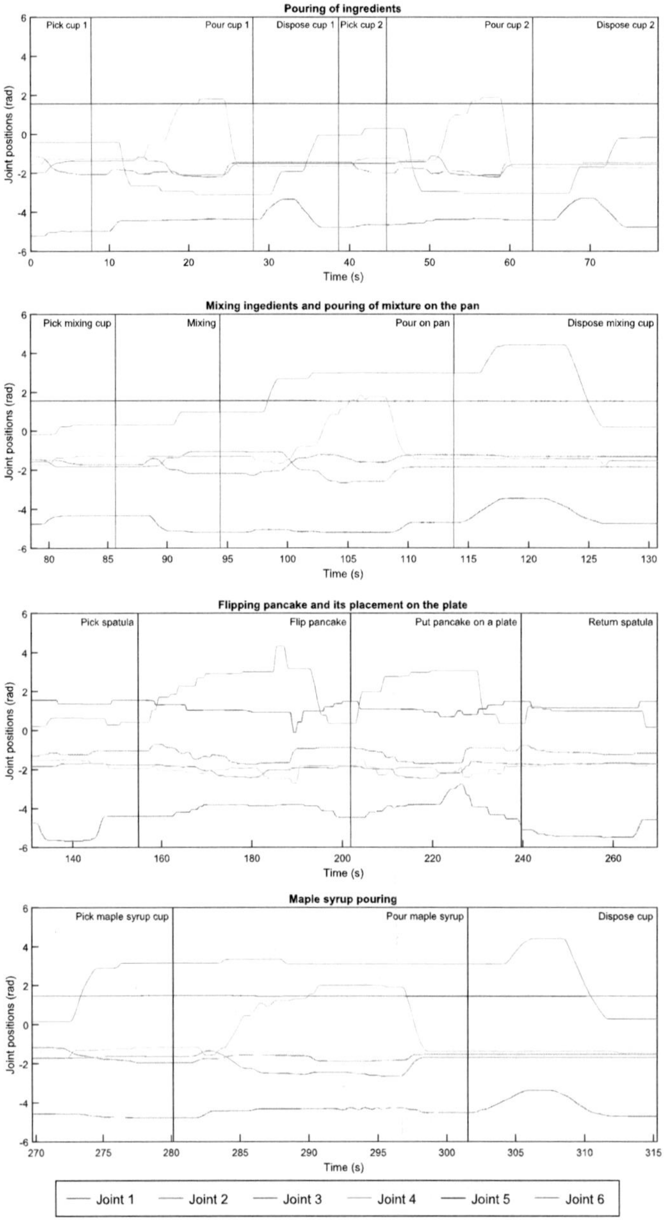

Fig. 5. Step by step visual illustration of experimental process represented alongside controllable parameters.

different robot poses. Pouring tasks require up to 5 way points with only two changes in robot pose and are thus easier to perform. In contrast, tasks like mixing and picking have simple trajectories with no changes in robot pose. Tasks which require significant robot displacement (cup disposal) due to large changes in position in cartesian space appear to be complex with 3–5 way points, but practically can be quickly hard-coded.

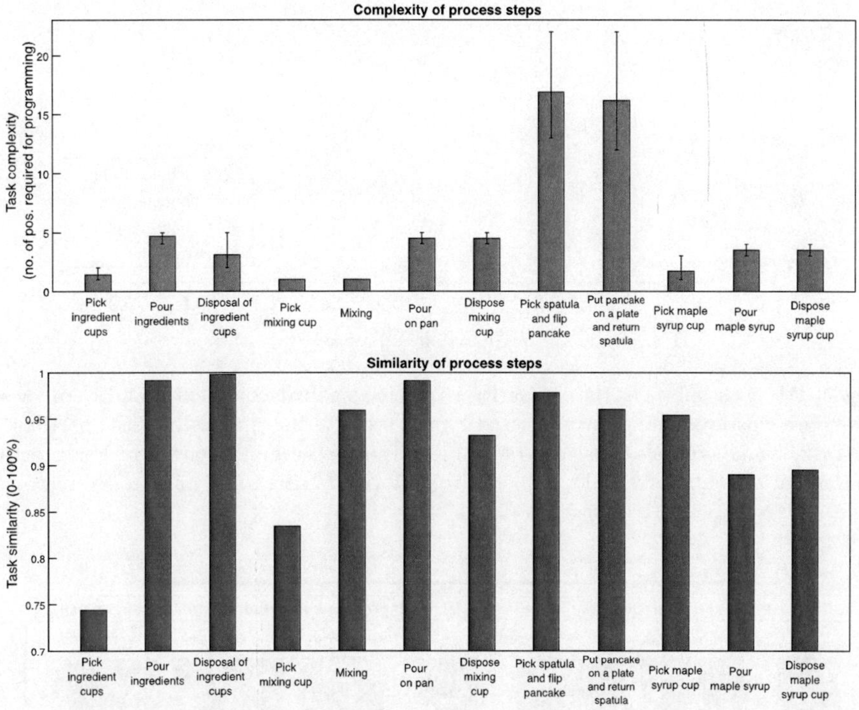

Fig. 6. Similarity and complexity metrics for the different sub-tasks.

From similarity aspect, most repetitious tasks are performed with reference to the fixed position (e.g., start pouring position) and have 95%+ similarity. Lowest similarity is observed for picking of ingredient cups since cup position and orientation are not constrained, with highest variability observed in cup orientations rather than position. Most complex sub-tasks such as flipping of pancakes experience 96%+ similarity since they require specific robot motions within the semi-structured environment. Sub-tasks which incorporate cyclical motion (maple syrup pouring) exhibit lower similarity (89%) since they cannot be easily repeated by humans during motion recording.

Combining these two metrics we normalize the magnitude by the maximum value recorded and perform unsupervised k-means clustering of the different tasks (Fig. 7). The clustering shows a similar distribution to that proposed in

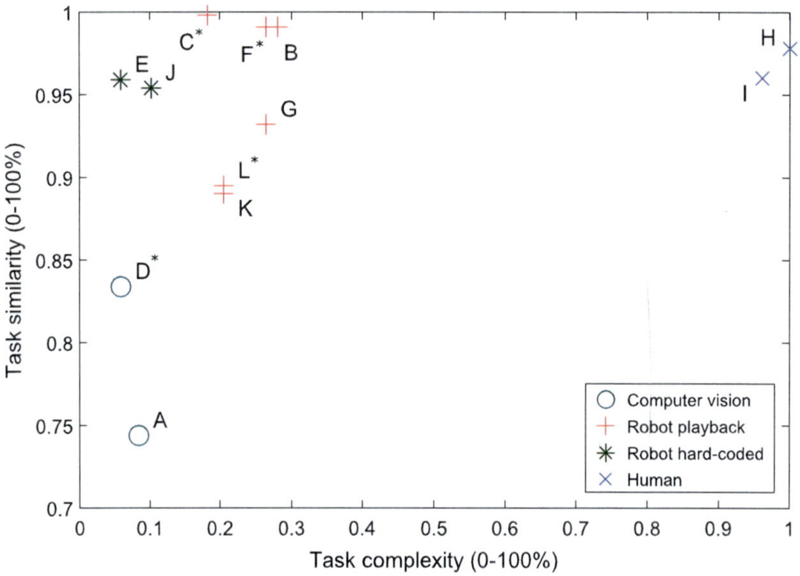

Fig. 7. For each sub-task the similarity is plotted against complexity and then these points are clustered using unsupervised k-means clustering. Labels indicate particular sub-task, while symbols illustrate recommended programming approach. For process demonstration, sub-tasks C, D, F and L are performed using robot hard-coded method.

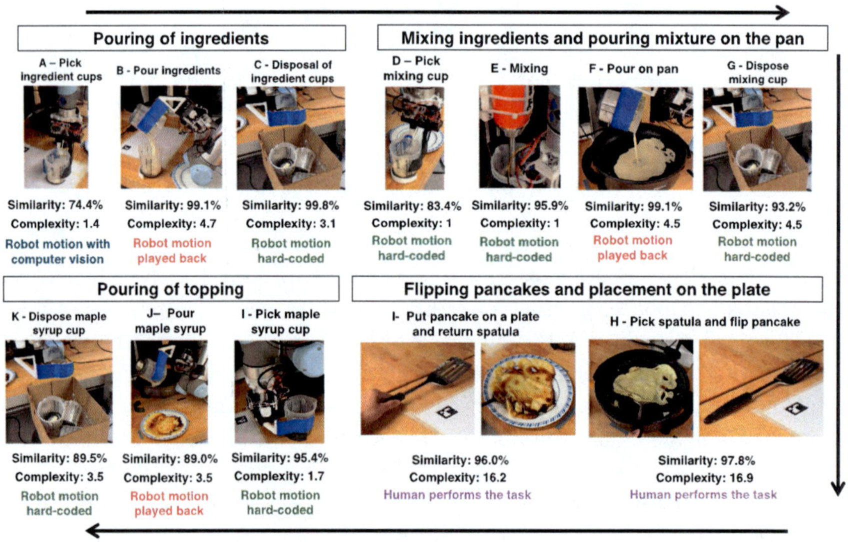

Fig. 8. Step by step visualization of automated pancake cooking process represented with similarity and complexity metrics alongside their method for task performing

Fig. 1 allowing the different tasks to labelled through this un-supervised method. The tasks in the vision cluster have low number of way points and low similarity, in this case up to two way points and up to 85% similarity. Hard-coded tasks cover a similar range of complexity but have a similarity higher than 95%. In contrast, robot playback is used for more complex tasks between 2–5 way points and similarity higher than 85%, while humans excel at tasks with highest complexity (10+ way points) and both high and low similarity tasks.

The clustering of the computer vision, hard-coded and playback clusters is much tighter than the clustering of sub-tasks for human action. This is due to the significantly higher complexity and range of complexity of the sub-tasks performed by a human.

3.2 Automation of the Process

Using the above metrics the process can be automated[1] using the methods identified. The steps of the pancake cooking process are shown in Fig. 8, highlighting the capabilities of the robot and human to perform these tasks. In two cases a change from the approach proposed by the clustering was made. This was for the cup disposal and picking of mixing cup sub-tasks. Whilst the similarity of 90%+ makes cup disposal suitable for both hard-coding and playback, the large change in robot position significantly increases complexity of this task, resulting in the clustering in the robot playback ground. However, hard-coding was ultimately chosen over robot playback since this particular task is repeated four times and shares multiple way points (e.g., disposal position). For the picking of the mixing cups which was labeled as a vision task, due to limitations of the field of view of the camera, this was instead performed with robot playback. These two cases reflect how the clustering offers useful insights into the method for each sub-task, however, there can still be practical considerations which may also affect the eventual task selected.

4 Discussion and Conclusion

In this paper we present metrics for analyzing human-taught robot trajectories to enable the appropriate approach for human–robot collaboration and for programming the robot. The two main metrics, complexity and similarity, enable us to segment the sub-tasks into four-different groups for the four different action types considered. Further work is required to better explore and optimize determining action type based on these metrics. Although developing methods to fully automate the programming and performance of tasks from human demonstration is the end goal for robotic automation, this is highly challenging. Our approach provides a mechanism for practically automating tasks and leveraging humans to perform tasks which robots would struggle.

[1] Video of the automated process is available on the link: https://bit.ly/3QutTG8.

The metrics and approach developed were evaluated on a specific task, the making of pancakes. Extending this and applying to other tasks would better explore the applicability of the approach. In addition, extending the different methods of robot programming to increasing methods that offer improved autonomy and robustness would increase the utility and potential of this approach.

Acknowledgements. This project was partially funded by the European Union's Horizon 2020 research and innovation programme under the Marie Skłodowska Curie grant agreement N° 945363.

References

1. Ajoudani, A., Zanchettin, A.M., Ivaldi, S., Albu-Schäffer, A., Kosuge, K., Khatib, O.: Progress and prospects of the human–robot collaboration. Auton. Robots **42**(5), 957–975 (2018)
2. Berndt, D.J., Clifford, J.: Using dynamic time warping to find patterns in time series. In: *KDD workshop*, vol. 10, pp. 359–370. Seattle, WA, USA (1994)
3. Billard, A., Kragic, D.: Trends and challenges in robot manipulation. Science, **364**(6446):eaat8414 (2019)
4. Bogue, R.: Robots in the laboratory: a review of applications. Indus. Robot: Int. J. (2012)
5. Chiang, C.J., Chirikjian, G.S.: Modular robot motion planning using similarity metrics. Auton. Robots **10**(1), 91–106 (2001)
6. Chua, P.Y., Ilschner, T., Caldwell, D.G.: Robotic manipulation of food products-a review. Indus. Robot: Int. J. (2003)
7. Dianatfar, Morteza, Latokartano, Jyrki, Lanz, Minna: Task balancing between human and robot in mid-heavy assembly tasks. Proc. CIRP **81**, 157–161 (2019)
8. Dragan, A.D., Bauman, S., Forlizzi, J., Srinivasa, S.S.: Effects of robot motion on human-robot collaboration. In *2015 10th ACM/IEEE International Conference on Human-Robot Interaction (HRI)*, pp. 51–58. IEEE (2015)
9. Evangelou, George, Dimitropoulos, Nikos, Michalos, George, Makris, Sotiris: An approach for task and action planning in human–robot collaborative cells using AI. Proc. Cirp **97**, 476–481 (2021)
10. Fonseca, E., Oliveira, I., Lobo, J., Mota, T., Martins, J., Au-Yong-Oliveira, M.: Kitchen robots: the importance and impact of technology on people's quality of life. In *World Conference on Information Systems and Technologies*, pp. 186–197. Springer (2019)
11. Junge, K., Hughes, J., Thuruthel, T.G., Iida, F.: Improving robotic cooking using batch Bayesian optimization. IEEE Robot. Autom. Lett. **5**(2), 760–765 (2020)
12. Kragic, D., Gustafson, J., Karaoguz, H., Jensfelt, P., Krug, R.: Interactive, collaborative robots: challenges and opportunities. In *IJCAI*, pp. 18–25 (2018)
13. Lei, Jie, Song, Mingli, Li, Ze-Nian., Chen, Chun: Whole-body humanoid robot imitation with pose similarity evaluation. Signal Process. **108**, 136–146 (2015)
14. Makris, Sotiris, Tsarouchi, Panagiota, Surdilovic, Dragoljub, Krüger, Jörg.: Intuitive dual arm robot programming for assembly operations. CIRP Ann. **63**(1), 13–16 (2014)
15. Ali Ahmad Malik and Arne Bilberg: Collaborative robots in assembly: a practical approach for tasks distribution. Proc. CIRP **81**, 665–670 (2019)

16. Malik, A.A., Bilberg, A.: Complexity-based task allocation in human–robot collaborative assembly. Indus. Robot: Int. J. Robot. Res. Appl. (2019)
17. Rusu, R.B., Gerkey, B., Beetz, M.: Robots in the kitchen: exploiting ubiquitous sensing and actuation. Robot. Auton. Syst. **56**(10), 844–856 (2008)
18. Villani, Valeria, Pini, Fabio, Leali, Francesco, Secchi, Cristian: Survey on human–robot collaboration in industrial settings: safety, intuitive interfaces and applications. Mechatronics **55**, 248–266 (2018)

Damping Design for Robot Manipulators

Tomás Coleman, Giovanni Franzese, and Pablo Borja(✉)

Department of Cognitive Robotics, Delft University of Technology, Delft, The Netherlands
{t.coleman,g.franzese}@tudelft.nl, pablo.borjarosales@plymouth.ac.uk

Abstract. This paper studies the tuning process of controllers for fully actuated manipulators. To this end, we propose a methodology to design the desired damping matrix—alternatively, the derivative gain of a PD controller—of the closed-loop system such that n second-order systems can approximate its behavior with a prescribed damping coefficient, where n denotes the degrees of freedom of the system. The proposed approach is based on the linearization of the closed-loop system around the desired configuration and is suitable for different control approaches, such as PD control plus gravity compensation, impedance control, and passivity-based control. Furthermore, we extensively analyze simulations and experimental results in a cobot.

Keywords: Damping coefficient · Performance · PD control · Impedance control

1 Introduction

Suppressing oscillations while prescribing desired stiffness values is of utmost importance in many modern robotic applications. For instance, in industrial applications, oscillations diminish the performance of robots that execute tasks involving high-speed motions [2]. Another example occurs in applications involving human-robot interaction, where robots often must be compliant enough to guarantee the safety of humans and correctly execute the task at hand [8]. Hence, choosing control gains that ensure an appropriate performance is a crucial step in designing controllers for robotic manipulators. Nevertheless, such gains are customarily selected through methods that are simple but not intuitive nor supported by an adequate analysis of the system's behavior. This ambiguity in the selection of the gains may lead to unexpected and undesirable behavior of the manipulator, which is especially important in situations where oscillations and overshoot jeopardize the success of the task. For instance, in applications involving human-robot interaction, overshoot might mean the difference between contact or no contact with the human or imply physical harm.

T. Coleman and G. Franzese: Equal contribution authorship.

P. Borja et al. (Eds.): HFR 2022, 26, pp. 74–89, 2023.
https://doi.org/10.1007/978-3-031-22731-8_6

In this paper, we propose a constructive methodology to design the damping matrix of the closed-loop system. To this end, we consider a control architecture that stabilizes the robot at the desired configuration. Then, we consider the port-Hamiltonian representation [10] of the closed-loop system and linearize it at the desired configuration. In particular, the proposed methodology is based on the simultaneous diagonalization of the inertia matrix and Hessian of the desired potential energy, yielding n second-order systems that are decoupled, where n equals the degrees of freedom of the system. Then, we design the damping coefficient for each decoupled system to prescribe the desired performance.

The proposed approach is similar to the one presented in [1]. However, two main differences with respect to the mentioned reference are:

(i) The desired potential energy is not required to be quadratic on the configuration error.
(ii) The forces Coriolis and centrifugal terms are not canceled.

The proposed approach is suitable for PD controllers, impedance control, and passivity-based controllers.

1.1 Notation

The symbol I_n denotes the $n \times n$ identity matrix; $\mathbf{0}$ is a vector or matrix whose entries are zeros; $\mathbb{1}$ is a vector whose elements equal one; $\partial f(x)/\partial x = \left[\partial f(x)/\partial x_1, \ldots, \partial f(x)/\partial x_n\right]^\top$; and $\mathtt{diag}(\cdot)$ denotes a diagonal matrix. When clear from the context, we omit the arguments of functions.

2 Proposed Methodology

2.1 Stabilization of Robotic Manipulators

A fully actuated and unconstrained robotic manipulator can be represented as follows

$$\begin{bmatrix} \dot{q} \\ \dot{p} \end{bmatrix} = \begin{bmatrix} \mathbf{0} & I_n \\ -I_n & \mathbf{0} \end{bmatrix} \begin{bmatrix} \dfrac{\partial H(q,p)}{\partial q} \\ \dfrac{\partial H(q,p)}{\partial p} \end{bmatrix} + \begin{bmatrix} \mathbf{0} \\ I_n \end{bmatrix} \tau; \quad H(q,p) := \frac{1}{2} p^\top M^{-1}(q) p + V(q), \quad (1)$$

where $q \in \mathbb{R}^n$ represents the joints configuration and $p \in \mathbb{R}^n$ the corresponding momenta; $\tau \in \mathbb{R}^m$ denotes the input vector (torques), $H : \mathbb{R}^n \times \mathbb{R}^n \to \mathbb{R}$ is the Hamiltonian (total energy) of the system; $M : \mathbb{R}^n \to \mathbb{R}^{n \times n}$ is the inertia matrix, which is positive definite; and $V : \mathbb{R}^n \to \mathbb{R}$ is the potential energy.

Theorem 1. *Consider the desired configuration $q_\star \in \mathbb{R}^n$, the desired potential energy $V_\mathrm{d} : \mathbb{R}^n \to \mathbb{R}$, and a positive definite matrix $D_\mathrm{d} \in \mathbb{R}^{n \times n}$. The control law*

$$\tau = \frac{\partial V(q)}{\partial q} - \frac{\partial V_\mathrm{d}(q)}{\partial q} - D_\mathrm{d} \dot{q} \quad (2)$$

stabilizes the system (1) *at the point* $(q_\star, \mathbf{0})$, *for any differentiable* $V_\mathsf{d}(q)$ *such that*

$$V_\mathsf{d}(q_\star) = 0 \quad \text{and} \quad V_\mathsf{d}(q) > 0,\ \forall\ q \in \mathbb{R}^n - \{q_\star\}. \tag{3}$$

Proof. From (3), we conclude that

$$H_\mathsf{d}(q_\star, \mathbf{0}) = 0 \quad \text{and} \quad H_\mathsf{d}(q,p) > 0,\ \forall\ (q,p) \in \mathbb{R}^n \times \mathbb{R}^n - \{(q_\star, \mathbf{0})\}. \tag{4}$$

Moreover, the closed-loop system takes the form

$$\begin{bmatrix} \dot{q} \\ \dot{p} \end{bmatrix} = \begin{bmatrix} \mathbf{0} & I_n \\ -I_n & -D_\mathsf{d} \end{bmatrix} \begin{bmatrix} \dfrac{\partial H_\mathsf{d}(q,p)}{\partial q} \\ \dfrac{\partial H_\mathsf{d}(q,p)}{\partial p} \end{bmatrix}; \quad H_\mathsf{d}(q,p) := \frac{1}{2} p^\top M^{-1}(q) p + V_\mathsf{d}(q). \tag{5}$$

Hence, the time derivative of $H_\mathsf{d}(q,p)$ is given by

$$\dot{H}_\mathsf{d} = \begin{bmatrix} \dfrac{\partial H_\mathsf{d}}{\partial q} \\ \dfrac{\partial H_\mathsf{d}}{\partial p} \end{bmatrix}^\top \begin{bmatrix} \mathbf{0} & I_n \\ -I_n & -D_\mathsf{d} \end{bmatrix} \begin{bmatrix} \dfrac{\partial H_\mathsf{d}}{\partial q} \\ \dfrac{\partial H_\mathsf{d}}{\partial p} \end{bmatrix} = -\left(\frac{\partial H_\mathsf{d}}{\partial p} \right)^\top D_\mathsf{d} \frac{\partial H_\mathsf{d}}{\partial p} \leq 0. \tag{6}$$

Accordingly, (4), together with (6), implies that $H_\mathsf{d}(q,p)$ qualifies as a Lyapunov function to prove the stability of the equilibrium point $(q_\star, \mathbf{0})$. For further details on Lyapunov stability, we refer the reader to [6]. Furthermore, to prove that the mentioned equilibrium is asymptotically stable—i.e., the robot converges to the desired configuration—note that

$$\dot{H}_\mathsf{d} = 0 \implies \frac{\partial H_\mathsf{d}}{\partial p} = M^{-1} p = \mathbf{0} \implies p = \mathbf{0} \implies \dot{p} = -\frac{\partial V_\mathsf{d}}{\partial q} = \mathbf{0} \implies q = q_\star.$$

Because $M^{-1}p = \dot{q}$, we conclude that $\dot{H}_\mathsf{d}$ equal to zero implies (q,p) equal to $(q_\star, \mathbf{0})$. Therefore, it follows from LaSalle's invariance principle [6] that the trajectories converge to the desired equilibrium $(q_\star, \mathbf{0})$.

□□□

Remark 1. A common choice for $V_\mathsf{d}(q)$ is

$$V_\mathsf{d}(q) = \frac{1}{2} \tilde{q}^\top K \tilde{q}, \tag{7}$$

where $K \in \mathbb{R}^{n \times n}$ is positive definite and $\tilde{q} := q - q_\star$. Moreover, the proposed selection yields

$$\tau = \frac{\partial V}{\partial q} - K\tilde{q} - D_\mathsf{d}\dot{q}, \tag{8}$$

which is a PD controller plus gravity compensation—where K is the gain of the proportional term and D_d corresponds to the gain of the derivative component.

Henceforth, we assume that $V_\mathsf{d}(q)$ is a twice differentiable function such that the following assumptions hold.

Assumption 1

$$\left(\frac{\partial V_{\mathrm{d}}(q)}{\partial q}\right)\Bigg|_{q=q_\star} = \mathbf{0} \quad \text{and} \quad \frac{\partial V_{\mathrm{d}}(q)}{\partial q} \neq \mathbf{0},\ \forall\, q \neq q_\star.$$

Assumption 2

$$\left(\frac{\partial^2 V_{\mathrm{d}}(q)}{\partial q^2}\right)\Bigg|_{q=q_\star} \succ 0.$$

We remark that Assumptions 1 and 2 are not necessary to prove stability. However, they are convenient to propose a constructive methodology to design D_{d}, as shown in Sect. 2.2. Moreover, the mentioned assumptions are not restrictive from a physical point of view. Indeed, they are satisfied by (7).

2.2 Damping Design

Lemma 1. *Consider two positive definite matrices, namely,* $N, P \in \mathbb{R}^{n\times n}$. *There exists an invertible matrix* $W \in \mathbb{R}^{n\times n}$ *such that*

$$W^\top N W = I_n \quad \text{and} \quad W^\top P W = \Sigma,$$

where $\Sigma \in \mathbb{R}^{n\times n}$ *is a diagonal matrix such that the elements of its main diagonal are positive.*

Proof. Because N is positive definite, its inverse exists and is positive definite. Hence, there exists a full rank matrix $\phi_{\mathrm{n}} \in \mathbb{R}^{n\times n}$ such that $N^{-1} = \phi_{\mathrm{n}}^\top \phi_{\mathrm{n}}$.[1] Moreover, the matrix $\phi_{\mathrm{n}} P \phi_{\mathrm{n}}^\top$ is positive definite. Thus, its singular value decomposition (SVD) yields

$$\phi_{\mathrm{n}} P \phi_{\mathrm{n}}^\top = U \Sigma U^\top,$$

where $U \in \mathbb{R}^{n\times n}$ is an orthogonal matrix, i.e., $U^\top = U^{-1}$. Select $W = \phi_{\mathrm{n}}^\top U$. Then, the proof is completed noting that

$$\begin{aligned} W^\top N W &= U^\top \phi_{\mathrm{n}} \phi_{\mathrm{n}}^{-1} \phi_{\mathrm{n}}^{-\top} \phi_{\mathrm{n}}^\top U = U^\top U = I_n; \\ W^\top P W &= U^\top \phi_{\mathrm{n}} P \phi_{\mathrm{n}}^\top U = U^\top U \Sigma U^\top U = \Sigma. \end{aligned}$$

□□□

To simplify the notation, we introduce the following matrix

$$\mathcal{K} := \left(\frac{\partial^2 V_{\mathrm{d}}}{\partial q^2}\right)\Bigg|_{q=q_\star}.$$

Moreover, we consider the full rank matrix $\phi_{\mathrm{m}} \in \mathbb{R}^{n\times n}$ such that

$$M^{-1}(q_\star) = \phi_{\mathrm{m}}^\top \phi_{\mathrm{m}}. \tag{9}$$

---

[1] The matrix ϕ_{n} can be obtained via Cholesky decomposition or singular value decomposition. For further details, we refer the reader to [5].

Accordingly, the SVD of $\phi_{\mathtt{m}}\mathcal{K}\phi_{\mathtt{m}}^{\top}$ is given by

$$\phi_{\mathtt{m}}\mathcal{K}\phi_{\mathtt{m}}^{\top} = U_{\mathtt{mk}}\Sigma_{\mathtt{mk}}U_{\mathtt{mk}}^{\top}, \tag{10}$$

where $U_{\mathtt{mk}} \in \mathbb{R}^{n\times n}$ is an orthogonal matrix and

$$\Sigma_{\mathtt{mk}} := \mathtt{diag}(\sigma_{\mathtt{mk}_1}, \ldots, \sigma_{\mathtt{mk}_n}) \succ 0. \tag{11}$$

Proposition 1. Consider the system (5) and the desired configuration $q_\star$, such that $V_{\mathtt{d}}(q)$ satisfies Assumptions 1 and 2. There exist $W_{\mathtt{mk}} \in \mathbb{R}^{n\times n}$, full rank, and $\Sigma_{\mathtt{D}} \in \mathbb{R}^{n\times n}$, diagonal positive definite, such that the selection

$$D_{\mathtt{d}} = W_{\mathtt{mk}}^{-\top}\Sigma_{\mathtt{D}}W_{\mathtt{mk}}^{-1} \tag{12}$$

ensures that the trajectories of (5) behave as the response of n second-order systems as they approach the equilibrium. Moreover, the damping coefficients of such second-order systems are determined by

$$\xi_i = \frac{\sigma_{\mathtt{D}_i}}{2\sqrt{\sigma_{\mathtt{mk}_i}}}, \qquad i \in \{1, \ldots, n\}, \tag{13}$$

where $\sigma_{\mathtt{D}_i}$ and $\sigma_{\mathtt{mk}_i}$ correspond to the main diagonal elements of $\Sigma_{\mathtt{mk}}$, defined in (11), and $\Sigma_{\mathtt{D}}$, respectively.

Proof. Define $M_\star := M(q_\star)$. Then, the linearization of (5) around the point $(q_\star, \mathbf{0})$ is given by $\dot{z} = \mathcal{A}z$, where

$$z := \begin{bmatrix} q - q_\star \\ p \end{bmatrix}, \quad \mathcal{A} := \begin{bmatrix} \mathbf{0} & M_\star^{-1} \\ -\mathcal{K} & -D_{\mathtt{d}}M_\star^{-1} \end{bmatrix}. \tag{14}$$

Consider the matrices

$$S_{\mathtt{d}} := \begin{bmatrix} I_n & \mathbf{0} \\ \mathbf{0} & M_\star^{-1} \end{bmatrix}, \quad \mathcal{A}_{\mathtt{d}} := \begin{bmatrix} \mathbf{0} & I_n \\ -M_\star^{-1}\mathcal{K} & -M_\star^{-1}D_{\mathtt{d}} \end{bmatrix}.$$

Note that $\mathcal{A}_{\mathtt{d}} = S_{\mathtt{d}}\mathcal{A}S_{\mathtt{d}}^{-1}$. Thus, $\mathcal{A}_{\mathtt{d}}$ is similar to $\mathcal{A}$, consequently their eigenvalues are the same. Furthermore, $\mathcal{A}_{\mathtt{d}}$ is a companion matrix of

$$L(\lambda) = I_n\lambda^2 + M_\star^{-1}D_{\mathtt{d}}\lambda + M_\star^{-1}\mathcal{K}.$$

Hence, the eigenvalues of $\mathcal{A}_{\mathtt{d}}$ are determined by the values $\lambda \in \mathbb{C}$ such that $L(\lambda)$ loses rank, i.e., its determinant equals zero [7]. Since $M_\star$ is full rank, we have the following chain of implications

$$\det(L) = 0 \iff \det(M_\star^{-1})\det(M_\star L) = 0 \iff \det\left(M_\star\lambda^2 + D_{\mathtt{d}}\lambda + \mathcal{K}\right) = 0. \tag{15}$$

Moreover, invoking the result of Lemma 1, there exists a full rank matrix $W_{\mathtt{mk}} \in \mathbb{R}^{n\times n}$ such that

$$W_{\mathtt{mk}}^{\top}\left(M_\star\lambda^2 + D_{\mathtt{d}}\lambda + \mathcal{K}\right)W_{\mathtt{mk}} = I_n\lambda^2 + W_{\mathtt{mk}}^{\top}D_{\mathtt{d}}W_{\mathtt{mk}}\lambda + \bar{\Sigma},$$

where $\bar{\Sigma}$ is a diagonal positive definite matrix. Hence, by choosing $D_{\mathtt{d}}$ as in (12), we obtain

$$\det(L) = 0 \iff \det\left(I_n\lambda^2 + \Sigma_{\mathtt{D}}\lambda + \bar{\Sigma}\right) = 0. \tag{16}$$

Note that $I_n\lambda^2 + \Sigma_{\mathtt{D}}\lambda + \bar{\Sigma}$ is a diagonal matrix. Hence, its determinant is zero if and only if one of its main diagonal elements equals zero. Consequently, (16) can be analyzed as n decoupled second-order systems. Furthermore, following the rationale in the proof of Lemma 1, we get that

$$W_{\mathtt{mk}} = \phi_{\mathtt{mk}}^{\top} U_{\mathtt{mk}} \quad \text{and} \quad \bar{\Sigma} = \Sigma_{\mathtt{mk}} \tag{17}$$

with $\phi_{\mathtt{m}}^{\top}$, $U_{\mathtt{mk}}$, and $\Sigma_{\mathtt{mk}}$ defined as in (9), (10), and (11), respectively. Therefore, the n second-order equations take the form

$$\lambda^2 + \sigma_{\mathtt{D}_i}\lambda + \sigma_{\mathtt{mk}_i} = 0. \tag{18}$$

Hence, equating (18) with the standard structure of a second-order system, i.e.,

$$\lambda^2 + 2\xi_i\omega_{\mathtt{n}_i}\lambda + \omega_{\mathtt{n}_i}^2 = 0,$$

we obtain (13). □□□

Remark 2. The model (1) is valid—considering appropriate coordinates—for the joint and task spaces. Consequently, the result of Proposition 1 is suitable to design the damping matrix in both spaces.

The proposed approach assumes that the stiffness matrix K is fixed for the specific purposes of the application at hand, i.e., the method is devised to compute $D_{\mathtt{d}}$ for a given K. However, the performance of the closed-loop system can be enhanced if K and $D_{\mathtt{d}}$ can be designed simultaneously. We illustrate this in Sect. 3.1.

Table 1. Different simulation cases

Case	C1	C2	C3	C4	C5	C6	C7	C8	C9	C10	C11
$q_\star$	$\frac{\mathbb{1}}{4}$	$\frac{\mathbb{1}}{4}$	$\frac{\mathbb{1}}{4}$	$\frac{\mathbb{1}}{4}$	$\frac{\mathbb{1}}{4}$	$\frac{\mathbb{1}}{2}$	$\frac{\mathbb{1}}{10}$	$\frac{\mathbb{1}}{4}$	$\frac{\mathbb{1}}{4}$	$\frac{\mathbb{1}}{4}$	$\frac{\mathbb{1}}{4}$
ξ	1	$\frac{\sqrt{2}}{2}$	1.5	$\xi_{\mathtt{d}}$	1	1	1	1	1	1	1
K	$K_{\mathtt{d}}$	$K_{\mathtt{d}}$	$K_{\mathtt{n}}$	$K_{\mathtt{d}}$	$K_{\mathtt{n}}$	$K_{\mathtt{d}}$	$K_{\mathtt{d}}$	$K_{\mathtt{m}}$	$\frac{K_{\mathtt{d}}}{4}$	$\frac{K_{\mathtt{d}}}{4}$	–

3 Results

We design the damping matrix of a *Franka Emika Panda* robot in joint space to validate the methodology. This robot has seven degrees of freedom given by the angles of seven revolute joints, denoted as q_i, with $i \in \{1, \ldots, 7\}$. We simulate the behavior of the closed-loop system considering different scenarios. Then, we perform two experiments: (i) we implement the proposed methodology to design $D_{\mathtt{d}}$, and (ii) we implement the controller provided in https://github.com/franzesegiovanni/franka_human_friendly_controllers.

3.1 Simulations

We consider the open-loop dynamics provided in [4].[2] Then, we simulate eleven different cases given in Table 1, where

$$\xi_{\mathtt{d}} = \left[0.5\ 0.7\ 0.7\ 1\ 1.3\ 1.3\ 1.5\right]; \qquad K_{\mathtt{d}} = \mathtt{diag}(80, 40, 80, 40, 80, 80, 20);$$
$$K_{\mathtt{m}} = 1000\phi_{\mathtt{m}}^{-1}\mathtt{diag}(2.5, 2.5, 2.5, 2.5, 3.6, 2.5, 10)\phi_{\mathtt{m}}^{-\top};$$

$$K_{\mathtt{n}} = \begin{bmatrix} 80 & 0 & 0 & 0 & 0 & 0 & 0 \\ 0 & 184 & 0 & -72 & 0 & 0 & 0 \\ 0 & 0 & 80 & 0 & 0 & 0 & 0 \\ 0 & -72 & 0 & 76 & 0 & 0 & 0 \\ 0 & 0 & 0 & 0 & 80 & 0 & 0 \\ 0 & 0 & 0 & 0 & 0 & 80 & 0 \\ 0 & 0 & 0 & 0 & 0 & 0 & 20 \end{bmatrix}.$$

For simulation purposes, we consider initial conditions $q_0 = \mathbf{0}$ in all the cases. Moreover, in all the cases—except for **C4**—we consider the same damping coefficient for the seven second-order equations (13).

For all the considered scenarios—except for **C10**—we compute $D_{\mathtt{d}}$ via the approach proposed in Sect. 2.2. In contrast to the rest of cases, in **C10** we consider $M(q_\star) = I_7$ to design $D_{\mathtt{d}}$. Hence, (13) can be rewritten as $\sigma_{\mathtt{d}_i} = 2\xi_i\sqrt{k_i}$ for $i \in \{1, \ldots, 7\}$. Thus, recalling that al the damping coefficients ξ_i are the same and K is a diagonal matrix[3] in **C10**, we have

$$D_{\mathtt{d}} = 2\xi K^{\frac{1}{2}}. \tag{19}$$

The expression (19) is common in practice to design $D_{\mathtt{d}}$. However, this approach is suitable only for diagonal stiffness matrices and completely neglects the inertia matrix of the system, which is one of the main sources of the nonlinear behavior of a robot. Henceforth, we refer to the expression (19) as the *naive approach* to design $D_{\mathtt{d}}$.

In all the cases—except for **C11**—we consider a control law of the form (8). Depending on the source, this control approach can be understood as a PD controller plus gravity compensation, an impedance control, or a particular class of passivity-based control strategy. For **C11**, we consider the control law (2) and the desired potential energy

$$V_{\mathtt{d}} = \sum_{i=1}^{7} \alpha_i \ln\left(\cosh(10\tilde{q}_i)\right) + \frac{1}{4}\tilde{q}_2^4,$$

[2] See https://github.com/marcocognetti/FrankaEmikaPandaDynModel.

[3] Namely, $K = \frac{1}{4}K_{\mathtt{d}}$.

with $\alpha_1 = \alpha_3 = \alpha_5 = \alpha_6 = 0.8$, $\alpha_2 = \alpha_4 = 0.4$, and $\alpha_7 = 0.2$. Thus,

$$\frac{\partial V_{\mathrm{d}}(q)}{\partial q} = \begin{bmatrix} 8\tanh(10\tilde{q}_1) \\ 4\tanh(10\tilde{q}_2) + \tilde{q}_2^3 \\ 8\tanh(10\tilde{q}_3) \\ 4\tanh(10\tilde{q}_4) \\ 8\tanh(10\tilde{q}_5) \\ 8\tanh(10\tilde{q}_6) \\ 2\tanh(10\tilde{q}_7) \end{bmatrix}.$$

We stress that the concept of stiffness matrix in impedance control is not valid for this case. Consequently, approaches as the one proposed in [1] are not suitable for this control law. Nevertheless, the method to design D_{d} proposed in Sect. 2.2 is still valid. Furthermore, the linearized systems obtained from **C1** and **C11** are the same.

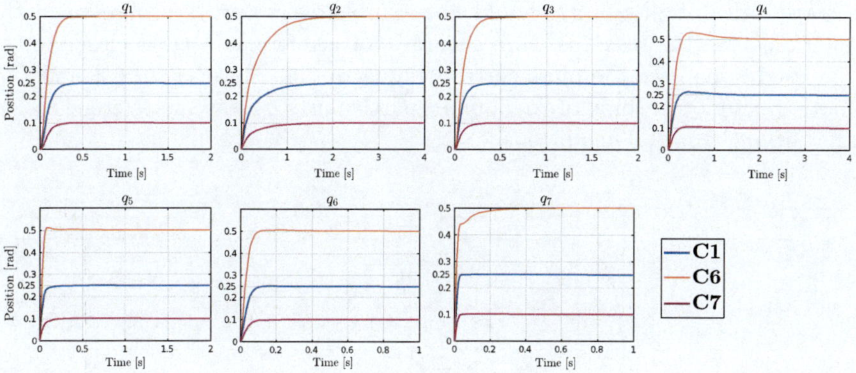

Fig. 1. Trajectories of the joints for a critically damped design and different references.

Limitations Due to Linearization The robotic manipulator used for simulations and experiments is a nonlinear system, while the proposed methodology to design D_{d} relies on linearizing the closed-loop system. Therefore, the performance analysis, i.e., the assignation of ξ, is valid only in a region close to the desired configuration. However, we underscore that the asymptotic stability of the desired equilibrium is guaranteed for the nonlinear model, i.e., the robot converges to the desired configuration independently of the validity of the linearized system using to design D_{d}. Unfortunately, the closed-loop system may exhibit different performance than the desired one for initial conditions far from the desired configuration. Nonetheless, in practice, the proposed method is expected to yield reasonable good behavior.

Figure 1 shows the plots of **C1**, **C6**, and **C7**. In these three cases D_{d} is desgined such that the seven second-order systems are critically damped. Thus,

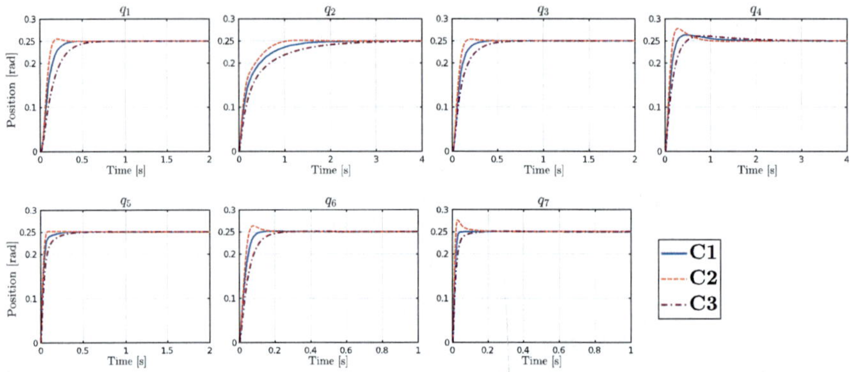

Fig. 2. Trajectories of the joints for different damping coefficients.

the trajectories of the joints should not exhibit overshoot. We observe in the simulations that the result is more accurate for a reference that is closer to the initial conditions—see the plots corresponding to **C6**. However, in **C1** and **C7**, we note overshoot in some of the joints—particularly q_4—as the reference is set farther from the initial conditions.

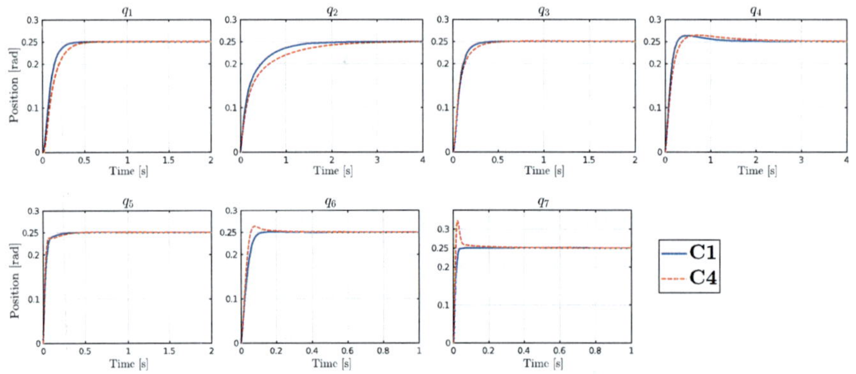

Fig. 3. Comparison between **C1** and **C4**. The damping coefficients ξ_i are designed independently in the latter case.

Choosing $\boldsymbol{\xi}$. The proposed approach permits assigning different damping coefficients to the seven second-order systems, which can be exploited to mitigate overshoot in their responses or increase convergence speed. In particular, we expect a faster convergence for $\xi < 1$ at expenses of some overshoot. Moreover, we expect that the response for $\xi > 1$ convegence slower than the one for $\xi = 1$. Figure 2 depicts the plots of **C1**, **C2**, and **C3**, where only ξ is modified. We observe that the convergence rates behave as predicted. However, in **C1**, which corresponds to seven critically damped systems, q_2 and q_5 behave as overdamped systems, while we appreciate overshoot in q_4—i.e., it behaves as an underdamped system.

While it is possible to design the damping coefficients independently, the intuition to propose them properly is not straightforward because the method changes the basis of the system—through the linear transformation $W_{\mathtt{mk}}$—to obtain the seven second-order systems. To illustrate this, in **C4**, we propose damping coefficients such that we could suppose three overdamped responses, three underdamped responses, and one critically damped response. Figure 3 shows the plots of **C1** and **C4**. For the latter, there is no critically damped response. Moreover, it is unclear how the proposed damping coefficients affect the response of each joint.

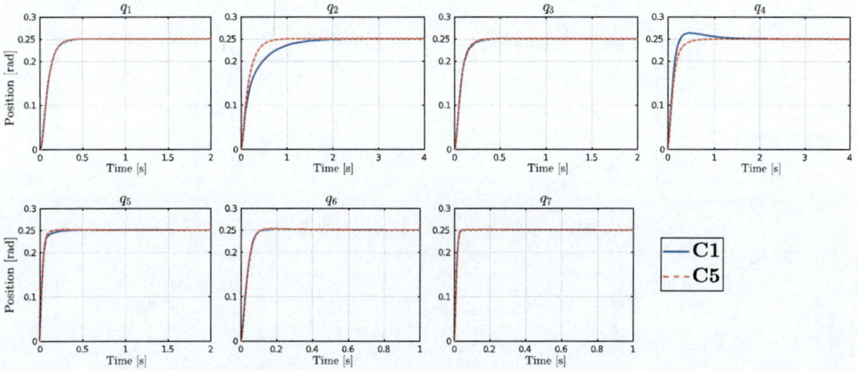

Fig. 4. Comparison between **C1** and **C5**, where the latter considers a non-diagonal stiffness matrix.

Non-diagonal K**.** In practice, the stiffness matrix is often proposed diagonal for simplicity—for instance [3]. Nevertheless, this approach is not convenient in some applications involving human-robot interaction, e.g., [9]. Moreover, the behavior of the joints is physically coupled. This coupling is the result of the non-diagonal structure of the mass inertia matrix. Hence, choosing K diagonal might not be physically intuitive.

Note that the simulations of **C1** show unexpected behaviors for q_2 and q_4. Whence, we can conjecture that the nonlinearities (physical couplings) have a greater impact on these coordinates than in the rest. To address this issue, in **C5**, we propose a stiffness matrix—namely, K_{n}—that couples q_2 and q_4. Then, in both cases, the damping matrix is obtained via the approach proposed in Sect. 2.2. Figure 4 shows the comparison between **C1** and **C5**, where the latter performs better than the former—especially for q_2 and q_4. While the design of the stiffness matrix is based on the system's physics, it does not stem from a formal argument, and further research on this topic is necessary.

If K is free, then the design of $D_{\mathtt{d}}$ can be simplified by choosing K based on $M(q_\star)$. In particular, the selection

$$K = \phi_{\mathtt{m}}^{-1}\Sigma_{\mathtt{K}}\phi_{\mathtt{m}}^{-\top}; \quad \Sigma_{\mathtt{K}} := \mathtt{diag}(\sigma_{\mathtt{K}_1}, \ldots, \sigma_{\mathtt{K}_n}) \succ 0$$

yields $W_{\mathtt{mk}} = \phi_{\mathtt{m}}^\top$ in (12). Hence, (13) reduces to

$$\xi_i = \frac{\sigma_{\mathtt{D}_i}}{2\sqrt{\sigma_{\mathtt{K}_i}}}, \qquad i \in \{1, \ldots, n\},$$

which is similar to the naive approach. Figure 5 depicts the comparison between **C1** and **C8**, where the stiffness matrix is proposed based on $M(q_\star)$. In the mentioned figure, we observe that the trajectories of all the joints, except for q_7, exhibit a better performance when the K is based on the inertia matrix.

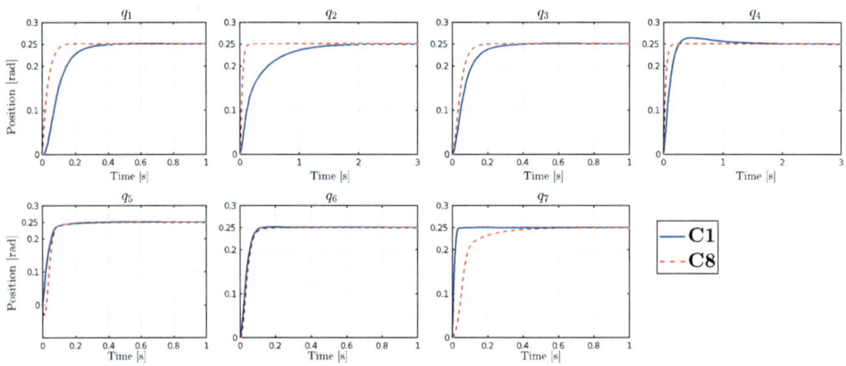

Fig. 5. Comparison between **C1** and **C8**, where the latter considers a stiffness matrix based on the inertia matrix of the system.

Proposed Method vs Naive Approach. The controllers designed by adopting the naive approach are suitable to stabilize the robot at its desired configuration because they satisfy the conditions provided in Sect. 2.1, i.e., the point $(q_\star, \mathbf{0})$ is a globally asymtotically equilibrium for the closed-loop system. An advantage of this method is its simplicity. On the other hand, an important disadvantage is the lack of intuition into how to prescribe a desired performance for the closed-loop system. Figure 6 illustrates the comparison between the damping design method proposed in Sect. 2.2 (**C9**) and the naive approach (**C10**) under the same conditions, where we observe that the proposed methodology outperforms the naive approach except for joints 2 and 4.

Linear vs Nonlinear Proportional Term. Some applications require desired potential energies that are not quadratic on the position error, e.g., [11]. Consequently, the gradient of the $V_{\mathtt{d}}(q)$ is nonlinear and the closed-loop system has no

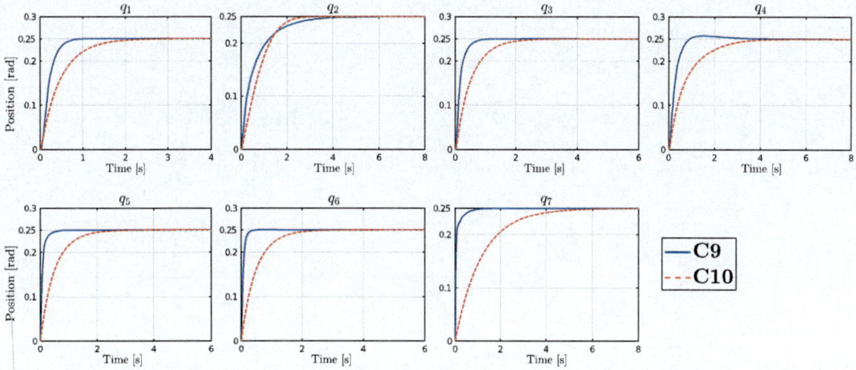

Fig. 6. Comparison between the proposed methodology (**C9**) and the naive approach (**C10**).

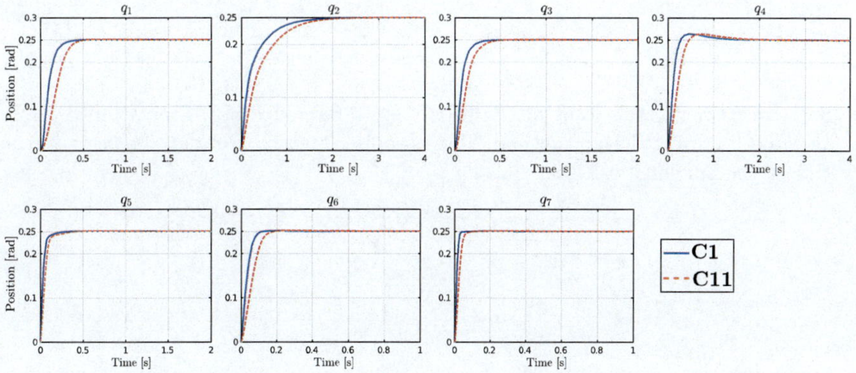

Fig. 7. Trajectories of a closed-loop system whose potential energy is quadratic on the position error (**C1**) and a closed-loop system whose potential energy is non quadratic on the position error (**C11**).

stiffness matrix. Nevertheless, since the proposed methodology is based on the linearization, it is still suitable to design the damping matrix for the closed-loop system. Figure 7 shows the trajectories of two systems whose linearizations are identical. However, one case considers a quadratic potential energy (**C1**) and the other corresponds to a non quadratic $V_{\mathrm{d}}(q)$ (**C11**).

3.2 Experimental Results

The procedure to validate the proposed methodology consists in calculating the damping matrix for for the closed-loop system (5) for a given desired potential energy of the form (7).[4] Then, we simultaneously change the desired angle of all

[4] The code to implement the proposed controller in the Panda robot is available at https://github.com/franzesegiovanni/franka_damping_design.

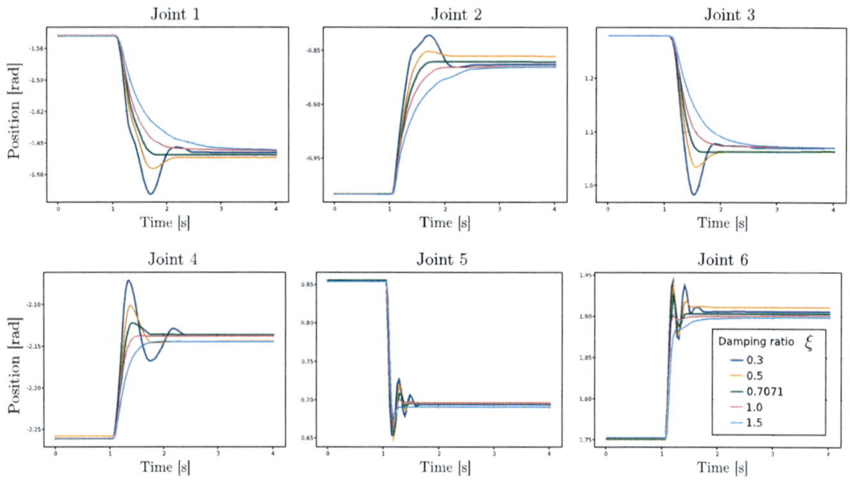

Fig. 8. Trajectory of each joint of the robot, for a reference of 0.1 rad from q_0, using the proposed damping design method. The damping matrix is updated at a frequency of 1 kHz at the current position.

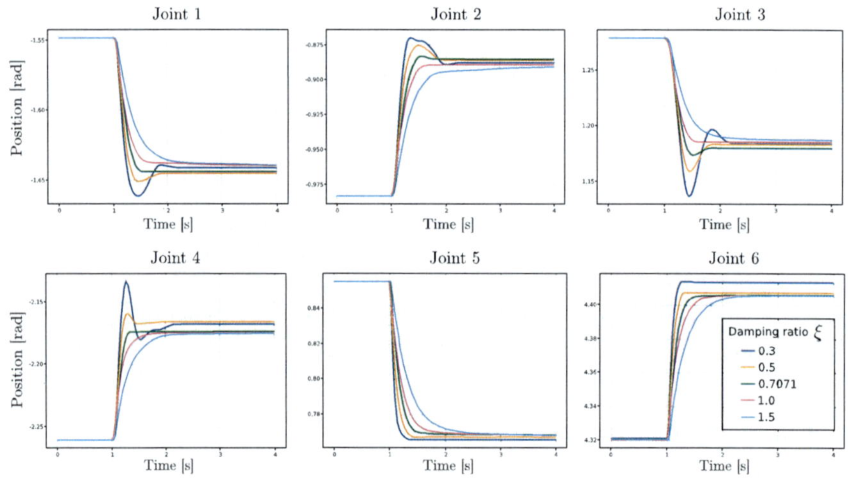

Fig. 9. Trajectory of each joint of the robot, for a reference of 0.1 rad from q_0, using the naive approach.

joints of the *Franka Emika Panda* robot. The initial configuration of the robot is given by

$$q_0 = \begin{bmatrix} -1.548 & -0.9832 & 1.2789 & -2.2607 & 0.8551 & 1.7509 & 0.1978 \end{bmatrix}^\top .$$

Moreover, the stiffness matrix considered for the experiments is given by

$$K = \texttt{diag}(100, 100, 100, 100, 100, 100, 8)$$

and the mass matrix is obtained using the functionality in the *Franka Emika Panda's* controller code base 'franka_ros'.[5]

The results in Fig. 8 show that, by adopting the proposed methodology, the variations in ξ_i affect the characteristic response of the system as expected, i.e., the system overshoots and oscillates more when the damping ratio decreases and when the damping ratio increases above $\xi_1 = 1$, the convergence is slower. We also observe that the steady-state error is affected by the selection of ξ. In particular, we omit the plot of q_7 because the response is too sensitive to provide a meaningful comparison.

The results of the same experiments, but carried out with naive approach, are shown in Fig. 9. It can be seen that the effect of selecting a specific damping ratio does not affect the performance in the expected manner in all the joints. This is especially apparent in the joints 4, 5, and 6 where even for low damping ratios, the system is drastically over damped. This highlights a major problem for the designer, as there is no correlation between the damping ratios and the characteristic behaviour of the joints. In this experiment, the joints 2 and 3 perform as expected, where the trajectories are underdamped, critically damped, and overdamped for $\xi = 0.7071$, $\xi = 1$, and $\xi = 1.5$, respectively. The mentioned behavior changes according to the robot's configuration, leading to the conclusion that the good performance at particular joints is because the assumption $M(q_\star) = I_7$ is an adequate approximation in that particular position and not by design.

The inconsistency of the naive approach is further highlighted in Fig. 10. Here, it can be more clearly seen that, using both methods, the joint is critically damped for $\xi = 1$. However, for $\xi = \sqrt{2}/2$, the response is underdamped (as expected) for the proposed methodology, while for the naive approach is still critically damped. Moreover, the naive approach results in a larger steady-state error for both damping ratios.

4 Conclusions

We have provided a method to design the damping matrix for fully actuated robotic arms, which is based on the linearization of the system and fits in different approaches, such as PD control, impedance control, and passivity-based control.

[5] See https://github.com/frankaemika/franka_ros.

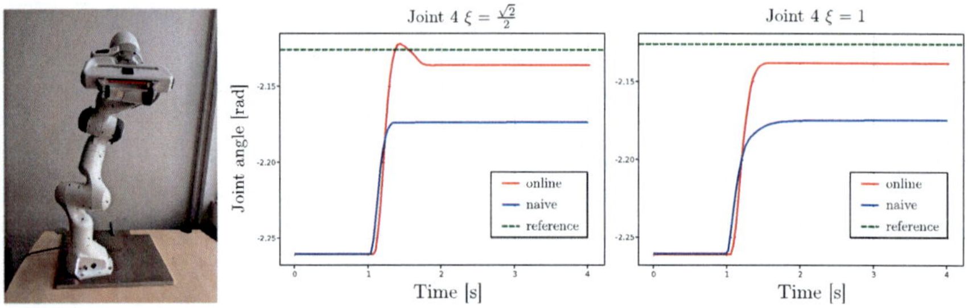

Fig. 10. Panda robot used for the experiments (left). Comparison of the trajectories of Joint 4 with $\xi = \sqrt{2}/2$ and $\xi = 1$ (right). Here, the plot corresponding to the proposed method is labeled as *online*.

We have shown that the proposed method is more consistent than the naive approach, which neglects the physical coupling given by the inertia matrix of the system. Moreover, we have provided some intuitive alternatives to design the stiffness matrix to enhance the performance of the closed-loop system. Finally, we have performed simulations and experiments to validate the methodology, supporting the theoretical claims.

Acknowledgments. The authors thank Cosimo Della Santina for the fruitful discussions during the elaboration of this manuscript.

Tomás Coleman's research is funded by the Netherlands Organization for Scientific Research project Cognitive Robots for Flexible Agro-Food Technology, grant P17-01.

Giovanni Franzese's research is funded by the European Research Council Starting Grant Teaching Robots Interactively (TERI), project reference 804907.

References

1. Albu-Schäffer, A., et al.: Cartesian impedance control of redundant robots: recent results with the DLR-light-weight-arms. In: 2003 IEEE International Conference on Robotics and Automation, vol. 3, pp. 3704–3709 (2003)
2. Albu-Schäffer, A., et al.: The DLR lightweight robot: design and control concepts for robots in human environments. Ind. Robot: Int. J. (2007)
3. Franzese, G., et al.: ILoSA: interactive learning of stiffness and attractors. In: 2021 IEEE/RSJ International Conference on Intelligent Robots and Systems (IROS), pp. 7778–7785. IEEE (2021)
4. Gaz, C., et al.: Dynamic identification of the Franka Emika panda robot with retrieval of feasible parameters using penalty-based optimization. IEEE Robot. Autom. Lett. **4**(4), 4147–4154 (2019)
5. Horn, R.A., Johnson, C.R.: Matrix Analysis. Cambridge University Press (2012)
6. Khalil, H.K.: Nonlinear Systems, 3rd edn. Prentice-Hall, New Jersey (2002)
7. Marcus, M., Minc, H.: A Survey of Matrix Theory and Matrix Inequalities, vol. 14. Courier Corporation (1992)
8. Peternel, L., et al.: Teaching robots to cooperate with humans in dynamic manipulation tasks based on multi-modal human-in-the-loop approach. Autonomous Robot. **36**(1), 123–136 (2014)

9. Prendergast, J.M., et al.: Biomechanics aware collaborative robot system for delivery of safe physical therapy in shoulder rehabilitation. IEEE Robot. Autom. Lett. **6**(4), 7177–7184 (2021)
10. van der Schaft, A.J., Jeltsema, D.: Port-Hamiltonian systems theory: an introductory overview. Foundations Trends Syst. Control **1**(2–3), 173–378 (2014)
11. Wesselink, T.C., Borja, P., Scherpen, J.M.A.: Saturated control without velocity measurements for planar robots with flexible joints. In: 2019 IEEE 58th Conference on Decision and Control (CDC), pp. 7093–7098. IEEE (2019)

Deformation-Aware Contact-Rich Manipulation Skills Learning and Compliant Control

Weiyong Si[1], Cheng Guo[2], Jiale Dong[3], Zhenyu Lu[1], and Chenguang Yang[1](✉)

[1] University of the West of England, and Bristol Robotics Laboratory, Bristol BS16 1QY, UK
cyang@ieee.org

[2] University of Bristol, and Bristol Robotics Laboratory, Bristol BS8 1TH, UK

[3] South China University of Technology, Guangzhou 510640, China

Abstract. In this paper, we study a vision-based reactive adaptation method for contact-rich manipulation tasks, based on the compliant control and learning from demonstration. For contact-rich tasks, the compliant control method is essential, especially when interacting with a deformable object with unknown properties, such as pizza dough. Learning from demonstration (LfD) provides a solution for this challenging task. However, the generalisation capabilities of LfD for deformable object manipulation tasks are still a challenging and opening issue, especially for unknown and dynamic tasks. Therefore, in this work, we investigate the vision and force-based perception feedback to enhance the generalisation of the LfD outcomes. The computer vision algorithm was developed to recognise the shape of the object and calculate the deviation between the desired shape and the current shape. The deviation of shape adjusts the parameters of learned primitive skills encoded by Dynamic Movement Primitives (DMPs). We adopt the pizza dough rolling task as the typical case to evaluate the performance of the proposed method. The shape and thickness of the dough can be made to the desired shape and thickness.

Keywords: Deformation-aware pizza dough rolling (DaPADR) · Learning from demonstration · Compliant control

1 Introduction

Collaborative robots (cobots) have gained much attention in both academic and industrial communities. In the past, collaborative robots, such as Baxter robot and Franka Emika Panda etc., have been employed in a number of fields, such as manufacturing, medical examinations, and home service [1]. Although the cobots

P. Borja et al. (Eds.): HFR 2022, 26, pp. 90–104, 2023.
https://doi.org/10.1007/978-3-031-22731-8_7

have been widely used in industry, it is still hard to interact with deformable objects, such as textile cloth, soft and deformable food in the kitchen and human tissue [2–4]. Manipulating soft and deformable objects is challenging due to the lack of an accurate model of the object and the unknown properties of the deformable objects. However, deformable objects, such as cloth, fruits and even our bodies, are very common in daily life, and interaction with deformable and soft objects is ubiquitous. Therefore, manipulating deformable objects by robots has gained much attention in both academic and industrial communities.

There is some literature on deformable object modelling and control, such as pizza dough manipulation [5–8]. Siciliano et al. has conducted comprehensive study on the pizza dough manipuation, including the perception, trajectory planning and controller design for pizza dough. For example, the modelling of the deformable object was studied, and optimization methods were investigated to generate trajectory for the robots in realtime [5]. The pizza dough manipulation by robots has attracted much attention. Stretching a dough with a rolling pin is a typical nonprehensile manipulation,[1] which has been studied in [5].

Recently, as machine learning methods have been adopted in the robotic field, various robot learning methods have been proposed to tackle the autonomous execution of robots. A number of literature on robot skill learning has been published [1]. Learning from demonstration (LfD) is one effective robot skill learning technology that relies on the human teachers' demonstrations [9]. Iterative learning and identification were investigated for the state monitoring [10]. In our previous work, LfD has been investigated for various manipulation tasks, such as writing characters, cutting vegetables/ fruit, assembly, and robot-assisted surgery [1]. In addition, the LfD based hierarchical framework was explored to learn the task of rolling pizza dough [11].

Vision-based trajectory planning and control methods, such as visual servoing, have been well studied in the industry [12]. However, pure vision-based planning and control method is hard to manipulate deformable objects autonomously. Due to the complex property of the deformable objects, it is impossible to plan the trajectories of robots in complex tasks manually in advance. Visual and haptic information provides real-time feedback for online planning and decision making of robots working on complex tasks under unknown conditions [13]. On the other hand, with developments in the field of machine learning, deep convolutional networks are outperforming traditional computer vision methods in many visual recognition tasks. Some new neural network architectures such as U-net [14,15], are capable of accurate segmentation of complex medical image including cardiac MR segmentation [14] and lung cancer nodule detection [15], that can reach near radiologist level performance. Thus the architecture of deep convolutional networks can also be used to predict the boundaries of flexible matter and help robots determine their workspace. These applications are widely seen to replace traditional computer vision which has been used to obtain more accurate, noise-free data.

[1] Note nonprehensile manipulation means manipulation without grasping, such as pushing, flipping, throwing, and squeezing.

To tackle with the challenging issue of autonomous manipulation of deformable objects, we proposed a learning and adaptive framework for compliant manipulation of deformable objects, as shown in Fig. 1. The contributions of this work can be summarized as follows: 1) Vision-based deformation-aware method was investigated for deformable object manipulation to achieve the target shape. Comparing our previous work [4], we investigated subgoal decision strategy based on shape recognition and deviation from the target shape. 2) In addition, to manipulate the deformable dough, a compliant controller was designed to execute the trajectories generated from LfD. We integrated the learning and compliant control for deformable objects in an unified framework. 3) We adopted pizza dough stretching tasks as evaluation case on a real robot; and the pizza dough's shape and thickness are the evaluation criteria.

The rest of this paper is organised as follows: Sect. 2 presents the preliminary knowledge of robot manipulator dynamics and compliant control. The proposed deformation-aware manipulation skill learning and complaint control method is also presented in Sect. 2. The experiment and results and performance analysis are presented in Sect. 3. Finally, we conclude the paper with a discussion about future work in Sect. 4.

2 Problem Formulation and Methodology

2.1 Vision-Based Deformation Recognition

The vision recognition is to detect the outline of the dough on the table and to attain the position of the dough edge in the robot's base coordinate system. The camera calibration is required to record the robot's working space before the experiment. As shown in Fig. 3., four ArUco markers are placed on the table with a certian distance from each other, and the object needs to be placed within this area and the dough also needs to stay in this workspace. With these markers, the camera's view plane (the camera's frame) can be mapped onto the table using a perspective transformation method. Also, the camera frame can be used to obtain an image of the dough within the working area. Once the image is obtained, the dough can be distinguished from the background using U-net and the final positions of the dough edges can be obtained using canny edge detection.

Perspective Transformation Perspective transformation is the projection of an image onto a new viewing plane, also known as projective mapping [16]. The transformation matrix required in the affine transformation of an image is a 2×3 two-dimensional plane transformation matrix, whereas the perspective transformation is essentially a three-dimensional transformation of space, where the three dimensional coordinates are projected onto a different viewing plane according to their sub-coordinate variance. In this experiment we need the diagram of the view plane where the table is located on the image.

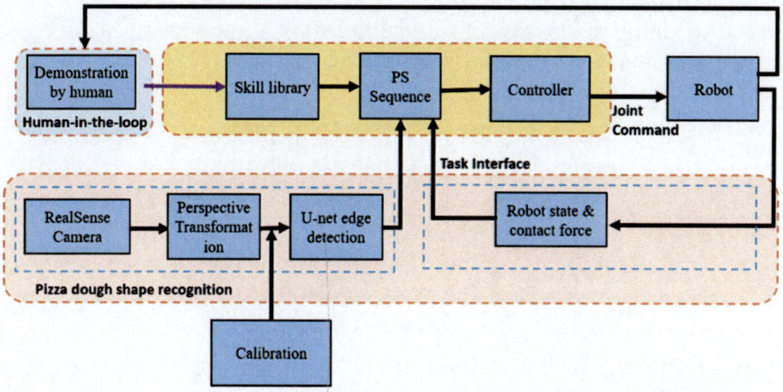

Fig. 1. The overall structure of the proposed framework.

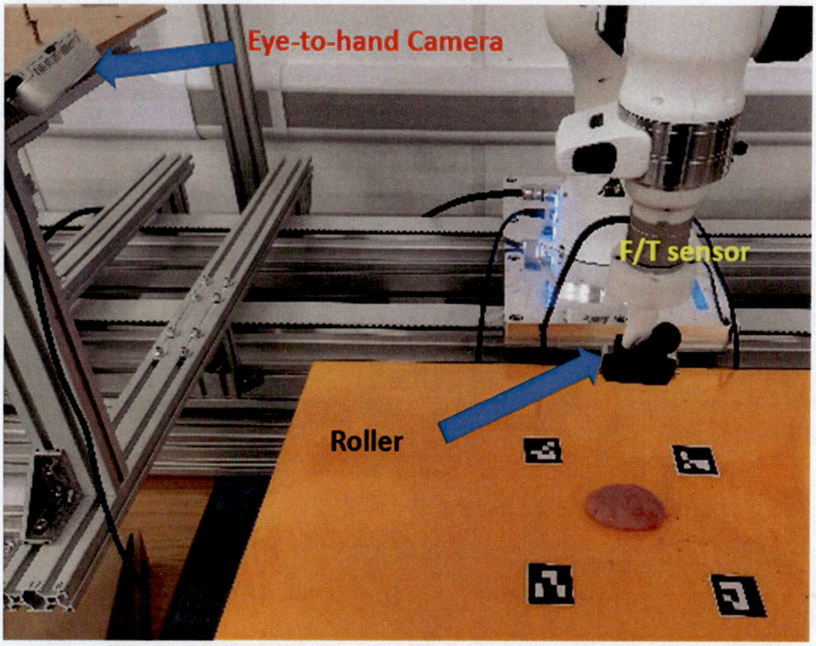

Fig. 2. The setup of deformation-aware stretching a pizza dough with a rolling pin. RealSense camera and F/T sensor were used to capture the deformation of shape and the contact force. A customised roller was used to stretch the pizza dough.

For a perspective transformation, it can be thought of a projection from a point on a quadrilateral onto the target quadrilateral after a linear transformation, where the shape of the target quadrilateral is a known condition. A general formula for a perspective transformation is as follows [17],

$$\begin{bmatrix} X' \\ Y' \\ \omega' \end{bmatrix} = \begin{bmatrix} a_{11} & a_{12} & a_{13} \\ a_{21} & a_{22} & a_{23} \\ a_{31} & a_{32} & a_{33} \end{bmatrix} \begin{bmatrix} u \\ v \\ 1 \end{bmatrix} \tag{1}$$

$$A = \begin{bmatrix} a_{11} & a_{12} & a_{13} \\ a_{21} & a_{22} & a_{23} \\ a_{31} & a_{32} & a_{33} \end{bmatrix} \tag{2}$$

where u and v are the pixels of the original image. New coordinates x, y after transformation by perspective is $x = \frac{X'}{\omega'}$ and $y = \frac{Y'}{\omega'}$.

The above equation shows that the coordinates under the new view plane can be expressed as [17]:

$$x = \frac{X'}{\omega'} = \frac{a_{11}u + a_{12}v + a_{13}}{a_{31}u + a_{32}v + a_{33}} = \frac{k_1u + k_2v + k_3}{k_7u + k_8v + 1} \tag{3}$$

$$y = \frac{Y'}{\omega'} = \frac{a_{21}u + a_{22}v + a_{23}}{a_{31}u + a_{32}v + a_{33}} = \frac{k_4u + k_5v + k_6}{k_7u + k_8v + 1} \tag{4}$$

To solve for the eight unknowns factors in this equation, it requires eight sets of equations. Four known coordinate points are the pixel positions read from ArUco markers in the original image and four coordinate points of the target image are set to $(0, 0), (0, 500), (500, 500), (500, 0)$. This is due to the fact that the corresponding work area captured by camera is a $12.5 * 12.5cm$ square on the table plane, which corresponds to $1/500 * 12.5cm$ per pixel on the image. The transformation matrix A is calculated for each pixel point in the original image, the new image is projected to give a top view of the dough on the table.

Image Segmentation Once the perspective-transformed image is obtained, the exact position of the dough needs to be obtained from this image. A deep learning approach with higher accuracy was chosen over traditional vision algorithms. The architecture of U-net was shown in Fig. 3, which was used to segment the background from the dough in this work. It consists of a systolic path (left side) and an extended path (right side). The systolic path follows the typical structure of a convolutional network like VGG19. It consists of two 3×3 convolution (padded convolutions) layers, and each layer followed by a rectified linear unit (ReLU) and a 3×3 convolution layer with stride 2 for downsampling. At each downsampling step, we keep the number of feature channels. Each step in the expansion path includes an upsampling of the feature map, followed by a 2×2 convolution ("up-convolution") that halves the number of feature channels, concatenates with the correspondingly cropped feature map in the contraction path, and performs two 3×3 convolutions, each followed by a ReLU. In the final

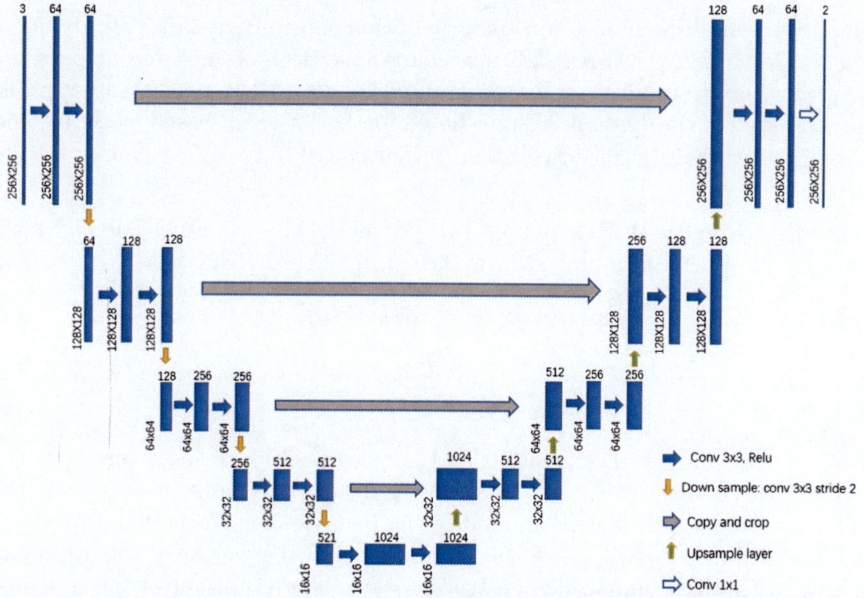

Fig. 3. U-net architecture. Each blue box corresponds to a multi-channel feature map. The number of channels is indicated at the top of the box. The dimensions of the feature map are in the lower left corner of the blue box. The arrows indicate the different operations.

layer, a 1×1 convolution is used to map each 64-component feature vector to the desired number of categories. 1×1 convolution classifies each pixel in the image thereby distinguishing the dough from the background. The output of the network is shown in Fig. 6(c) and (f).

The network is trained using a stochastic gradient descent using the input photos and their accompanying segmentation maps. Before being fed into the network, the input pictures and segmentation maps are downsized to 256 × 256 to match the network's input. Unlike the network mentioned in the U-net paper, the output of this network has exactly the same size as the input image, which ensures that every pixel can be classified. The advantage is acquiring the image with a segmentation of the original input image, which better improves the accuracy of dough recognition.

Finally, by performing canny edge detection on the generated dough mask image, the positions corresponding to edge pixel are transformed into positions in the robot coordinate system, thus completing the recognition of dough contours.

2.2 Primitive Skills Modelling

DMPs was proposed to model the primitive behavior by roboticist [18]. In our previous work, we have proposed various unified and compact representation of the primitive motion skills by dynamic systems. For example, we studied the unified representation of position and orientation of primitive skills [19]. Dynamic system (DS)-based primitive skill modelling have been studied recently, as the convergence and stability can be guaranteed. Due to the reactive and stable characteristics of the dynamic system-based method, the DS-based method could be used to deal with the safety-critical scenarios [20].

Dynamic Movement Primitives (s) We model the primitive skills by position and orientation in a unified formulation in Cartesian space,

$$\tau \dot{z} = \alpha_z(\beta_z(g^p - p) - z) + f_p(x) \tag{5}$$

$$\tau \dot{p} = z \tag{6}$$

$$\tau \dot{x} = -\alpha x \tag{7}$$

where g^p and p are target position and current position respectively, z is the velocity of dynamic system, β_z and α_z are designed parameters. $f_p(x)$ is the nonlinear term, which is used to modify the behaviour of the second-order system. The definition of $f_p(x)$ will be given later and the weight of this term can be learned through human demonstration data. The evolution of the dynamic system is determined by the canonical system, Eq. 7, where the τ can tune the duration of the dynamic system, and the α is a positive parameter. The canonical system is shared by the position and orientation dynamic system to coordinate the translation and rotation simutaneously.

The orientation skill can also be encoded by the DMPs, however, the skill model is adopted by the quaternion-based form [21]. In our previous work, we encoded the translation and orientation simultaneously by DMPs [19]. The details of the model can refer [19], we define the model here to facilitate understanding,

$$\tau \dot{\eta} = \alpha_z \left(\beta_z 2 \log(g^o * \bar{q}) - \eta\right) + f_o(x) \tag{8}$$

$$\tau \dot{q} = \frac{1}{2} \eta * q \tag{9}$$

where η is the angular velocity, g^o and q are the target angle and current angle encoded by quaternion respectively. $\bar{q}$ represents the quaternion conjugation, and $*$ denotes the quaternino product, details can refer to [21]. β_z and α_z are designed parameters, which can be consistent with the position DMPs, $f_o(x)$ is the nonlinear term, which is used to modify the behaviour of the second-order system.

The nonlinear force term is used to define the transient behavior of the dynamic system. This term usually consists of a set of nonlinear basis function (RBFs), which can be given by,

$$f_p = \frac{\sum_{k=1}^{N} w_k^p \Phi_k(x)}{\sum_{k=1}^{N} \Phi_k(x)} x \tag{10}$$

$$f_o = \frac{\sum_{k=1}^{N} w_k^o \Phi_k(x)}{\sum_{k=1}^{N} \Phi_k(x)} x \tag{11}$$

$$\Phi_k(x) = \exp(-h_k(x - c_k)^2) \tag{12}$$

where N is the number of the RBFs, and the x is the phase variable determined by the canonical system. w_k^p and w_k^o are the weights of the ith RBFs, $i \in [1, N]$ and h_k and c_k are the centres and widths of the RBFs, respectively.

We generate the complex trajectory by merging the individual primitive skill (PS), encoded by DMPs for the complex and contact-rich task with multiple-step, as shown in Algorithm 1. The multiple primitive skill are encoded by behavior tree [22], the DMP-based primitive skill was modelled as execution nodes. The target position and orientation of each primitive skills are calculated by shape recognition and desired shape and thickness. In this work, we only consider circular desired shape, four position and four orientation skills are needed. The four primitive skills are approaching, rolling forward, rolling back and back origin. For the circular shape, the primitive skill library with four primitive skills can generate trajectory for the roller. However, the vision-based trajectory generation algorithm can also be employed to different shapes. Different shape only needs to modify the control flow nodes in behavior tree.

Algorithm 1 Vision-based Trajectory Generation

while (!achieve desired shape and thickness) **do**
    starting point P_{st}, direction dir = vision feedback()
    while (current position != end point) **do**
        PS1.run(current position, P_{st}), PS2.run(current direction, dir)
    end while
    while (current position != end point) **do**
        PS3.run(current position, P_{st}), PS4.run(current direction, dir)
    end while
    while (current position != end point) **do**
        PS5.run(current position, P_{st}), PS6.run(current direction, dir)
    end while
    while (current position != end point) **do**
        PS7.run(current position, P_{st}), PS8.run(current direction, dir)
    end while
end while

2.3 Compliant Control

The dynamic of the robot manipulator needs to be modelled for the compliant controller. For a serial robot manipulator, the dynamic equation in joint space can be given,

$$M(q)\ddot{q} + C(q, \dot{q})\dot{q} + G(q) + \tau_f(\dot{q}) + d_u = \tau_c + J^T(q)F_e \tag{13}$$

where $M(q)$ is the positive-definite and symmetric matrix; $C(q, \dot{q})$ represents the Coriolis and centrifugal term; $G(q)$ is the gravity torque; $\tau_f(\dot{q})$ is the friction torque and d_u represents the unmodelled terms. In this work, we ignored these terms. q is the joint angle, $\dot{q}$ and $\ddot{q}$ are the velocity and acceleration, respectively. F_e is the interaction force between end-effector and environment. τ_c is the control input, and $J^T(q)$ is the Jacobian transpose.

We model the manipulation skill in the task space to improve the generalization capability of the learned skills for different robots. In addition, encoding the manipulation skills in the task space is straightforward to learn versatile manipulation, including translation and orientation skills etc. Therefore we derive the controller in the task space to achieve the learned skills. The relationships between position, velocity and acceleration of the end-effector and joint variables are described as,

$$x(t) = f(q) \tag{14}$$

$$\dot{x}(t) = J(q)\dot{q} \tag{15}$$

$$\ddot{x} = \dot{J}(q) + J(q)\ddot{q} \tag{16}$$

where $x(t)$ is the position in the task space, $\dot{x}(t)$ and $\ddot{x}(t)$ represent the velocity and acceleration of the end-effector in task space. $f(q)$ is the feedforward kinematic relationship.

Considering the above equations and (1), we can derivate the dynamics of the robot manipulator in task space,

$$\Lambda_p(q)\ddot{x} + B(q,\dot{q})\dot{x} + G_p(q) + T_f(\dot{q}) + D = T_c + F_e \tag{17}$$

where $\Lambda_p(q)$ is the inertial matrix in task space, $B(q,\dot{q})$ account for Coriolis and centrifugal term, $G_p(q)$ is the gravity force, $T_f(\dot{q})$ is the friction force,[2] D is the unmodelled force. T_c is the control input, and F_e is the interaction force, so we can get T_c,

$$T_c = A_d\ddot{x} + D_d(\dot{x} - \dot{x}_d) + K_d(x - x_d) \tag{18}$$

where A_d , D_d and K_d are the desired mass, damping and stiffness. The x_d and $\dot{x}_d$ are the desired position and velocity, respectively. The impedance is determined based on the desired interaction behaviour, and the properties of the environment. The stability of impedance controller can refer to [23].

3 Experiment

We carry out pizza dough stretching with a rolling pin to evaluate the performance of the proposed framework on a 7 DoFs cobot. A RGBD (RealSense) camera was used to capture the deformation of the object and a computer vision algorithm was used to calculate the difference between the real shape and the desired shape. A customised roller was used to stretch the pizza dough.[3] The

[2] Note we ignore the friction force in practice.

[3] Note the pizza dough is not as large as the real pizza dough, because the roller is not large enough. The proposed method can be employed for big dough if we adopt a large roller.

setup of the experiment can be seen in Fig. 2. We defined evaluation metrics for dough rolling, as shown in Fig. 4. The red outline is the desired shape, and we defined two axes (the yellow lines) to judge whether the desired shape is achieved. During the dough rolling, we adjust the orientation of the roller based on the error along these two axes.

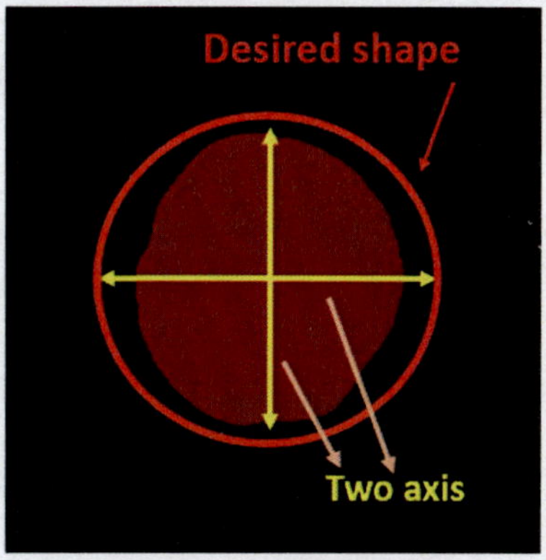

Fig. 4. The evaluated metric of dough rolling. The red outline is the desired shape, and we defined two axes (the yellow lines) to judge whether the desired shape is achieved. During the dough rolling, we adjust the orientation of the roller based on the error along these two axes.

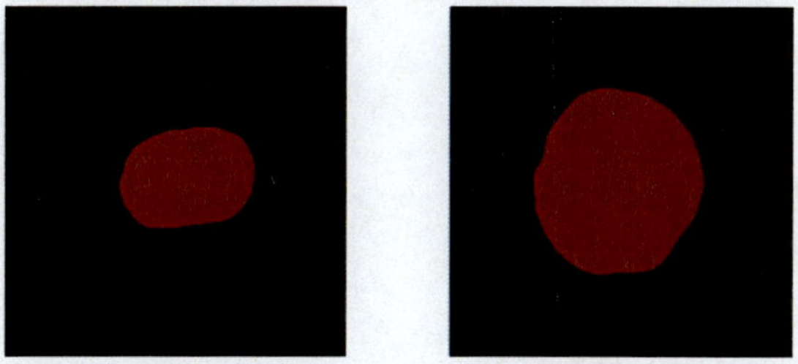

Fig. 5. The result of deformation recognisation and segmentation. The left one is the original shape of dough, and the right one is the final shape of dough.

3.1 The Result of Deformation Segmentation

As shown in Figs. 5 and 6, the changing of the shape of the dough is shown during the rolling process. We test the image recognition and segmentation accuracy by U-net. We collected 20 real images during the dough rolling, and humans

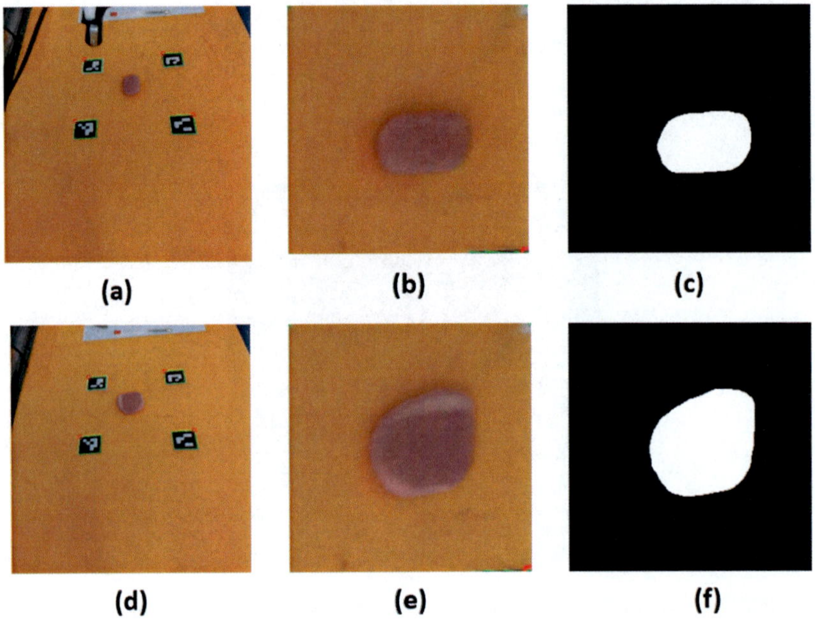

Fig. 6. Snapshot of dough stretching. (a)–(c) are the initial shape of the dough. (d)–(f) are the shape after stretching. (a) is the original image from camera. (b) is the middle image after neural networks processing. (c) is the segmentation of the dough. The machine vision algorithm can detect and calculate the deformation of dough in real time.

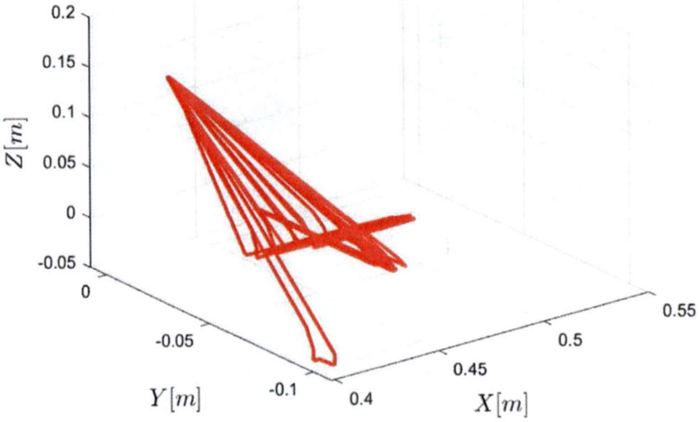

Fig. 7. The trajectory in 3D dimention.

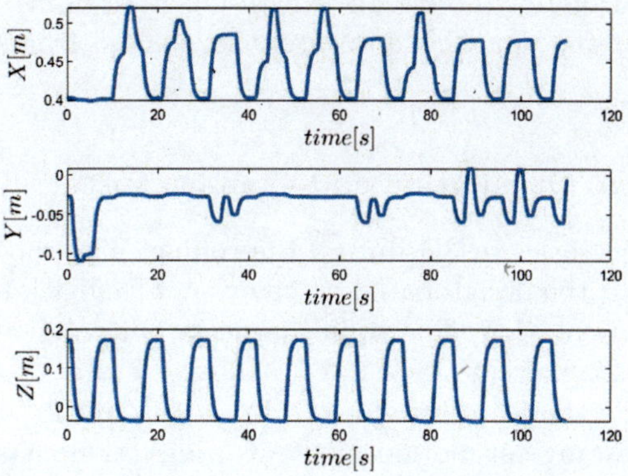

Fig. 8. The position trajectory along X, Y and Z.

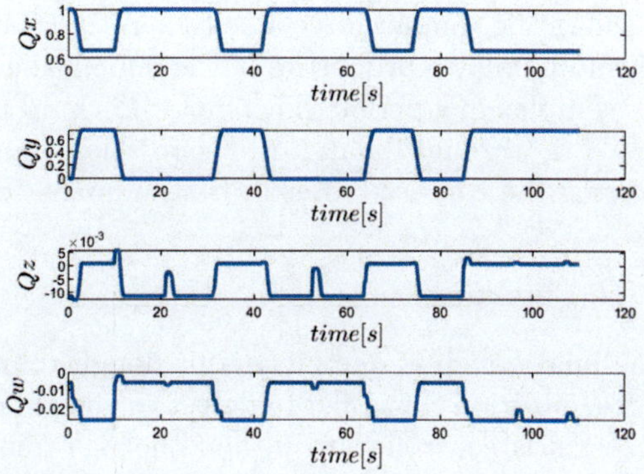

Fig. 9. The orientation trajectory represented by quaternion.

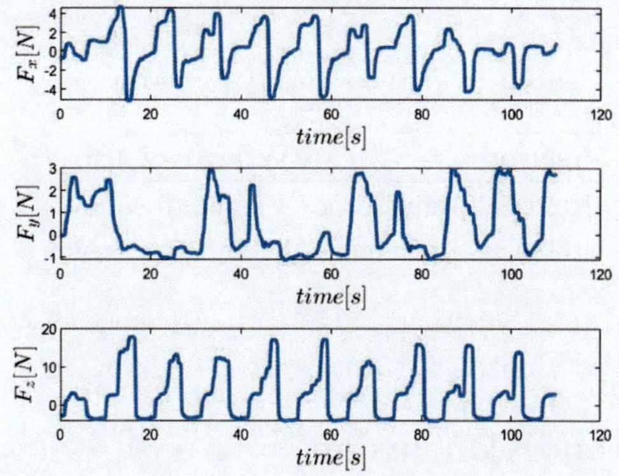

Fig. 10. The contact force along X, Y and Z.

labelled the real outline of the dough. The segmentation error is less than 1% pixel, meaning that the dough is accurately segmented from the image by the model.

3.2 The Position, Orientation and Contact Force

Figure 7 is the trajectories in 3D during the rolling. The robot approaches the dough and then rolls the dough based on the visual feedback to adjust the start point and rolling orientation. The rolling process is iteractively changing until the desired shape is achieved.

In Fig. 8, position trajectories along X, Y, and Z during the rolling is presented. The robot's range of motion in the X-Y direction changes as the dough is deforming and stretching.

Figure 9 is the orientation trajectories along X, Y, and Z during the rolling. In this experiment, we can see the robot keep changing between two directions, which are based on the direction of movement of the dough stretched in different axes. Figure 10 shows the contact force along X, Y, and Z during the rolling. Due to the complex and unknown properties of the dough, the change of contact force along X-Y-Z is complex in a period. Therefore, it is hard to control the contact force accurately. LfD through bilateral teleoperation provides a promising approach to transferring the contact-rich skill from humans to robots.

3.3 The Result of Thickness

The thickness of the final dough is one metric for dough rolling by robots. We adopt the hybrid force/ position controller to detect the height of the final dough. The F_z along the Z-axis is controlled to 0, and the X-Y motion trajectory is controlled by teleoperation. The height along the Z-axis was recorded, and the variance of thickness can be measured.

We compare the performance of human rolling by teleoperation and autonomous rolling by the robot, and the results are shown in Table 1.

Table 1. Flatness quality performance.

Category	Variance of height
Human teleoperation	3 mm ± 0.3mm
Robot autonomous	3.3 mm ± 0.2 mm

4 Discussion and Conclusion

In this paper, we proposed a deformation-aware method for the manipulation of deformable objects. The perspective transformation and U-net method were used to recognise the object and segment the object from the background. Based on the target shape and size of the object, we calculate the desired task variables (desired position and orientation). The realtime trajectory was generated from the learned skill based on human demonstration. And compliant controller was designed to track the desired trajectory. Finally, we evaluate the shape of the final dough by calculating the error along the two axes of the dough. And the variance of the dough in the Z axis reflects the thickness quality. The result shows that the shape error is bellow 5% and the variance of the thickness is 0.3mm. However, the softness of the dough varies a lot, and we will investigate the variable impedance control and iterative learning method to overcome the challenge in the future.

Acknowledgments. This work was supported in part by the H2020 Marie Skłodowska-Curie Actions Individual Fellowship under Grant 101030691.

References

1. Si, W., Wang, N., Yang, C.: A review on manipulation skill acquisition through teleoperation-based learning from demonstration. Cognit. Comput. Syst. **3**(1), 1–16 (2021)
2. Matas, J., James, S., Davison, A.J.: Sim-to-real reinforcement learning for deformable object manipulation. In: Conference on Robot Learning. PMLR, pp. 734–743 (2018)
3. Hu, Z., Sun, P., Pan, J.: Three-dimensional deformable object manipulation using fast online gaussian process regression. IEEE Robot. Automat. Lett. **3**(2), 979–986 (2018)
4. Si, W., Guan, Y., Wang, N.: Adaptive compliant skill learning for contact-rich manipulation with human in the loop. IEEE Robot. Automat. Lett. **7**(3), 5834–5841 (2022)
5. Kim, J.-T., Ruggiero, F., Lippiello, V., Siciliano, B.: Planning framework for robotic pizza dough stretching with a rolling pin. In: Robot Dynamic Manipulation. Springer, pp. 229–253 (2022)
6. Gutiérrez-Giles, A., Ruggiero, F., Lippiello, V., Siciliano, B.: Closed-loop control of a nonprehensile manipulation system inspired by the pizza-peel mechanism. In: 18th European Control Conference (ECC). IEEE 2019, pp. 1580–1585 (2019)
7. Petit, A., Lippiello, V., Fontanelli, G.A., Siciliano, B.: Tracking elastic deformable objects with an rgb-d sensor for a pizza chef robot. Robot. Auton. Syst. **88**, 187–201 (2017)
8. Satici, A.C., Ruggiero, F., Lippiello, V., Siciliano, B.: A coordinate-free framework for robotic pizza tossing and catching. In: 2016 IEEE International Conference on Robotics and Automation (ICRA). IEEE, pp. 3932–3939 (2016)
9. Si, W., Wang, N., Li, Q., Yang, C.: A framework for composite layup skill learning and generalizing through teleoperation. Front. Neurorobotics 16 (2022)
10. Lu, Z., Wang, N., Yang, C.: A novel iterative identification based on the optimised topology for common state monitoring in wireless sensor networks. Int. J. Syst. Sci. **53**(1), 25–39 (2022)
11. Figueroa, N., Ureche, A.L.P., Billard, A.: Learning complex sequential tasks from demonstration: a pizza dough rolling case study. In: 2016 11th ACM/IEEE International Conference on Human-Robot Interaction (HRI). IEEE, pp. 611–612 (2016)

12. Espiau, B., Chaumette, F., Rives, P.: A new approach to visual servoing in robotics. IEEE Trans. Robot. Autom. **8**(3), 313–326 (1992)
13. Pairet, È., Ardón, P., Mistry, M., Petillot, Y.: Learning generalizable coupling terms for obstacle avoidance via low-dimensional geometric descriptors. IEEE Robot. Automat. Lett. **4**(4), 3979–3986 (2019)
14. Oktay, O., Schlemper, J., Folgoc, L.L., Lee, M., Heinrich, M., Misawa, K., Mori, K., McDonagh, S., Hammerla, N.Y., Kainz, B. et al.: Attention u-net: learning where to look for the pancreas. arXiv preprint arXiv:1804.03999 (2018)
15. Ronneberger, O., Fischer, P., Brox, T.: U-net: convolutional networks for biomedical image segmentation. In: International Conference on Medical Image Computing and Computer-Assisted Intervention. Springer, pp. 234–241 (2015)
16. Mezirow, J.: Perspective transformation. Adult Educ. **28**(2), 100–110 (1978)
17. Hartley, R., Zisserman, A.: Multiple view Geometry in Computer Vision, 2nd edn. Cambridge University Press, New York, NY, USA (2003)
18. Ijspeert, A.J., Nakanishi, J., Hoffmann, H., Pastor, P., Schaal, S.: Dynamical movement primitives: learning attractor models for motor behaviors. Neural Comput. **25**(2), 328–373 (2013)
19. Si, W., Wang, N., Yang, C.: Composite dynamic movement primitives based on neural networks for human-robot skill transfer. Neural Comput. Appl. 1–11 (2021)
20. Billard, A., Mirrazavi, S., Figueroa, N.: Learning for Adaptive and Reactive Robot Control: A Dynamical Systems Approach. MIT Press (2022)
21. Ude, A., Nemec, B., Petrić, T., Morimoto, J.: Orientation in cartesian space dynamic movement primitives. In: 2014 IEEE International Conference on Robotics and Automation (ICRA). IEEE, pp. 2997–3004 (2014)
22. Saveriano, M., Piater, J.: e & i Elektrotechnik und Informationstechnik **137**(6), 309–315 (2020). https://doi.org/10.1007/s00502-020-00816-7
23. Hogan, N.: Impedance control: an approach to manipulation: Part i-theory (1985)

Designing Robots with the Context in Mind- One Design Does Not Fit All

Ela Liberman-Pincu[1](✉), Elmer D. van Grondelle[2], and Tal Oron-Gilad[1]

[1] Ben-Gurion University of the Negev, Beer-Sheva, Israel
elapin@post.bgu.ac.il, orontal@bgu.ac.il
[2] Delft University of Technology, Delft, The Netherlands
e.d.vangrondelle@tudelft.nl

Abstract. Robots' visual qualities (VQs) impact people's perception of their characteristics and affect users' behaviors and attitudes toward the robot. Recent years point toward a growing need for Socially Assistive Robots (SARs) in various contexts and functions, interacting with various users. Since SAR types have functional differences, the user experience must vary by the context of use, functionality, user characteristics, and environmental conditions. Still, SAR manufacturers often design and deploy the same robotic embodiment for diverse contexts. We argue that the visual design of SARs requires a more scientific approach considering their multiple evolving roles in future society. In this work, we define four contextual layers: the domain in which the SAR exists, the physical environment, its intended users, and the robot's role. Via an online questionnaire, we collected potential users' expectations regarding the desired characteristics and visual qualities of four different SARs: a service robot for an assisted living/retirement residence facility, a medical assistant robot for a hospital environment, a COVID-19 officer robot, and a personal assistant robot for domestic use. Results indicated that users' expectations differ regarding the robot's desired characteristics and the anticipated visual qualities for each context and use case.

Keywords: Context-driven design · Visual qualities · Socially assistive robot

1 Introduction

1.1 Socially Assistive Robots

Recent years indicate a growing need for Socially Assistive Robots (SARs) [1–3]. Examples of SARs exist in the para-medical field [4, 5] and for domestic use, taking care of the elderly or people with disabilities [6, 7], or helping with children [8, 9]. Since SAR types have functional differences, we expect the user-robot interaction to vary by use context, functionality, user characteristics, and environmental conditions [10, 11]. Yet, our market research revealed that SAR manufacturers often design and deploy the same robotic embodiment for diverse contexts (see Sect. 2.1, Table 2). Most studies in the field of SARs' appearance evaluate users' perceptions of existing off-the-shelf SARs [12–15]. Few studies looked at isolated visual qualities using designated SARs [16].

P. Borja et al. (Eds.): HFR 2022, 26, pp. 105–119, 2023.
https://doi.org/10.1007/978-3-031-22731-8_8

The lack of design research, standards, or a consistent body of knowledge in this field forces designers to start from scratch when designing new robots [17, 18]. Thus, the design of SARs requires a more scientific approach considering their evolving roles in future society. Technological products, even innovative and cutting-edge ones, often fail in the market when the design does not evoke the desired human cognitive response and action, does not apply to environmental conditions, or leads to unrealistic expectations [19–21]. When developing new SARs, the focus is mainly on guaranteeing functionality and safety. Aesthetics and robot look are part of the design process but not necessarily context-specific, i.e., one design fits all. Hekkert and van Dijk (2011) [22] suggest the designer should begin by defining a vision for the context and desired interaction of a new product to set the most appropriate visual qualities (VQs) and come up with a suitable solution for particular design problems.

In this paper, we first classify SARs by their use contexts by outlining four contextual layers: the domain in which the SAR exists, its physical environment, intended users, and role. For example, a robot for the para-medical field, supporting non-professional older adults in their private homes for physical exercises. Then, we examine how potential users perceive SARs in context by delving into the robots' essential characteristics and desired VQs. We used an online questionnaire to collect participants' expectations of robot characteristics by contexts of use and their related VQs. We then analyze this data to evaluate users' perceptions of each robot's desired character and the factors affecting the participant's selection of VQs. Finally, we compared the findings with previous work to form design tools to support user- and interaction-centered designs for diverse tasks and use cases of SARs.

1.2 Initial Mapping of Visual Qualities Perceptions

Previously we evaluated the effect of three VQs for SARs: body structure, outline, and color, on users' perception of the SAR's characteristics [23]. We have empirical findings on how isolated VQs impact people's perception of its characteristics: friendly, childish, innovative, threatening, old-fashioned, massive, elegant, medical, and the robot's gender, as presented in Table 1. For example, to achieve the perception of a friendly SAR, a designer should consider using A-shape or hourglass structure and avoid V-shape, choose light colors (e.g., a combination of white and blue), and avoid dark colors.

Table 1. VQs' effect on self-designed SAR characteristics. Dark boxes represent significant effects (adapted from Liberman Pincu et al. [23]).

	Friendly	Childish	Innovative	Threatening	Old-fashioned	Massive	Elegant	Medical	Robot sex
Structure	■		■	■		■	■		■
Color	■	■		■		■	■	■	■
Outline			■		■		■		

Significance level $p<.05$

These links between SAR characteristics and VQs provide designers with an initial mapping for selecting the VQs most suitable to the robot's role and its desired characteristics, increasing the possibility of aligning with user expectations, at least for the initial encounters with the SAR.

2 Deconstruction of Contexts Layers- Domain, Environment, Users, and Role.

To further enhance the design guidelines to assist designers in the design process of a new SAR and to align these characteristics with the relationship models [24], we now evaluate user expectations in different contexts of use. We map the relevant and desired characteristics for different SARs by a deconstruction process parsing into four contextual layers: Domain, Environment, Users, and Role. The following sections details each layer.

2.1 Domains

Our literature survey and market research lead to seven popular domains for SARs: Healthcare (including Eldercare and Therapy), Educational, Authority (including Security), Companion, Home assistance, Business, and Entertainment. Table 2 provides examples for each domain.

Table 2. Common domains in the market and the literature

Domain	Commercial examples	References
Healthcare	Temi (Robotemi), Pepper (Softbank), NAO (Softbank), Misty (Misty Robotics), QTrobot (LuxAI)	[2, 4, 5]
Educational	Pepper (Softbank), NAO (Softbank), Qtrobot (LuxAI), Buddy (Blue Frog Robotics)	[25–27]
Authority	Knightscope (Knightscope), Cobalt (cobalt robotics)	[28–31]
Companion	Temi (Robotemi), Misty (Misty Robotics), Buddy (Blue Frog Robotics), Aido (Aido)	[32–34]
Home assistance	Misty (Misty Robotics), Aido (Aido)	[6, 35, 36]
Business	Temi (Robotemi), Pepper (Softbank), NAO (Softbank), Cobalt (cobalt robotics), Buddy (Blue Frog Robotics)	[37, 38]
Entertainment	Pepper (Softbank), NAO (Softbank), Buddy (Blue Frog Robotics), Aido (Aido)	[39, 40]

2.2 Physical Environments

SARs are intended for varied environments. The basic level refers to the robot's intended physical location: indoor or outdoor. This classification affects many engineering decisions considering environmental conditions such as light, noise, humidity, dust, and surface conditions (floor, carpet, grass, etc.). The second level refers to privacy: Personal (i.e., home or private office); Semi-public, meaning there are different users all familiar with the robot (e.g., workplace, assisted living residence, etc.); or public, meaning there are multiple users, some are passersby interacting with the robot for the first time. Figure 1 illustrates the levels of physical environments.

2.3 Users

Users can be classified by demographic information, like gender, age, or culture, or by their needs, abilities, and disabilities (cognitive and physical) [40, 41]. In addition, users can be professional (trained to work with the robot, e.g., a trained nurse working with a medical robot) or non-professional [42] (e.g., a hotel guest interacting with a receptionist robot), as well as random (occasional passersby) or familiar with the robot (regularly interacting in the workplace or elsewhere but not as part of their professional work, e.g., a security robot placed at a building entrance). Figure 2 illustrates a classification of users.

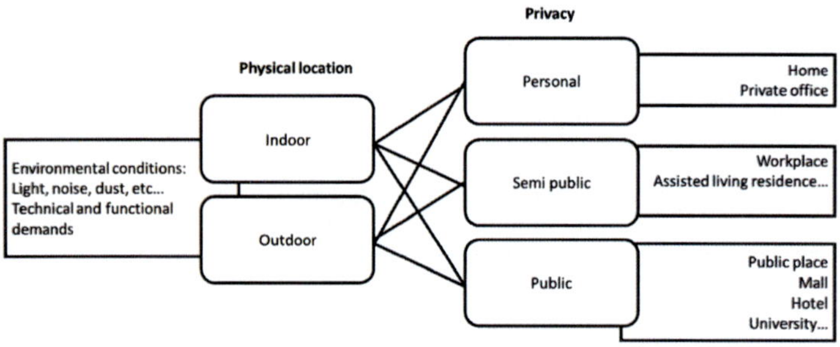

Fig. 1. The two levels of physical environments (physical location and privacy).

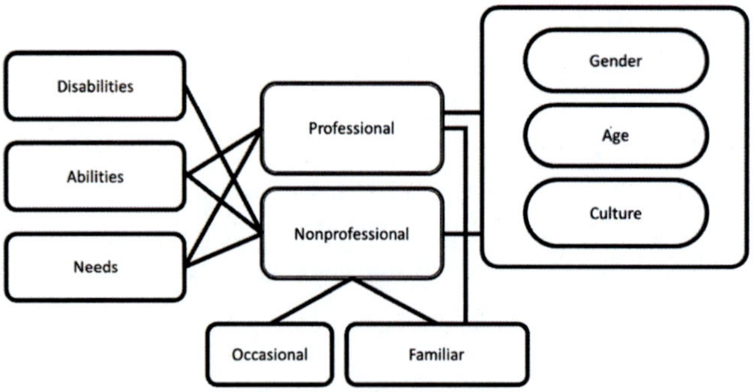

Fig. 2. Classification of users by their demographics, characteristics, and familiarity with the robot.

2.4 Roles and Tasks

The human-robot relationship links to the robot's role and tasks, answering questions such as: Is this robot here to help me? In what way? Different human-robot relationship theories suggest classifying relationships by hierarchy [43]; Should I obey the robot? Who supervises who? Who leads the interaction?

Different role categorizations are found in the literature; Onnasch and Roesler (2021) [44], for example, classified eight abstract roles for various application domains: Information exchange, Precision (e.g., robots for micro-invasive surgery), Physical load reduction, Transport (transport objects from one place to another), Manipulation (the robot physically modifies its environment), Cognitive stimulation, Emotional stimulation, Physical stimulation. Abdi et al. [45] identified five roles of SAR in elder care: affective therapy, cognitive training, social facilitator, companionship, and physiological therapy.

For our model, we used the eight roles based on Onnasch and Roesler (2021) [44], excluding *precision* (which is less related to social relationships) that was replaced with *regulation*: Information exchange, Physical load reduction, Transport, Manipulation, Cognitive stimulation, Emotional stimulation, Physical stimulation, and Regulation.

Each role is classified into a three-level hierarchy of human-robot relationships (robot-led interaction, equal or human-led interaction). For example, in the *physical stimulation* section, we can have a training robot at the *robot-led interaction* level, a teammate at the *equal* level, and a physical therapy robot at the *human-led interaction* level. Figure 3 illustrates the hierarchy of relationships.

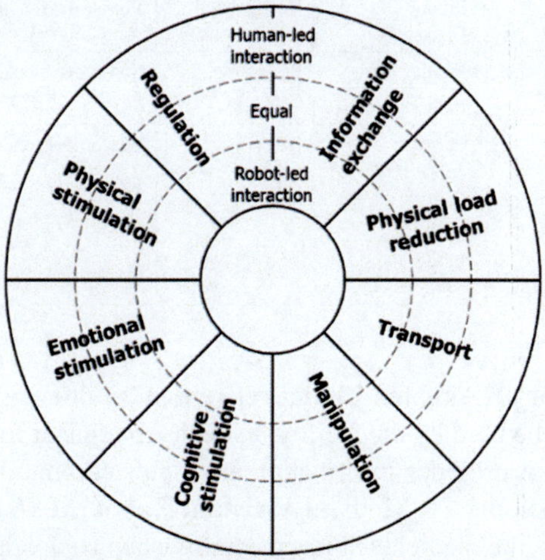

Fig. 3. Eight SAR roles (adapted from Onnasch and Roesler (2021)) by three hierarchy levels of leadership (robot-led interaction, equal or human-led interaction).

3 Evaluating Users' Expectations and Design Perceptions

3.1 Aim and Scope

Based on potential users' evaluations, this study aims to define the appropriate characteristics for robots in different contexts of use. The outcomes will be used together with previous findings [23, 46] to form tools and guidelines for designers and manufacturers.

3.2 Four Use Cases

To apply the deconstruction layers of design, we defined four SAR use cases that differ by their contextual layers: a service robot for an Assisted Living/retirement residence facility (ALR), a Medical Assistant Robot (MAR) for a hospital environment, a Covid-19 Officer Robot (COR), and a Personal Assistant Robot (PAR) for home/domestic use. The following paragraph details each case. Table 3 summarizes them.

Table 3. Four SAR use cases

	Domain	Environment		Users		Role
ALR	Business	An assisted living residence facility	Semi-public	Older adults	Non-professional	Information exchange
			Indoor			Human-led interaction
MAR	Healthcare	Hospital	Public	Medical crews	Professional	Information exchange/ Transport
			Indoor	Hospitalized and caregivers	Non-professional	Human-led interaction/equal
COR	Authority	Public places	Public	Passersby	Non-professional	Regulation
			Indoor/outdoor			Robot-led interaction
PAR	Home assistance	Home	Personal	Diverse	Non-professional	Physical load reduction/Cognitive stimulation/Emotional stimulation
			Indoor			Human-led interaction/equal

A service robot for an Assisted Living/retirement residence facility (ALR) aims to roam the lobby and be used by the facility residents to register for various classes and activities. In addition, it provides information and helps communicate (via video calls and chats) with staff members. **A Medical Assistant Robot (MAR) for a hospital environment** aims to assist the medical team, especially when social distancing is required. Through it, the medical team can communicate in video calls with isolated patients and bring equipment, food, and medicine into patients' rooms. **A COVID-19 Officer Robot (COR)** aims to ensure passersby comply with Covid-19 restrictions like social distancing or wearing a face mask. **A Personal Assistant Robot (PAR) for home/domestic use** seeks to assist users with daily tasks, recommend activities at home and outside, and remind them of their duties and appointments. The robot allows users to watch videos, listen to music, play, and have video chats with family and friends.

3.3 Evaluation Method and Online Questionnaire Design

Using Qualtrics, we designed an online questionnaire where participants were exposed to one of the four use cases. First, they were asked to define the robot's desired characteristics by marking relevant words out of a word bank. The word bank contained twelve words based on previous studies related to SARs' perception [47–49] and that were found relevant to our four use cases: innovative, inviting, cute, elegant, massive, friendly, authoritative, aggressive, reliable, professional, intelligent, and threatening. Following Benedek & Miner's product reaction cards (2002) [50], we followed a similar procedure to our previous studies; however, in this case, participants were not reacting to a design but an idea. In addition, they had the option of adding their own words. Following, they were asked to select three types of VQs: body structure, outline, and color scheme from a set of options, by the alternative that, in their opinion, best expresses the desired characteristics that they have chosen. Figure 4 illustrates the questionnaire design.

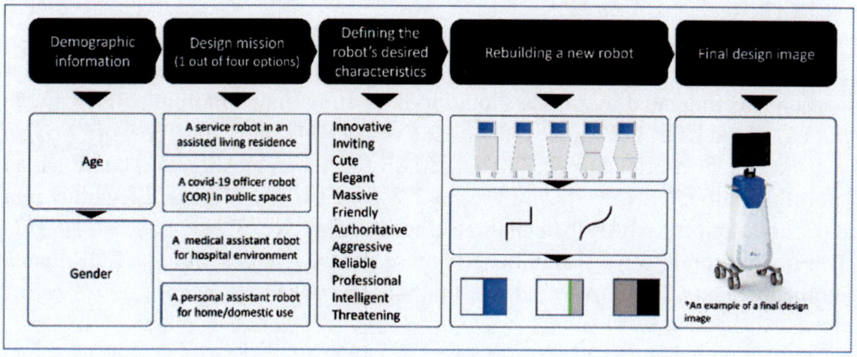

Fig. 4. Questionnaire design.

The online questionnaire was distributed using social media and snowball distribution between November 2021 to March 2022 (via posts on Facebook and WhatsApp). In total, we collected data from 228 adult respondents. Table 4 Summarizes the respondents' demographics by use case.

4 Results

4.1 The Effect of Context on Users' Expectations

We used Chi-square tests of independence to evaluate the effect of the context and the participants' demographic information on their selections of desired characteristics. Participants were grouped into three age groups: up to 29, ages 30–49, and 50 and above. Three participants preferred not to indicate gender; hence in our evaluations of gender effect, $N = 225$ instead of $N = 228$. The words *massive*, *aggressive*, and *threatening* were selected by less than 2% of the participants and therefore were excluded from our analysis.

Table 4. Summary of the respondents' data for each case study

Case study	Gender			Age	Total
	Other	Males	Females		
ALR	1	29	24	M = 38.7, SD = 17.5	54
MAR	1	24	23	M = 35.3, SD = 12.2	48
COR	–	30	25	M = 37.7, SD = 15.5	55
PAR	1	39	31	M = 43.0, SD = 18.0	71
Total	3	122	103	M = 39.0, SD = 16.4	228

Results confirm that users have different expectations regarding the robot's characteristics suitable for each context of use. Table 5 presents the top three words selected by participants for each context. Figure 5 illustrates the chosen words for each context in a radar chart.

We have found statistically significant relations between the contexts and four describing words: *Inviting, Friendly, Elegant,* and *Authoritative.* For example, 78% of the participants indicated that ALR should look inviting, much more than PAR (56%), MAR (54%), and COR (42%), X^2 (3, $N = 228$) = 14.8742, $p < .01$. The word *Friendly* is more likely to be ascribed to ALR (81%), MAR (81%), and PAR (75%) but less likely to be attributed to COR (55%), X^2 (3, $N = 228$) = 13.1663, $p < .01$. Elegant is more suitable for describing a PAR than all three other contexts, X^2 (3, $N = 228$) = 11.5077, $p < .01$. Finally, authoritative is significantly more suitable for describing a COR than all three other contexts, X^2 (3, $N = 228$) = 44.4546, $p < .01$.

Table 5. Top three words selected by participants for each context and their rate.

Case study	Most selected word	2nd word	3rd word
ALR	Friendly (81%)	Inviting (78%)	Reliable (67%)
MAR	Friendly (81%)	Professional (67%)	Reliable (67%)
COR	Professional (67%)	Reliable (60%)	Authoritative (58%)
PAR	Friendly (75%)	Reliable (69%)	Professional (65%)

In addition, we have found that the participants' demographic data (gender and age) affect their selections. Female participants were significantly more likely to select the words *Cute* (X^2 (1, $N = 225$) = 6.68, $p < .01$) and *Friendly* (X^2 (1, $N = 225$) = 10.813, $p < .01$). Though a Chi-square test of independence showed that there were no significant associations between gender and the selection of the word *Innovative,* male participants were more likely to select it (48%) than female participants (37%), (X^2 (1, $N = 225$) = 2.503, $p = .11$. The words *Aggressive* and *Threatening* were selected only by male participants.

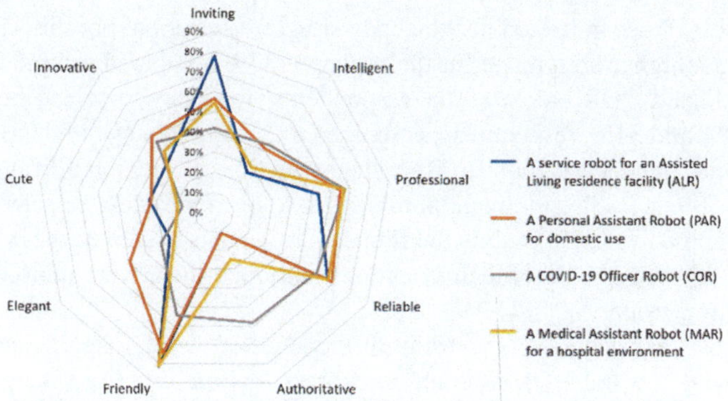

Fig. 5. Users' assigned characteristics by the context of use.

User age category was found to affect the desire for Innovative robots significantly; younger participants (up to 29) selected this character more frequently (53%) than mid-age (ages 30–49) (42%) and older (50 and above) participants (26%), the relationship between these variables was significant, X^2 $(2, N = 228) = 10.78, p < .01$. Furthermore, results indicate two worth noting yet, not statistically significant trends. Implying that age has a positive correlation with selecting the word *Cute* and a negative correlation with choosing the word Professional (i.e., older participants tend to desire cuter, less professional-looking robots). Table 6 summarizes the factors affecting the participants' selection of words.

Table 6. Factors affecting the participants' selection of words.

	Innovative	Inviting	Cute	Elegant	Friendly	Authoritative
Context		■		■	■	■
Gender			■		■	
Age	■					

Significance level p<.05

4.2 Participants' Selection of Visual Qualities

After selecting the characteristics of the SAR, participants were asked to choose the most suitable visual qualities for the context of use they had. Some VQs were selected more frequently than others regardless of the use context or other factors. For example, most respondents (79%) preferred rounded edges over chamfered ones. Only 10% of the respondents chose the dark color scheme, and most participants (49%) preferred the white color scheme. The two most selected structures were the Hourglass (27%) and the A shape (26%).

The context of use impacted just the body structure selection. For the ALR, respondents showed a higher preference for the A shape (37% compared to 26% in the overall data). PAR and COR increased the respondents' tendency to select the Hourglass structure (32% and 31%, respectively, compared to 27% in the overall data).

Participants' gender was found to affect their selection of colors significantly, X^2 (2, N = 225) = 7.939, $p < .05$; the male participants were more likely to select the white and blue combination (54%), while the female participants preferred the white option. Participants' age did not affect their selections, although there were minor differences among the three groups.

Participants' expectations (according to the selected words) significantly affected their selection of several VQs. Wanting to express *Inviting* increased the participants' probability of selecting a rounded outline, X^2 (1, N = 228) = 7.946, $p < .01$, and decreased the participants' probability of selecting the *Dark* color scheme, X^2 (2, N = 228) = 5.537, $p = .06$. To express *Cute* participants selected using the *white* color, X^2 (2, N = 228) = 16.03, $p < .01$, and were more likely to select the *A shape* structure, X^2 (4, N = 228) = 13.23, $p < .05$. Participants who wished to express *Elegant* showed a tendency to select chamfered outline significantly more than in the general population of the study (36% compared to 21% in the overall data), X^2 (1, N = 228) = 23.26, $p < .01$. Wanting to express *Friendly* increased the participants' probability to select the *Hourglass* structure, X^2 (4, N = 228) = 9.5, $p < .05$, the *Rounded* outline X^2 (1, N = 228) = 6.43, $p < .05$, and the *White and blue* color combination, X^2 (2, N = 228) = 8.15, p < .05. Table 7 concludes our findings.

5 Discussion and Future Work

SARs are becoming more prevalent in everyday life [1–3], establishing different kinds of relationships [43]. The body of knowledge in human-robot interaction keeps growing to ensure that these robots follow human' social norms and expectations [10, 11, 51]. However, knowledge is limited regarding the design research of SARs [17, 18]. Most research in the field focuses on evaluating users' perceptions of existing off-the-shelf SARs [12–15]. In this work, we sought to define what these expectations are. Using an online questionnaire, we collected data from 228 respondents regarding their expectations from SARs in four use cases differ in their four layers of context. Results confirm that users have different expectations regarding the robot's characteristics that are suitable for each context. However, in most cases (excluding COR), the top selected word was *Friendly* (see Sect. 4.1 and Table 5).

Further, when asked to select VQs that best express these expectations, we found that participants' demographic data significantly affected their selections (see Sect. 4.2 and Table 7). The four final designs (formed by looking at most users' selections) are almost similar and differ only by the structure; all designs are rounded-edged white robots. Figure 6 presents the four designed robots by use case.

We then combined our findings of this study with our previous empirical findings [23] to set up design guidelines to assist designers in creating a new SAR according to its context. Figure 7 presents desired characteristics in each context according to our recent findings; the table on the right presents design suggestions for each characteristic.

Table 7. Participants' selection of visual qualities.

	Structure (Five levels)	Outline (Two levels)	Color (Three levels)
Overall	A shape (26%) Diamond (18%) Hourglass (27%) Rectangle (10%) V shape (19%)	Rounded (79%) Chamfered (21%)	Dark (10%) White (49%) White and blue (41%)
Assisted Living service robot (ALR)	A Shape (37%)	Rounded (78%)	White (46%)
Personal assistant robot (PAR)	Hourglass (32%)	Rounded (85%)	White (46%)
COVID-19 Officer Robot (COR)	Hourglass (31%)	Rounded (75%)	White (47%)
Medical Assistant Robot (MAR)	–	Rounded (77%)	White (56%)
Male	–	Rounded (79%)	*White and blue (54%)*
Female	–	Rounded (80%)	*White (50%)*
Up to the age of 29	Hourglass (31%)	Rounded (80%)	White and blue (54%)
Ages 30–49	–	Rounded (75%)	White and blue (48%)
50 and above	A Shape (30%)	Rounded (83%)	White (53%)
Inviting	–	**Rounded****	Not Dark
Cute	*A shape**	–	**White****
Elegant	–	**	–
Friendly	*Hourglass**	*Rounded **	*White & blue**

Italic represent a significant level of $p < .05$; Bold represent a significance level of $p < .01$.

For example, in designing a new service robot for an assisted living residence facility (ALR), designers should inspire for a friendly, inviting, and reliable look; hence, they may consider choosing a rounded, white and blue, a-shaped design.

The results align with previous studies that found that design preferences are more related to users' personal preferences. There is no consensus among **users** regarding the appropriate appearance for SARs. Hence, specifically for the case of a personal robot (PAR), the design should allow users to make adjustments using mass customization [52]. In addition, these results may indicate that the participatory design of SARs should be done carefully using two-way evaluations, allowing users to express themselves without relying on them to make design decisions. Participatory design outcomes should be assessed with other users using evaluation tools like Microsoft reaction cards [50].

This research, however, is subject to several limitations. First, participants were not asked to justify their selections; hence, we could not track their intentions. Second, the participants could only select three VQs: structure, outline, and color, with a closed set

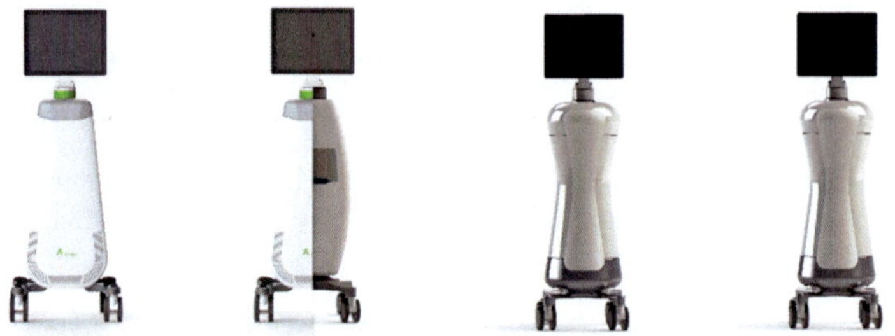

a service robot for an Assisted Living/retirement residence facility (ALR)	Medical Assistant Robot (MAR) for a hospital environment	a COVID-19 Officer Robot (COR)	a Personal Assistant Robot (PAR) for domestic use
A Shape (37%) Rounded (78%) White (46%)	A Shape (27%)/Diamond (25%) Rounded (77%) White (56%)	Hourglass (31%) Rounded (75%) White (47%)	Hourglass (32%) Rounded (85%) White (46%)

Fig. 6. Four options of robots design by use case (by looking at the majority of users' selections). See Table 7 for the full details.

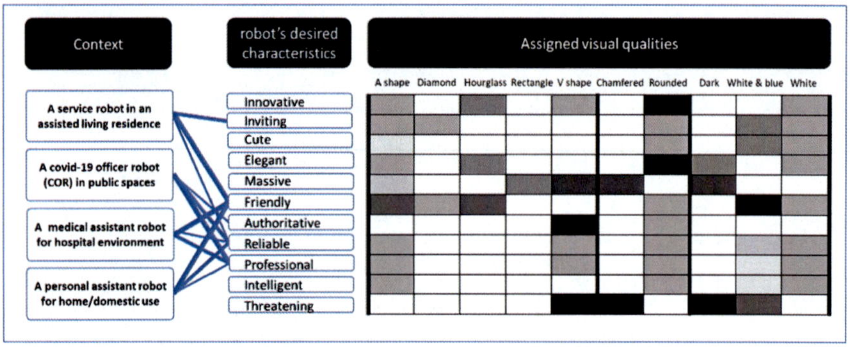

Fig. 7. The desired characteristics in each context according to our recent findings; the table on the right presents design suggestions for each characteristic

of options. Therefore, all outcomes share the exact proportions and dimensions. The robot's height affects user perception [13]; in our previous study evaluating the effect of the COVID-19 officer robot's appearance, some participants mentioned it should be taller [28]. Our subsequent studies will explore stakeholders' perceptions as well as the effect of culture. Such findings will provide further support for the design process of new SARs depending on their context of use, their intended role, and users. And form design guidelines for future SARs.

Acknowledgment. This research was supported by the Ministry of Innovation, Science and Technology, Israel (grant 3-15625), and by the Ben-Gurion University of the Negev through the Helmsley Charitable Trust, the Agricultural, Biological and Cognitive Robotics Initiative, the W. Gunther

Plaut Chair in Manufacturing Engineering and by the George Shrut Chair in Human Performance Management.

References

1. Kodate, N., et al.: Home-care robots—attitudes and perceptions among older people, carers and care professionals in Ireland: a questionnaire study. Health Soc. Care Community (2021). https://doi.org/10.1111/hsc.13327
2. Chita-Tegmark, M., Scheutz, M.: Assistive robots for the social management of health: a framework for robot design and human–robot interaction research. Int. J. Soc. Robot. **13**(2), 197–217 (2021)
3. Zachiotis, G.A., Andrikopoulos, G., Gornez, R., Nakamura, K., Nikolakopoulos, G.: A survey on the application trends of home service robotics. In: 2018 IEEE International Conference on Robotics and Biomimetics (ROBIO), pp. 1999–2006. IEEE (2018)
4. Aymerich-Franch, L., Ferrer, I.: Socially assistive robots' deployment in healthcare settings: a global perspective. arXiv e-prints, arXiv-2110 (2021)
5. Tavakoli, M., Carriere, J., Torabi, A.: Robotics, smart wearable technologies, and autonomous intelligent systems for healthcare during the COVID-19 pandemic: an analysis of the state of the art and future vision. Adv. Intell. Syst. 2000071 (2020)
6. Smarr, C.A., Fausset, C.B., Rogers, W.A.: Understanding the potential for robot assistance for older adults in the home environment. Georgia Institute of Technology (2011)
7. Papadopoulos, I., Koulouglioti, C., Lazzarino, R., Ali, S.: Enablers and barriers to the implementation of socially assistive humanoid robots in health and social care: a systematic review. BMJ Open **10**(1), e033096 (2020)
8. Guneysu, A., Arnrich, B.: Socially assistive child-robot interaction in physical exercise coaching. In: 2017 26th IEEE International Symposium on Robot and Human Interactive Communication (RO-MAN), pp. 670–675. IEEE (2017)
9. Cagiltay, B., Ho, H.R., Michaelis, J.E., Mutlu, B.: Investigating family perceptions and design preferences for an in-home robot. In: Proceedings of the Interaction Design and Children Conference, pp. 229–242 (2020)
10. Caudwell, C., Lacey, C., Sandoval, E.B.: The (Ir) relevance of Robot Cuteness: an exploratory study of emotionally durable robot design. In: Proceedings of the 31st Australian Conference on Human-Computer-Interaction, pp. 64–72 (2019)
11. Onnasch, L., Roesler, E.: A Taxonomy to structure and analyze human–robot interaction. Int. J. Soc. Robot. 1–17 (2020)
12. Lazar, A., Thompson, H.J., Piper, A.M., Demiris, G.: Rethinking the design of robotic pets for older adults. In: Proceedings of the 2016 ACM Conference on Designing Interactive Systems, pp. 1034–1046 (2016)
13. Wu, Y.H., Fassert, C., Rigaud, A.S.: Designing robots for the elderly: appearance issue and beyond. Arch. Gerontol. Geriatr. **54**(1), 121–126 (2012)
14. von der Pütten, A., Krämer N.: A survey on robot appearances. In: 2012 7th ACM/IEEE International Conference on Human-Robot Interaction (HRI), pp. 267–268. IEEE
15. Reeves, B., Hancock, J.: Social robots are like real people: First impressions, attributes, and stereotyping of social robots. Technol. Mind Behav. **1**(1) (2020)
16. Björklund, L.: Knock on wood: does material choice change the social perception of robots? (2018)
17. Sandoval, E.B., Brown, S., Velonaki, M.: How the inclusion of design principles contribute to the development of social robots. In: Proceedings of the 30th Australian Conference on Computer-Human Interaction, pp. 535–538. (2018)

18. Hoffman, G.: Anki, Jibo, and Kuri: What We Can Learn from Social Robots That Didn't Make It. IEEE Spectrum (2019)
19. Fairbanks, R.J., Wears, R.L.: Hazards with medical devices: the role of design. Ann. Emerg. Med. **52**(5), 519–521 (2008)
20. Bartneck, C.: Why do all social robots fail in the market?. [Podcast]. https://doi.org/10.17605/OSF.IO/7KFRZ ISSN 2703–4054 (2020)
21. Bhimasta, R.A., Kuo, P.Y.: What causes the adoption failure of service robots? A Case of Henn-na Hotel in Japan. In: Adjunct Proceedings Of The 2019 Acm International Joint Conference on Pervasive and Ubiquitous Computing and Proceedings of the 2019 ACM international symposium on wearable computers, pp. 1107–1112 (2019)
22. Hekkert, P., Van Dijk, M.: ViP-Vision in design: A guidebook for innovators. BIS publishers (2011)
23. Liberman-Pincu, E., Parmet, Y., Oron-Gilad, T.: Judging a socially assistive robot (SAR) by its cover; The effect of body structure, outline, and color on users' perception. arXiv preprint arXiv:2202.07614. (2022)
24. Liberman-Pincu, E., van Grondelle, E.D., Oron-Gilad. T.: Designing robots with relationships in mind- Suggesting two models of human- socially assistive robot (SAR) relationship. In: Proceedings of 2021 HRI '21 Companion, March 8–11, 2021, Boulder, CO, USA. ACM, New York., NY, USA, 5 pages. https://doi.org/10.1145/3434074.3447125
25. Rosanda, V., Istenic Starcic, A.: The robot in the classroom: a review of a robot role. In: International Symposium on Emerging Technologies for Education, pp. 347–357. Springer, Cham (2019)
26. Chin, K.Y., Hong, Z.W., Chen, Y.L.: Impact of using an educational robot-based learning system on students' motivation in elementary education. IEEE Trans. Learn. Technol. **7**(4), 333–345 (2014)
27. Belpaeme, T., Kennedy, J., Ramachandran, A., Scassellati, B., Tanaka, F.: Social robots for education: a review. Sci. Robot. **3**(21), eaat5954 (2018)
28. Liberman-Pincu, E., David, A., Sarne-Fleischmann, V., Edan, Y., Oron-Gilad, T.: comply with me: using design manipulations to affect human-robot interaction in a covid-19 officer robot use case. Multimodal Technol. Interact. **5**(11), 71 (2021)
29. Espinas, M.F.C., Roguel, K.M.G., Salamat, M.A.A., Reyes, S.S.G.: Security Robots vs Security Guards
30. Agrawal, S., Williams, M.A.: Would you obey an aggressive robot: a human-robot interaction field study. In: 2018 27th IEEE International Symposium on Robot and Human Interactive Communication (RO-MAN), pp. 240–246. IEEE (2018)
31. Geiskkovitch, D.Y., Cormier, D., Seo, S.H., Young, J.E.: Please continue, we need more data: an exploration of obedience to robots. J. Hum. Robot Interact. **5**(1), 82–99 (2016)
32. Robinson, H., MacDonald, B., Kerse, N., Broadbent, E.: The psychosocial effects of a companion robot: a randomized controlled trial. J. Am. Med. Dir. Assoc. **14**(9), 661–667 (2013)
33. Borenstein, J., Pearson, Y.: Companion robots and the emotional development of children. Law Innov. Technol. **5**(2), 172–189 (2013)
34. Gasteiger, N., Loveys, K., Law, M., Broadbent, E.: Friends from the future: a scoping review of research into robots and computer agents to combat loneliness in older people. Clin. Interv. Aging **16**, 941 (2021)
35. Yamazaki, K., Ueda, R., Nozawa, S., Kojima, M., Okada, K., Matsumoto, K., ... Inaba, M.: Home-assistant robot for an aging society. Proceedings of the IEEE **100**(8), 2429–2441 (2012)
36. Fuentes-Moraleda, L., Diaz-Perez, P., Orea-Giner, A., Munoz-Mazon, A., Villace-Molinero, T.: Interaction between hotel service robots and humans: a hotel-specific service robot acceptance model (sRAM). Tourism Manage. Perspect. **36**, 100751 (2020)

37. Rosete, A., Soares, B., Salvadorinho, J., Reis, J., Amorim, M.: Service robots in the hospitality industry: an exploratory literature review. In: International Conference on Exploring Services Science, pp. 174–186. Springer, Cham (2020)
38. Kwak, S.S., Kim, M.S.: User preferences for personalities of entertainment robots according to the users'psychological types. Bull. Japan. Soc. Sci. Des. **52**(4), 47–52 (2005)
39. Bogue, R.: The role of robots in entertainment. Ind. Robot: Int. J. Robot. Res. Appl. (2022)
40. Flandorfer, P.: Population ageing and socially assistive robots for elderly persons: the importance of sociodemographic factors for user acceptance. Int. J. Popul. Res. **2012** (2012)
41. Cortellessa, G., Scopelliti, M., Tiberio, L., Svedberg, G.K., Loutfi, A., Pecora, F.: A Cross-Cultural Evaluation of Domestic Assistive Robots. In: AAAI fall symposium: AI in eldercare: new solutions to old problems, pp. 24–31 (2008)
42. Raigoso, D., Céspedes, N., Cifuentes, C.A., Del-Ama, A.J., Múnera, M.: A survey on socially assistive robotics: clinicians' and patients' perception of a social robot within gait rehabilitation therapies. Brain Sci. **11**(6), 738 (2021)
43. Prescott, T.J., Robillard, J.M.: Are friends electric? The benefits and risks of human-robot relationships. Iscience **24**(1), 101993 (2021)
44. Onnasch, L., Roesler, E.: A taxonomy to structure and analyze human–robot interaction. Int. J. Soc. Robot. **13**(4), 833–849 (2021)
45. Abdi, J., Al-Hindawi, A., Ng, T., Vizcaychipi, M.P.: Scoping review on the use of socially assistive robot technology in elderly care. BMJ Open **8**(2), e018815 (2018)
46. Liberman-Pincu, E., Oron-Gilad. T.: Impacting the perception of socially assistive robots-evaluating the effect of visual qualities among children. In: Proceedings of the 30th IEEE International Conference on Robot and Human Interactive Communication (RO-MAN) August 8–12, 2021—Vancouver, BC, CA (Virtual Conference) (2021)
47. Bartneck, C., Kulić, D., Croft, E., Zoghbi, S.: Measurement instruments for the anthropomorphism, animacy, likeability, perceived intelligence, and perceived safety of robots. Int. J. Soc. Robot. **1**(1), 71–81 (2009)
48. Carpinella, C.M., Wyman, A.B., Perez, M.A., Stroessner, S.J.: The robotic social attributes scale (RoSAS) development and validation. In: Proceedings of the 2017 ACM/IEEE International Conference on Human-Robot Interaction, pp. 254–262 (2017)
49. Kalegina, A., Schroeder, G., Allchin, A., Berlin, K., Cakmak, M.: Characterizing the design space of rendered robot faces. In: Proceedings of the 2018 ACM/IEEE International Conference on Human-Robot Interaction, pp. 96–104 (2018)
50. Benedek, J., Miner, T.: Product reaction cards. Microsoft, July, 29 (2002)
51. Saunderson, S., Nejat, G.: How robots influence humans: a survey of nonverbal communication in social human–robot interaction. Int. J. Soc. Robot. **11**(4), 575–608 (2019)
52. Liberman-Pincu, E., Oron-Gilad, T.: Exploring the effect of mass customization on user acceptance of socially assistive robots (SARs). In: Proceedings of the 2022 ACM/IEEE International Conference on Human-Robot Interaction, pp. 880–884 (2022)

Estimation of Whole-Body Muscular Activation from an Optimal Set of Scarce Electromyographic Recordings

Marco Baracca[1(✉)], Giuseppe Averta[2], and Matteo Bianchi[1]

[1] Research Center "E. Piaggio" and Department of Information Engineering, University of Pisa, Pisa, Italy
marco.baracca@phd.unipi.it

[2] Department of Control and Computer Engineering of the Politecnico di Torino, Torino, Italy

Abstract. Monitoring workers' status is crucial to prevent work-related musculoskeletal disorders and to enable a safe human-robot interaction. This is typically achieved relying on muscle activation recordings, commonly performed via wearable electromyographic EMG sensors. However, to properly acquire whole-body muscular status, a large number of sensors is needed. This represents a limitation for a real deployment of wearable acquisition systems, due to cost and wearability constraints. To overcome this problem, we propose a solution to provide a reliable muscles estimation from a limited number of EMG recordings. Our method exploits the covariation patterns between muscles activation to complement the recordings coming from a reduced set of optimally placed sensors, minimizing the estimation uncertainty. Using a dataset of EMG data recorded from 10 subjects, we demonstrate that it is possible to reconstruct the temporal evolution of 10 whole-body muscles with a maximum normalized estimation error of 13%, using only 7 EMG sensors.

Keywords: Ergonomics · Human motion control · EMG · Optimal sensing

1 Introduction

Despite the increasing importance that human robot collaboration has gained for increasing the quality of human work and decreasing the risk of injuries [1], more than half of workers in the European Union (EU) report disturbances in carrying out daily tasks, which often result in chronic Work-related MusculoSkeletal Disorders (WMSDs) [6]. The most commonly identified issues are muscles-related pain, especially at the back, upper limb and lower limb level [17]. WMSDs affect workers with different ages and types of occupations, with high costs for companies and the healthcare systems [7,13]. A possible way to reduce this problem is to monitor the biomechanical state of the worker during the day through wearable technologies [1].

To solve this problem, many works in literature provide solutions for the assessment of ergonomic indexes [11,15], fatigue quantification during working activities [4,8,16] and to enhance human-robot interaction [12]. Most of these approaches are based on

P. Borja et al. (Eds.): HFR 2022, 26, pp. 120–130, 2023.
https://doi.org/10.1007/978-3-031-22731-8_9

recording of muscular activities to estimate fatigue and joints overload, relying on the usage surface ElectroMyoGraphic (sEMG) sensors. However, for a real deployment of this technology in large scale, it is important to recall that EMG sensors are usually expensive and, given the large number of muscles present in the human body, the amount of sensors needed to obtain a complete measure of the biomechanical state could be very large, with direct implications on costs and wearability of the sensing setup. Therefore, to increase the usability and acceptability of sEMG sensing systems, it is important to optimize the number and the placement of the sensor elements, without penalising the quality of the sensing.

A possible way to tackle the problem of dimensionality reduction is to exploit the concept of motor synergies. Introduced in [9], motor synergies are a set of strategies performed by the nervous system to handle the complexity of human body. This concept opens to the possibility to analyze the functional coordination of elemental motor variables targeting the execution of a given tasks. At the muscular level, a seminal work is [5], where authors found co-activation patterns during motion, usually named muscle synergies. From an observability point of view, the concept of synergies can be associated to the possibility to estimate the overall kinematic-muscular state in spite of a reduced amount of sensory information, exploiting the commonalities shared by the motor units. In [2], the authors presented a framework to estimate human arm biomechanical state (sEMG signals and joint trajectories) during daily-living activities from a reduced number of measurements. This approach relies on a Minimum Variance Estimation (MVE) algorithm combined with a representation based on functional Principal Component Analysis (fPCA) to exploit a dataset of recorded movements as *a priori* knowledge used to complement the whole-state estimation from a reduced set of optimal signals that minimize the estimation uncertainty in spite of scarce sensory information. However, the authors focused only on the application of this approach for the estimation of the human upper limb motion.

In this work, we validate the feasibility to generalize this approach to the estimation of the whole-body muscular state. To do this, we used a dataset containing movements performed during industrial tasks. Data were split in *a priori* and validation set. The first was used to identify the optimal sEMG placement to minimize estimation uncertainty, while the second was used to assess the performances of our method on data that were not used to build the a-priori knowledge. Interestingly, our results demonstrate that, starting from a set of 10 muscles, we can remove up to 3 sensors without substantially worsening the estimate obtained.

The paper is organized as follows: we first summarize the theoretical foundations of this method; then, we report the methods and the implementation of the optimization procedure on the whole-body dataset; finally, we discuss the results obtained in the reconstruction phase.

2 Theoretical Framework

As introduced in the previous section, the goal of this work is to provide an estimation of muscle activation signals using a limited number of sensor elements. To achieve this results we have choosen the Minimum Variance Estimation (MVE) approach to use the knowledge of *a priori* information provided by a set of data recorded in advance. However, to exploit the MVE approach, we need to represent movements in a static domain instead of the temporal domain. The strategy used in this work to solve the problem is the one presented in [2]. It relies on 3 different phases: i) an encoding phase to translate the signal from the time domain to a static representation obtained via functional Principal Component Analysis; ii) an estimation phase, in which MVE is used to estimate the missing information; and iii) a decoding phase to re-obtain the temporal evolution of state from the static representation.

2.1 Encoding Phase

The encoding phase is based on functional Principal Component Analysis (fPCA), an extension of Principal Component Analysis suitable to manage time-series. In a nutshell, given a dataset of time-varying data, fPCA extracts a basis of function ordered by importance (where the importance is represented by the explained variance of the dataset itself).

Considering a sEMG signal $m(t)$, its linear functional decomposition can be defined as:

$$m(t) \simeq \bar{m} + S_0(t) + \sum_{i=1}^{s_{max}} \alpha_i S_i(t), \tag{1}$$

where $\bar{m}$ is the average of the signal, $S_0(t)$ is the average temporal muscular activation profile through the whole dataset, $S_i(t)$ is the ith functional Principal Component (fPC) and α_i is the weight associated to the element $S_i(t)$.

The first component of the basis of function $S_1(t)$ can be extracted from the dataset by solving the following problem:

$$\max_{S_1} \sum_{j=1}^{R} \left(\int S_1(t) m_j(t) dt \right)^2 \tag{2}$$

subject to

$$||S_1(t)||_2^2 = 1. \tag{3}$$

The other components $S_i(t)$ can be computed as:

$$\max_{S_i} \sum_{j=1}^{R} \left(\int S_i(t) m_j(t) dt \right)^2 \tag{4}$$

subject to

$$\begin{gathered} ||S_i(t)||_2^2 = 1 \\ \int_0^{t_{end}} S_i(t) S_k(t) dt = 0, \forall k \in \{1, ..., i-1\} \end{gathered} \tag{5}$$

For practical details on the implementation, the interested reader could refer to [14].

This decomposition can be applied to each recorded muscle to express the temporal activation profile into a set of weights. Given a set of M muscles, we can define an extended static state x_e representing a single trial as:

$$x_e = \left[\bar{m}_1 \; \alpha_{1,1} \; \dots \; \alpha_{1,k} \mid \bar{m}_2 \; \alpha_{2,1} \; \dots \; \alpha_{2,k} \mid \dots \mid \bar{m}_M \; \alpha_{M,1} \; \dots \; \alpha_{M,k} \right]^T. \tag{6}$$

2.2 Estimation Phase

The extended state defined in (6) permits to apply MVE for the estimation of temporal signals. In this section, we briefly summarize the principle behind the approach and we refer the interested reader to [3] for more details.

Considering a set of d sensors to measure the state of the system $x \in \mathbb{R}^l$, we can assume a linear relation between the state and the output and define the vector of measures $y \in \mathbb{R}^d$ as:

$$y = Hx + v \tag{7}$$

where $H \in \mathbb{R}^{d,l}$ is a full row rank measurement matrix and v is the measurement noise. In the case where $d < l$ we have an infinite number of solutions defined as:

$$x = H^{\dagger} y + N_h \xi \tag{8}$$

where $H^{\dagger}$ is the pseudo-inverse of H, N_h represents the null space of H and $\xi \in \mathbb{R}^{l-d}$ is a free vector of parameters. Usually, the most common solution is to use the pseudo-inverse which returns the least-squared solution. However, this is not always the best solution in terms of error between the the real state and its estimation.

The solution proposed with the MVE is to exploit the covariation patterns between the elements of the state to reduce the estimation error. The information used is organized in a covariance matrix P_0 defined as:

$$P_0 = \frac{(X - \bar{x})(X - \bar{x})^T}{N - 1} \tag{9}$$

where $\bar{x}$ is a matrix whose columns contain the average μ_0 of X. Given P_0, the best estimate $\hat{x}$ of x is the vector which solves the following optimization problem:

$$\hat{x} = \text{argmin} \frac{1}{2}(x - \mu_0)^T P_0^{-1}(x - \mu_0) \tag{10}$$

Assuming that v is a zero mean Gaussian noise with covariance matrix R, the solution of (10) can be found in closed form as:

$$\hat{x} = (H^T R^{-1} H + P_0^{-1})^{-1}(H^T R^{-1} y + P_0^{-1} \mu_0) \tag{11}$$

We can also define the *a posteriori* covariance matrix as:

$$P_P = (H^T R^{-1} H + P_0^{-1})^{-1} \tag{12}$$

This matrix is important because it returns information regarding the uncertainty of the associated state estimation and can be used to select what element of the state has to be measured to maximize the information. To solve the problem we can set the following minimization:

$$H_{opt} = \underset{H}{\text{argmin}}\ \sigma_{max}(P_P(H)) \tag{13}$$

to find the optimal selection matrix H_{opt} and consequently the best set of muscle to be recorded.

2.3 Decoding Phase

After we computed the estimated extended state $\hat{x}_e$ we want to re-obtain the temporal evolution of the different signals represented by it. This step can be easily done using (1) to combine the original fPCs and obtain the reconstructed signal in the time domain.

3 Optimal Sensor Setup

3.1 Dataset

To validate our approach to whole-body muscles estimation we used the dataset available in [10], which consists of kinematic and muscular data of subjects performing different industrial-like tasks. The three tasks taken into account are: 1) lifting and lowering boxes with different weights from shelves at different heights; 2) drilling and 3) painting with a lightweight tool. Every subject involved in this dataset perfomed three repetition for each task condition. For the purpose of our study, we took into account only the post-processed muscular data of the right-handed subjects (10 out of 12). Muscular activation were recorded with a sEMG system (Delsys Trigno Wireless platform) and the signal processing consists in: normalization w.r.t. the maximum voluntary contraction, filtering with a 2nd order Butterworth low-pass filter (cut-off frequency 2 Hz), and rectification. The list of the represented muscles can be found in Table 1. Given that each task is represented by a different number of movements, we randomly discarded part of the trials to create a task-balanced dataset.

Table 1. List of muscles recorded during task execution.

Index	Muscle
1	Anterior Deltoid
2	Posterior Deltoid
3	Biceps Brachii
4	Triceps Brachii
5	Trapezius Descendens
6	Erector Spinae
7	Gluteus Maximus
8	Rectus Femoris
9	Biceps Femoris
10	Tibialis Aterior

3.2 Sensor Optimization

Given the particular structure of the extended state, the matrix H is composed of block of k-dimensional diagonal matrices. To preserve this structure, we chose the genetic algorithm as method to solve the optimization problem, where each candidate solution is an N-element vector containing the N indexes of the muscles to be measured. In this way, given the single individual of the population, the matrix H can be easily computed and, combining it with the *a priori* knowledge, we can obtain the *a posteriori* covariance matrix P_P through (12). The optimal EMG sensor selection was identified minimizing the Shatten p-norm of P_P defined as:

$$\|P_P\|_p := \left(\sum s_n^p(P_P)\right)^{\frac{1}{p}} \tag{14}$$

The minimization of this cost function leads to the minimization of the maximum singular value of P_P, and consequently to the minimization of the uncertainty of estimation. The genetic algorithm was implemented using the software Matlab (Population size = 150, Max number of generations = 200, Elite Count = 10).

To verify the stability of our estimation procedure, the dataset was split in 10 groups, one for each subject, and a k-fold validation was implemented. For each iteration, one of the subjects was selected as a validation set, while the remaining ones was used to build the *a priori* knowledge. For each *a priori* dataset, we performed the fPCA to extract the principal components and the associated weights. Then, weights were used to compute the average μ_0 and the *a priori* covariance matrix P_0.

We performed the optimization for each fold with different number of sensors used (from 1 to 9). The results obtained, in terms of best values of the cost function reached, are reported in Fig. 1.

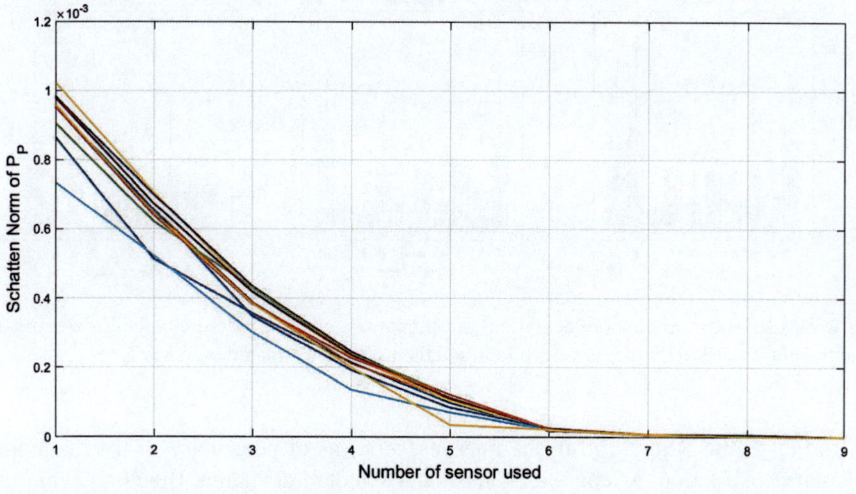

Fig. 1. Shatten norm obtained for different subjects with different number of EMG sensors used

We can observe that we can remove up to 3 sensors without noteworthy deterioration of the estimation uncertainty. These 3 cases (usage of 7, 8 and 9 sensors respectively) return also a stable muscle selection through different subjects used as validation set.

We have evaluated also the stability of the representation obtained with fPCA across different subjects. To do this, we performed the dot product to assess the similarity between fPCs of the same order and of the same degree of freedom. We obtained median values ranging from 0.998, for the first order component, to 0.952 for the fifth component. More details are reported in Table 2.

Table 2. Median, 25th and 75th percentile of similarity index between functional components for each fPC order.

fPC order	**0**	**1**	**2**	**3**	**4**	**5**
Median	0.9947	0.9980	0.9937	0.9858	0.9765	0.9524
75th percentile	0.9978	0.9996	0.9989	0.9970	0.9947	0.9899
25th percentile	0.9817	0.9908	0.9646	0.9148	0.8420	0.5962

4 Results

To validate the reliability of our approach, we calculated the differences between the real recorded signal and the estimation for all the validation sets. We used as metric the Root Mean Square Error (RMSE). This procedure was repeated separately for the case with 7, 8 and 9 sensors used.

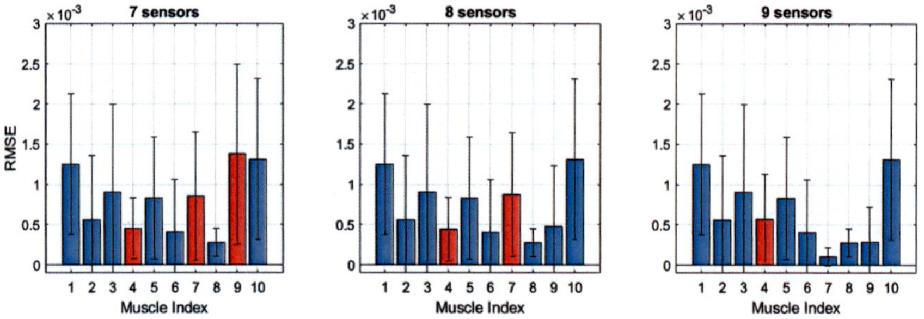

Fig. 2. RMSE (mean ± standard deviation) of optimal setup with different number of sensors. In blue the muscles directly measured while in red the muscles estimated with MVE

In Fig. 2, the RMSE for all the muscles, in terms of percentage of the maximum voluntary contraction, is reported as mean and standard deviation. Blue bars represent the muscles directly measured, while the red ones stand for the muscles estimated by the MVE. It is possible to observe that the estimation error made by the MVE is similar to the one introduced by the fPCA decomposition in measured muscles (which is the best achievable results given a fixed number of components used).

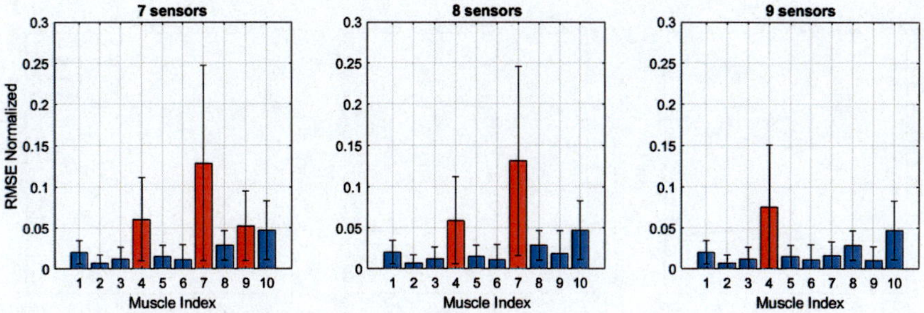

Fig. 3. RMSE normalized (mean ± standard deviation) with the maximum activation range for each muscles. In blue the muscles directly measured while in red the muscles estimated with MVE

We also normalized the RMSE of each muscle with the maximum range reached on the dataset to compare the error with the muscle activation level of the tasks analyzed. The results in term of mean and standard deviation are represented in Fig. 3. In this case, there is a difference between recorded and estimated muscles. However, we can see that, even in the worst case, the average estimation error for the muscles is always below 13%.

However, these results can only inform on the average error of the reconstruction. To verify the reconstruction accuracy in time, we also compared the reconstructed temporal profile with the real data in time (see Figs. 4 and 5). It is possible to notice that both the values and the shape obtained are very similar to the reference even with a modest sized sensory information available.

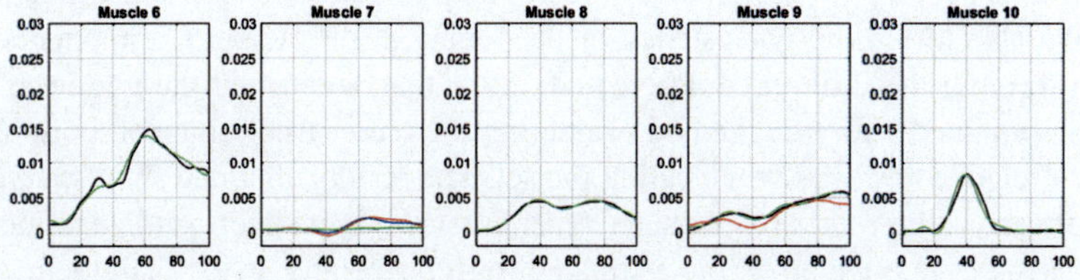

Fig. 4. Comparison between the real signal (black) and the MVE output using 7 (red), 8 (blue) and 9 (green) EMG sensors during a lifting action. The activation level is normalized with the Maximum Voluntary Contraction of each muscle. The estimated signals are Muscle 4 (for red, blue and green lines), Muscle 7 (for red and blue lines) and Muscle 9 (red line)

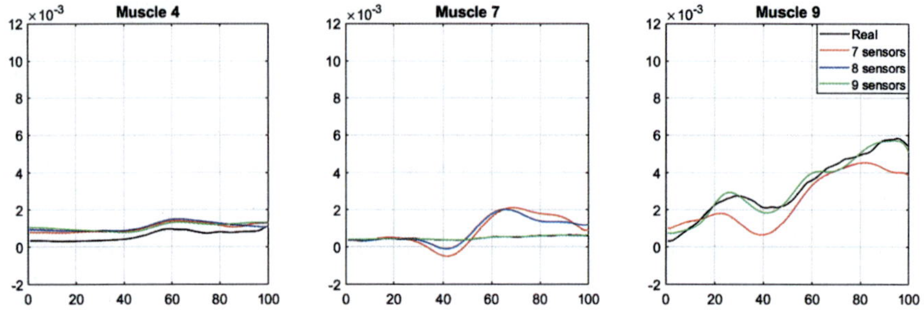

Fig. 5. Detail on the estimated muscles for action represented in Fig. 4

5 Discussions and Conclusions

Improve worker's body sensorization is a key factor to enhance human-robot interaction and assess its condition performing daily working activities. Obtaining a good estimation of the biomechanical state of the worker's body is essential to prevent injuries and ailments that - in the long term - can lead to chronic conditions that may affect the worker's daily life. One of the data necessary to carry out this assessment is that concerning the muscular activity which allows to monitor the state of fatigue of the worker. However recording all the muscles present in human body can be expensive and this type of measurement can be affected by different sources of noise. Furthermore, a system with a large number of sensors would create impairments to the worker in his activities, discouraging its use.

To reduce the number of sensor we took the Minimum Variance Estimation approach, which exploit the information contained in a *a priori* dataset to compensate for noisy or missing measurement, and we applied it to estimate whole body muscle activity. To do this, we used a dataset containing a set of working-like movement performed by different subjects.

Given the small size of the dataset, we performed a k-fold validation using each subject as validation data. For each *a priori* set we performed an optimization to find the optimal sensor setup for each number of sensor elements possible. We found that, starting from a set of 10 sensors, we can remove up to 3 sensors without worsening the output. We also tested the consistency of the functional Principal Components representation through the different fold created. After that, we tested the goodness of the estimation obtained with this method evaluating the reconstruction error, both in absolute and in relative terms, for each validation set. The results obtained show that, in term of absolute error, the estimated muscles reach a level of precision similar to measured ones. In terms of relative error, the estimated muscles behave worst than the measured one. However only the Gluteus Maximum reaches an error higher than 10% (13.1%), while for the other 2 muscles the relative error is about 6%.

Despite the good result obtained, we believe that there is room of improvement for this work. The first step is to increase the number of muscles taken into account in order to obtain a more fine grained assessment of the entire body. However this approach requires a number of trials contained in the *a priori* dataset much larger of the dimension of the extended state and it is necessary to gather a new dataset containing an higher number of recorded movements. With a sufficient number of trials in the algorithm can be integrated also the kinematic data (as done in a previous work on human arm) to achieve a complete biomechanical estimation necessary for ergonomic assessment. Another point to be developed is the extension of this approach for real-time estimates. In fact, based on the fPCA decomposition to move from the time domain to the weight domain, it is necessary to have recorded the complete movement in time. An idea could be to iteratively perform this approach in real time using the movement recorded up to that moment. However, the feasibility of this solution is still under investigation.

Acknowledgements. This paper is part of a project that has received funding from European Union's Horizon 2020 Research and Innovation Program under Grant Agreement No. 101017274 (DARKO), and Grant Agreement No. 871237 (SOPHIA). The research leading to these results has received partial funding also from the Italian Ministry of Education and Research (MIUR) in the framework of the CrossLab project (Departments of Excellence), and in the framework of PRIN (Programmi di Ricerca Scientifica di Rilevante Interesse Nazionale) 2017, with the project TIGHT: Tactile InteGration for Humans and arTificial systems, grant number 2017SB48FP.

References

1. Ajoudani, A., Albrecht, P., Bianchi, M., Cherubini, A., Del Ferraro, S., Fraisse, P., Fritzsche, L., Garabini, M., Ranavolo, A., Rosen, P.H., Sartori, M., Tsagarakis, N., Vanderborght, B., Wischniewski, S.: Smart collaborative systems for enabling flexible and ergonomic work practices [industry activities]. IEEE Robot. Autom. Mag. **27**(2), 169–176 (2020). https://doi.org/10.1109/MRA.2020.2985344
2. Averta, G., Iuculano, M., Salaris, P., Bianchi, M.: Optimal reconstruction of human motion from scarce multimodal data. IEEE Trans. Human Mach. Syst. (2022)
3. Bianchi, M., Salaris, P., Bicchi, A.: Synergy-based hand pose sensing: reconstruction enhancement. Int. J. Robot. Res. **32**(4), 396–406 (2013)
4. Cifrek, M., Medved, V., Tonković, S., Ostojić, S.: Surface EMG based muscle fatigue evaluation in biomechanics. Clin. Biomech. **24**(4), 327–340 (2009)
5. D'Avella, A., Saltiel, P., Bizzi, E.: Combinations of muscle synergies in the construction of a natural motor behavior. Nat. Neurosci. **6**(3), 300–308 (2003)
6. De Kok, J., Vroonhof, P., Snijders, J., Roullis, G., Clarke, M., Peereboom, K., van Dorst, P., Isusi, I.: Work-related musculoskeletal disorders: prevalence, costs and demographics in the EU. Eur. Agency Saf. Health Work **1** (2019)
7. James, S.L., Abate, D., Abate, K.H., Abay, S.M., Abbafati, C., Abbasi, N., Abbastabar, H., Abd-Allah, F., Abdela, J., Abdelalim, A., et al.: Global, regional, and national incidence, prevalence, and years lived with disability for 354 diseases and injuries for 195 countries and territories, 1990–2017: a systematic analysis for the global burden of disease study 2017. Lancet **392**(10159), 1789–1858 (2018)
8. Jebelli, H., Lee, S.: Feasibility of wearable electromyography (EMG) to assess construction workers' muscle fatigue. In: Advances in Informatics and Computing in Civil and Construction Engineering, pp. 181–187. Springer (2019)
9. Latash, M.L., Scholz, J.P., Schöner, G.: Toward a new theory of motor synergies. Motor Control **11**(3), 276–308 (2007)
10. Lorenzini, M., Kim, W., Ajoudani, A.: Human Kino-Dynamic Measurements Dataset for Factory-like Activities (2021). https://doi.org/10.5281/zenodo.5575139

11. Lorenzini, M., Kim, W., Ajoudani, A.: An online multi-index approach to human ergonomics assessment in the workplace. IEEE Trans. Human Mach. Syst. (2022)
12. Peternel, L., Fang, C., Tsagarakis, N., Ajoudani, A.: A selective muscle fatigue management approach to ergonomic human-robot co-manipulation. Robot. Comput. Integr. Manuf. **58**, 69–79 (2019)
13. Punnett, L., Wegman, D.H.: Work-related musculoskeletal disorders: the epidemiologic evidence and the debate. J. Electromyogr. Kinesiol. **14**(1), 13–23 (2004)
14. Ramsay, J.O.: Functional Data Analysis. Wiley Online Library (2006)
15. Ranavolo, A., Ajoudani, A., Cherubini, A., Bianchi, M., Fritzsche, L., Iavicoli, S., Sartori, M., Silvetti, A., Vanderborght, B., Varrecchia, T., et al.: The sensor-based biomechanical risk assessment at the base of the need for revising of standards for human ergonomics. Sensors **20**(20), 5750 (2020)
16. Rocha, V.d.A., do Carmo, J.C., Nascimento, F.A.d.O.: Weighted-cumulated s-EMG muscle fatigue estimator. IEEE J. Biomed. Health Inform. **22**(6), 1854–1862 (2017)
17. Russo, F., Di Tecco, C., Fontana, L., Adamo, G., Papale, A., Denaro, V., Iavicoli, S.: Prevalence of work related musculoskeletal disorders in Italian workers: is there an underestimation of the related occupational risk factors? BMC Musculoskelet. Disord. **21**(1), 1–16 (2020)

Exploration of Human-Robot Collaboration Methods for Iterative Design Optimization of Paper Airplanes

Nana Obayashi(✉), Kai Junge, and Josie Hughes

Create Lab, EPFL, 1015 Lausanne, Switzerland
nana.obayashi@epfl.ch, kai.junge@epfl.ch, josie.hughes@epfl.ch
https://www.epfl.ch/labs/create

Abstract. To develop engineered systems with enhanced functionality, improving their design is essential. Robots and humans each have specialized competence advantageous to design. Robots excel in terms of fabrication reliability and observation of trends in data whereas humans have prior knowledge and experience to inform decision making. By enabling robots and humans to collaborate on design problems, we may be able to achieve the best of both worlds. We explore this concept for the design problem of paper airplanes with an objective to reach a target distance and present various collaboration mechanisms for iterative design, test, and evaluation processes. We demonstrate that with the optimal collaboration technique where the robot provides initial design exploration followed by human exploitation, a target-reaching design is found quicker than other collaborative or non-collaborative methods. This work presents an initial exploration of collaborative human-robot design methods which could inform future methodologies for increasingly complex design problems.

Keywords: Human-robot collaboration · Design optimization · Human-in-the-loop

1 Introduction

Engineering design is required for every object, structure or robot that is created and is fundamental to the functionality, user perception and economic potential [3]. Developing optimal designs of a physical structure is complex as the design space can be vast, hard to quantify, costly to fabricate/evaluate, and the objective function challenging to formulate [14]. The process can also be non-linear, requiring iterative phases of design, testing and evaluation [16]. Humans have a great capacity for design, showing creativity and ability to use previous experience and learnt knowledge to influence and inform the design of a structure. The human brain allows us to internally simulate physical situations providing us with a powerful onboard simulator and optimizer [1]. However, computational and robotics tools offer potential for significant advances in design and

P. Borja et al. (Eds.): HFR 2022, 26, pp. 131–145, 2023.
https://doi.org/10.1007/978-3-031-22731-8_10

experimental validation, particularly in the domain of robot design [2,6]. This includes both in automating the fabrication, design and evaluation of engineered structures [13], and also intelligent learning based design optimization [4,17]. By understanding the relative advances of humans and robots, it may be possible to develop collaborative human-robot design methodologies.

While the technologies for fully automated design exploration and fabrication are still in their infancy, there is significant potential for human-in-the-loop design optimization and fabrication. However, we need to develop frameworks and approaches for successful collaborative design and fabrication between robots and humans that leverage the advantages of their respective skills. Previous work has explored the automated induction of design heuristics from human design data [10], and the learning of CAD construction from human data set [15]. There has also been AI-assisted design teams [12] which have shown improved agility and performance. While these novel approaches highlight the potential of human-robot design collaboration they have not yet explored the addition of robots to the physical making process where a more iterative approach is required for design.

To explore this problem we focus on the design of paper airplanes, more specifically their wing planform. This is a design space which is complex and nonlinear and driven by fluid-structure interactions [7]. These interactions are hard to accurately capture in simulation, and thus can be better explored through physical design prototype iteration. We utilize a robotic system that can automatically design, fabricate and optimize these airplanes using model-free approaches to explore optimal approaches to human-robot design collaboration.

We hypothesize that by sharing the design and fabrication and performing it collaboratively, the performance, or speed by which an optimal design can be found can be improved with respect to baselines of human or robot only design. Specifically, we propose a number of different iterative collaborative methods that seek to identify the key benefits of humans and robots during the design process. These shift the role of the human and robot during the iterative process to either exploration or exploitation. This allows us to explore the optimal use of human and robot input and how this can be best combined.

Utilizing a robot system for paper airplane prototyping and evaluation [9], we investigate how it can be used to explore collaborative design. By deploying shared human-robot collaborative design, we demonstrate that a collaborative approach where the robot performs initial exploration and the human follows up with a period of exploration allows for the required airplane performance to be reached quicker in comparison to either robot or human designing alone. In addition, sharing the physical fabrication offers a trade-off in terms of total time, reliability, and repeatability of the experimental process.

In the remainder of this paper we present the overall design problem, different collaborative modalities that are investigated, and the experimental setup. Experimental results for exploring shared fabrication and evaluation process (i.e.: 'making' or 'throwing' the airplane) are first given, followed by the collaborative design optimization experiments. These provide some initial insights into potential collaborative methods between human and machine, however are preliminary in nature. We conclude with a discussion regarding potential insights this can offer for scale experiments.

2 Methods

To investigate and leverage the collaborative capabilities of humans and robots, we focus on the complex system of paper airplane design. While a paper airplane wing can commonly be defined by simple parameters such as wing span, chord lengths, and sweep angle to name a few, the resulting flight behavior and design space is highly complex and nonlinear. The resulting behavior of a launched paper airplane can often be stochastic and driven by factors arising from the environment and also from the structure itself [9]. Unlike conventional aircraft wings, a paper airplane does not have an airfoil, and it can be difficult to precisely fabricate certain desired geometric properties out of paper (i.e. parameters such as wing dihedral and trim). Furthermore, the airplane may also be subject to high aeroelastic deformations through the flight. These are all factors that make simulations or numerical analysis of paper airplanes challenging, compared to those for rigid aircraft wings.

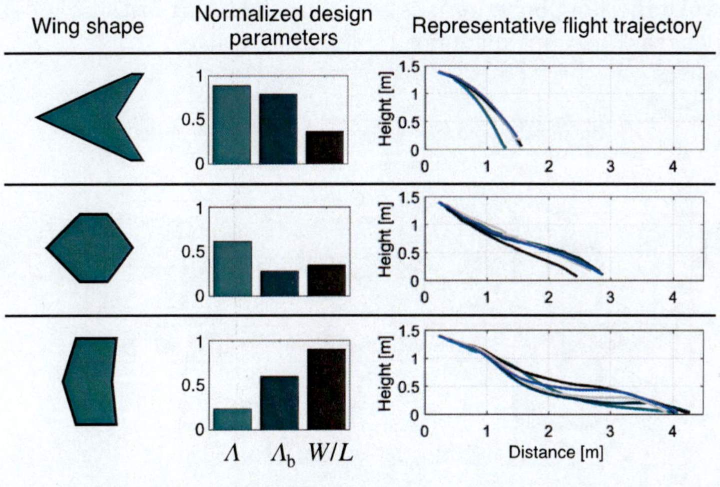

Fig. 1. Selection of three example parameterized wing designs and their representative flight trajectories. This illustrates that the wing parameters are heavily coupled and their behaviors are difficult to intuitively predict.

To illustrate the complexity of this system, three example paper airplane wing designs and their performance obtained from an automated data capture setup using cameras, are shown in Fig. 1. Each wing is defined by three geometric parameters: front sweep angle, Λ, back sweep angle, Λ_b, and width-to-length ratio, W/L (Fig. 5a). For each of the three wings, its normalized geometric parameters are shown. Within some design constraints, there are infinite combinations of the geometric parameters, and these parameters are closely coupled to create a resulting wing shape. Furthermore, changing the geometric parameters even slightly will lead to different flight trajectories, as will be evident in the sections to follow.

Designing a paper airplane to fly a given distance is a challenging design problem and requires physical evaluation and iteration. Humans often have excellent intuition in design while a machine can explore a design space in an unbiased, systematic manner. In throwing the paper airplane, a human's throwing motion leads to variable launching conditions while a robot can infinitely repeat the same throwing motion. These benefits and shortcomings in both humans and robots make this system of paper airplanes an interesting problem for an iterative design process and a perfect avenue to explore human-robot collaboration. In this section we present first the collaborative methods for fabricating and evaluating (or throwing) the paper airplane performance, and then the collaborative methods for selecting new wing shapes. Figure 2 graphically presents the collaborative nature of the paper airplane optimization problem. The optimization is conducted iteratively, meaning in each iteration, a particular airplane design is sampled from the feasible design space, which is then evaluated by fabricating and launching the airplane. For this problem, the objective function is the target distance, while the stopping criteria is achieving the target distance (given some tolerance) or reaching a maximum of 15 design iterations.

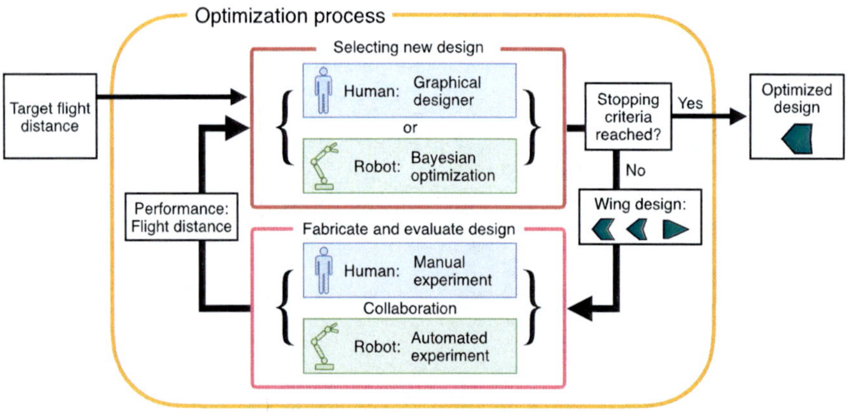

Fig. 2. Summary of the proposed design optimization approach of collaboration between human and robot in the sampling and evaluation process to iteratively accomplish a design objective.

2.1 Design Collaboration Types

Approaches for Collaborative Fabrication and Evaluation The experimental process of fabricating and evaluating the paper airplane can be roughly divided into two tasks: 'making' (cutting the paper and attaching the holder) and 'throwing' of the paper airplane. Dividing the process into these two tasks highlight the need of different skills which enable the relative advantages of the robot and human. There are four possible combinations of sharing (or not sharing) these tasks between the robot and the human, as shown in Fig. 3. To evaluate these different approaches, we can explore the reliability and repeatability of the 'making' and 'throwing' of the airplanes by considering the speed to complete these tasks and by repetitively recording the flight performance of a single design.

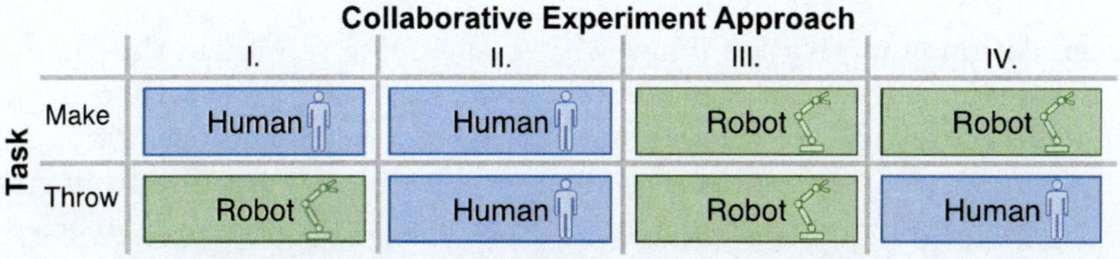

Fig. 3. Four possible approaches of sharing the experimental tasks of 'making' and 'throwing' the paper airplane between human and robot.

Approaches for Collaborative Wing Design For each iteration of the design, a wing design is selected from the design space either by the robot using Bayesian optimization [11], or by a human using an interactive wing designer GUI (Fig. 5b). In both cases, the designer has access to previous designs and their evaluations (flight trajectories and distances). Throughout the optimization process, the robot and human can collaborate by having some iterations designed by the robot and others by the human. In this experiment, we investigate five collaborative approaches as show in Fig. 4, and specified as such:

1. **Human Only:** Human designs for all iterations.
2. **Robot Only:** Robot designs for all iterations.
3. **Human-Robot:** Human designs first four wings, then the robot continues.
4. **Robot-Human:** Robot designs first four wings, then a human continues.
5. **Alternate:** Robot and human alternates between two design of wings, with the robot starting the process.

We often make generalized assumptions that humans are good at learning from experience and that they have good intuitions. On the other hand, the Bayesian optimization algorithm is good at evenly exploring the design space. Furthermore, we must also consider the limitations in human's attention span

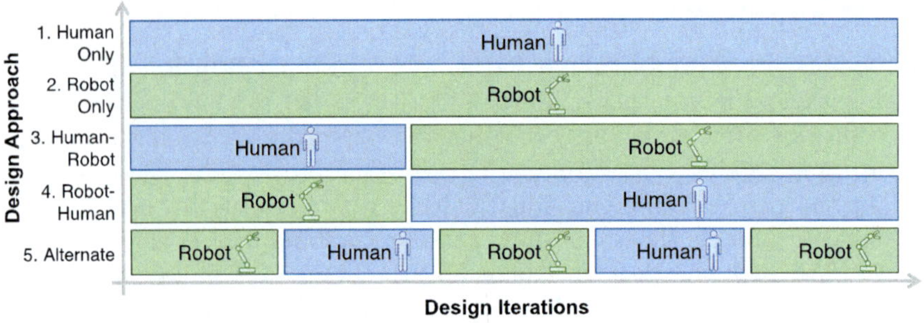

Fig. 4. Visual representation of five design approaches used in the iterative design experiment of paper airplanes consisting of various combination of human-robot collaboration.

and degradation in motivation if met with unsatisfying results. In the same way, the optimization algorithm could find a local minimum rather than a global minimum. By investigating whether or not to collaborate and altering who starts the experiment, we can explore the relative advantages and disadvantages of abilities possessed by humans and robots. Through these design approaches we investigate the significance of exploration and exploitation for a complex design space by robots and humans.

2.2 Design of a Wing

The parameterized paper airplane wing is shown in Fig. 5a. The wing designs in this work have a fixed surface area of $200\,\text{cm}^2$ and defined using three parameters: front sweep angle, Λ, back sweep angle, Λ_b, and width-to-length ratio, W/L. The ranges for the design parameters are as follows: $\Lambda = [0°, 74°]$, $\Lambda_b = [14°, 152°]$, $W/L = [0.25, 1.71]$. The wing is cut out of an A4-sized $160\,\text{g/m}^2$ DCP paper using a CO_2 laser cutter, to which a 3D-printed holder is attached (Fig. 5b).

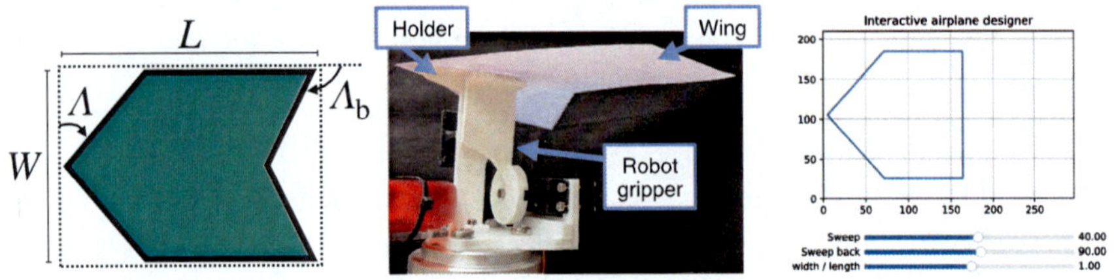

a) Wing parameterization b) Paper airplane held by the robot c) Interactive GUI for wing design

Fig. 5. a) Parameterization of a paper airplane wing, b) Fabricated paper airplane held by the robot about to be placed on the launcher, c) Interactive GUI used by a human paper airplane designer with three adjustable geometric parameters.

Two methods exist in choosing the three geometric parameters (Λ, Λ_b, W/L) that define a new wing: 1) using an interactive GUI (Fig. 5c) by a human or 2) using Bayesian optimization sampling [5]. Bayesian optimization can capture the stochastic and probabilistic nature of this design problem and allows for a sequential decision making much like how a human iteratively samples new designs [8]. Furthermore, the design space of the wing has been reduced to just three parameters, making Bayesian optimization a suitable algorithm. An exploration ratio of 0.7 is chosen from heuristics and trial and error for the given design problem. This ratio has room to be tuned further in future works. The interactive GUI contains slider bars for the three parameters and the visualization of the wing shape will be updated as the human designer adjusts the parameters. In both methods, the combination of the geometric parameters are subject to constraints, such as the design having to fit on an A4-sized paper, having a fixed surface area, and the root chord of the wing having to be larger than the tip chord.

2.3 Experimental Setup

The robotic experimental setup for the making and launching of the paper airplanes is shown in Fig. 6a where it follows the process summarized in Fig. 6b to conduct the experiment. The robotic setup ensures the fabrication and launch of the airplane to be repeatable, fully autonomously.

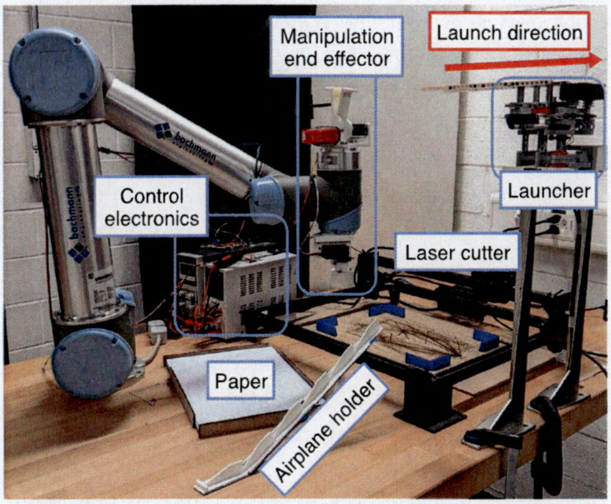

a) Robot paper airplane experimental setup

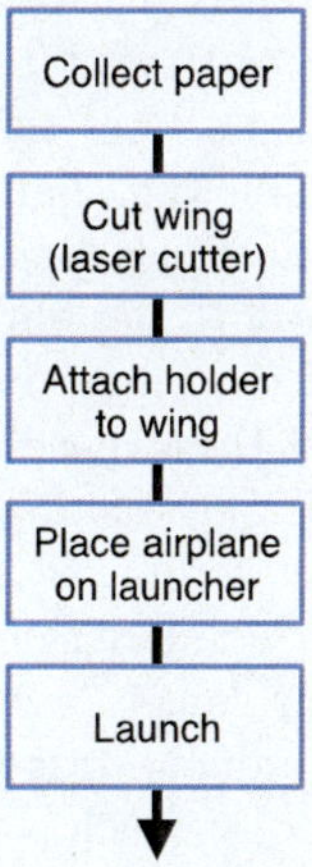

b) Paper airplane experiment process

Fig. 6. a) Robotic experimental setup for paper airplane fabrication and launching, b) Process of making and launching a paper airplane, starting from paper collection ending with launching.

In obtaining a data point of any single paper airplane design, each design is flown five times using the robotic setup and the resulting distances are averaged. A single human tester completed the task sharing for the collaborative fabrication and evaluation. For the collaborative wing design, the objective of the optimization is a target flight distance. Five total optimization experiments are conducted with two additional validation experiments. A new human tester without knowledge of the airplane performance conducted each experiment. All of the participants had an engineering background.

3 Results

In this section, the results for collaborative fabrication and evaluation methods are first presented. This is followed by exploration of the different collaborative design methods for a given target flight distance. As a final demonstration, the optimal human-robot technique is tested on a new target flight distance with two different human designers in comparison to the robot-only baseline.

3.1 Collaborative Fabrication and Evaluation of Paper Airplanes

First, the different divisions of tasks in fabrication and evaluation between a human and a robot was evaluated in terms of time (Fig. 7a) and repeatability (Fig. 7b). In the 'make' task, a human can cut out a pre-drawn wing from a piece of paper and attach a holder in approximately 100 s, which is half the time required by a robot. The bottleneck of the robot in the 'make' task is the amount of time it takes for a laser engraver to cut through the paper. From visual inspection, the paper airplane created by the robot and the human are comparable (i.e.: both have high repeatability). In the 'throw' task of picking up the fabricated paper airplane and throwing it, a human and robot takes about 10 s and 20 s per throw, respectively which are comparable results within the scale of the whole process. We can then conclude that in terms of increasing speed of the experimental process, either 1) the human should complete both tasks, or 2) the human should 'make' and the robot should subsequently 'throw'.

Next we investigate the performance repeatability of the paper airplane for the four combinations of the task-sharing between the human and robot. The paper airplane is launched five times for each collaborative approach. We observe two prominent trends. Firstly, the robot is able to throw the airplane more consistently which is visible from the box plot in Fig. 7b. The variance in distance of a human throw is approximately 10 times higher than that of the robot (in the order of $0.1\,\mathrm{m}^2$ compared to $0.01\,\mathrm{m}^2$). This illustrates the consistency of the robot, which is critical for a precise measurement. Secondly, the robot is able to throw the airplane for a longer distance compared to the human. The maximum distance reached by human throws are lower than the minimum distance reached by the robot throws.

From these task-sharing experiments, we conclude that when the human 'makes' the paper airplanes and the robot 'throws' them, it is the optimal app-

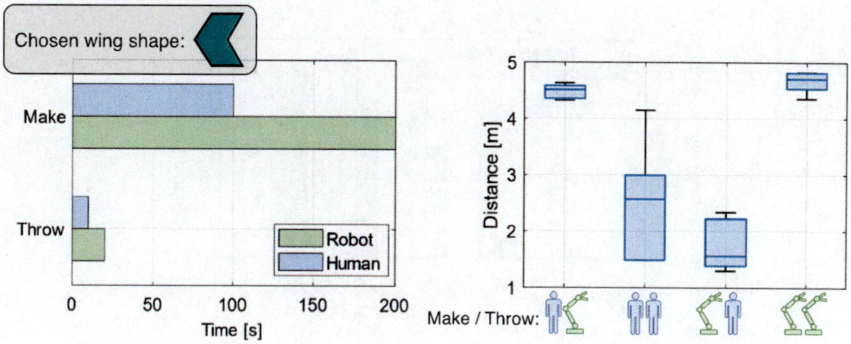

a) Task completion time comparison between human and robot

b) Repeatability comparison between human and robot

Fig. 7. a) Comparison of the time it takes for a human and robot to individually 'make' and 'throw' a paper airplane, b) Flight distance and variance of five launches of the same paper airplane where the 'make' and 'throw' were accomplished by either a human or a robot as indicated.

roach with respect to both speed and repeatability. This human-robot collaborative approach is used for all following experiments.

3.2 Collaborative Wing Design of Paper Airplanes

The proposed approaches in human-robot collaboration for designing paper airplane wing shapes given in Sect. 2.1 are investigated. In particular, we pay attention to how the human and the robot either explores or exploits a design in order to attain an objective. The designer is given an objective of designing a paper airplane that flies a target distance of 5 m with a 20 cm tolerance and variance less than 0.1 m^2 and is allowed a maximum of 15 design iterations. This target distance was chosen to be close to the maximum distance that the airplanes within the design space can achieve. Each design is flown five times to capture the true system behavior of the airplanes. The average and variance of five flight distances are taken as the representative distance and error displayed by a wing design. Figure 8 gives a general performance overview of the proposed design approaches, showing the error or closeness to the target distance achieved at the stopping iteration. The approaches are ordered from left to right in the best to worst performing in the figure. The **Robot-Human** approach achieved a design that landed closest to the target and with the least number of iteration. This was followed by the **Human Only** approach. Only these two approaches achieved the target within the chosen tolerance and the maximum 15 iterations. The other three approaches, **Human-Robot**, **Alternate**, and **Robot Only** did not achieve the target within the maximum allowed design iterations.

Figure 9 shows the design progression and distance achieved at each iteration using the five aforementioned collaborative design approaches. The error bars

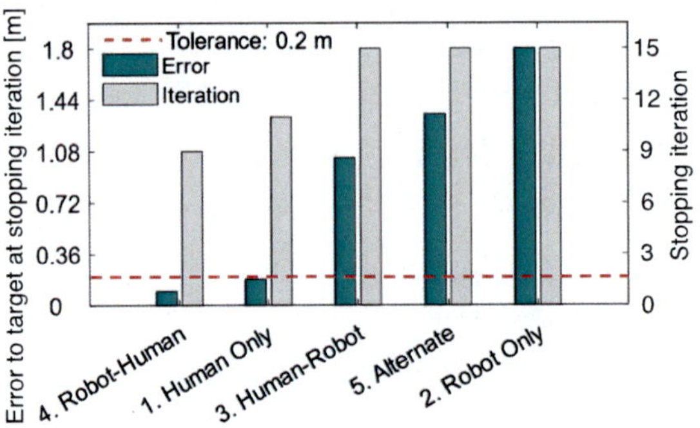

Fig. 8. Deviation from the target distance (5 m) of the average flight distance recorded at the stopping iteration. Results from each collaborative design approaches are ordered by best performance from left to right.

show the variance in landing distance for the five flights performed for each design. The results from each of the collaboration approaches are now discussed.

1. Human Only The second best performance was seen when the human alone iteratively designed the wings. The human designer designed a wing that reached a distance of 4.82 m with a variance of 0.069 m^2 in 11 iterations. The human designer initially tested visibly varying wing shapes exploring the design space but in the later iterations returned to similar high performing wings. Humans have a dynamic exploration ratio whether it is by visual intuition or from systematic guesses.

2. Robot Only The approach with the robot alone did not reach the target within the allowed number of iterations. The Bayesian optimization initially explored various designs visually 'unintuitive' to humans. At iteration 4, it reached a maximum average flight distance of 4.18 m. After exploring about six more designs in shape variety, from iteration 12 to 15, the robot exploits shapes similar to that of iteration 4. However, the robot got stuck at a local minima and could not achieve the objective.

3. Human-Robot In this approach, the human designer began the optimization by iteratively designing four wings. Without prior intuition of the system, the human designer explored more 'conventional' paper airplane wing designs (i.e.: designs that resemble a fighter jet). Beyond iteration 4, the robot took over to complete the iterative design. However, the robot explored very little and again stayed in a local minimum with the wing shape similar to one designed by the human. Although also dependent on the exploration ratio of the Bayesian optimization, the robot's performance depends strongly on the initial 'seeds' of the iterative process. The human seeds show limited diversity and exploration of the design space, sticking to more intuitive shapes.

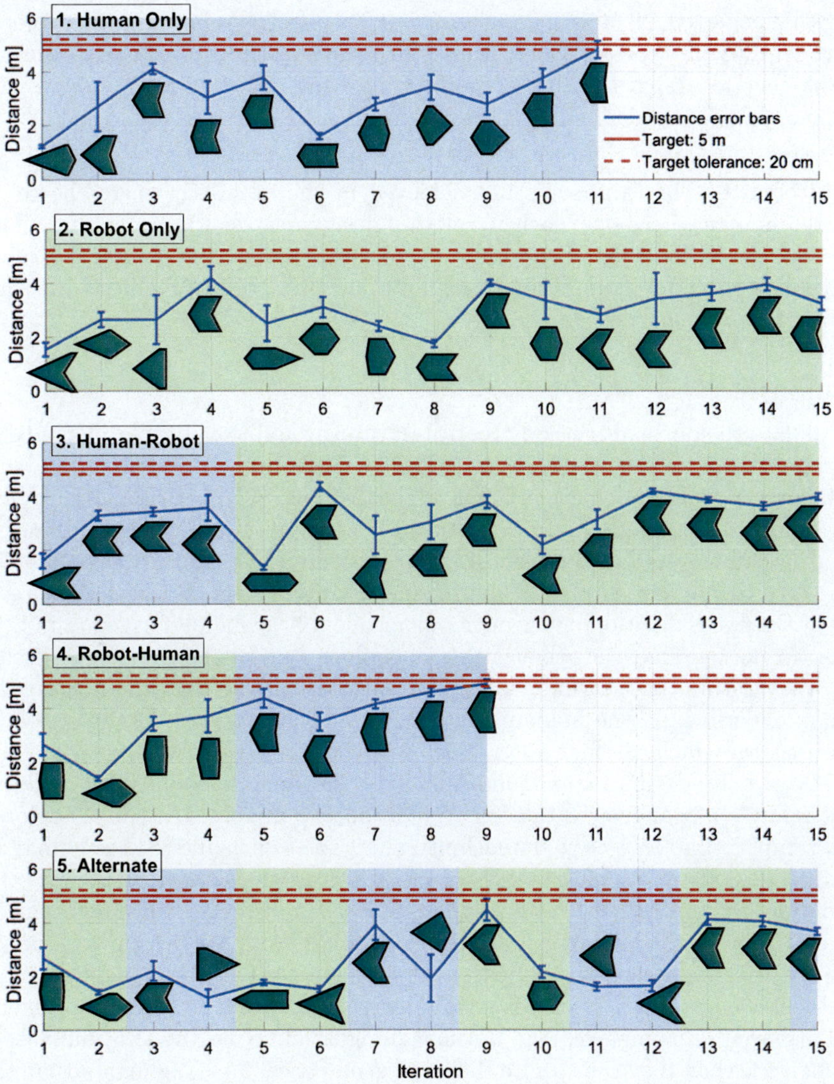

Fig. 9. Wing shape progression and the achieved distance and error of five flights each, throughout the iterations for five different design approaches. Target distance of 5 m with a 20 cm tolerance is shown. The background of the figure is colored blue for design carried out by a human and colored green for design carried out by Bayesian optimization.

4. Robot-Human The best performance was seen when the robot started the optimization by iteratively designing four wings, and handing it off to a human designer. In only 5 iterations following that of the robot, the human designed a wing that achieved 4.9 m with a variance of $0.016\,\text{m}^2$. The robot initially explored two very different types of wings, from which the human received early intuition on the general design of a better performing wing. We therefore observe that humans are good at exploiting a design given 'good' initial examples.

5. Alternate The target was not achieved in the approach where the robot and the human alternates, each iteratively designing two wings per turn. This approach seems challenging and impractical as neither the human nor the robot has enough room for exploration or exploitation. As a result, the design neither reaches a global nor local minima.

3.3 Validation Experiment

To validate the optimal method (**Robot-Human**) additional experiments comparing this method to a robot-only baseline have been performed. The goal of the process was to achieve a target flight distance of 3 m with a 20 cm tolerance, and variance less than $0.1\,\text{m}^2$ within 15 design iterations. The collaborative robot-human method was performed by two individuals who have not previously completed any paper airplane experiments. Figure 10 shows the design iteration of the two validation experiments alongside a baseline performed only by a robot. In the baseline, the robot is able to explore, but fails to exploit the design effectively within the maximum allowed iterations. For the **Robot-Human** validation experiments, the human was able to find an optimal design in 5 or 8 iterations (of which the first 4 iterations were performed by the robot).

The two validation experiments highlight the human's ability to exploit a small sample of designs. In the first validation experiment, the robot explored *good* samples that were close to reaching the target, which allowed the human to produce the target-reaching design in only 1 iteration. Whereas, in the second validation experiment, the robots explored *bad* samples that did not reach the target. However, the human was able to extrapolate from *bad* designs to produce a target-reaching design in the subsequent 4 iterations. This result highlights and validated the effectiveness of this collaborative method for this design problem.

Furthermore, comparing the optimal designs found by the two humans, it can be seen that they have quite different geometries. This highlights that the role of design optimization is complex, as there are multiple 'good' solutions.

4 Discussion and Conclusion

Developing mechanisms of collaboration and friendly interactions between robots and humans is imperative for robots taking on many critical tasks in society. Engineering design is often an overlooked task, but one where effective human-robot collaboration has the potential for significant impact as efficiency and personalization is increasingly desired. Focusing on the design of paper airplanes, we

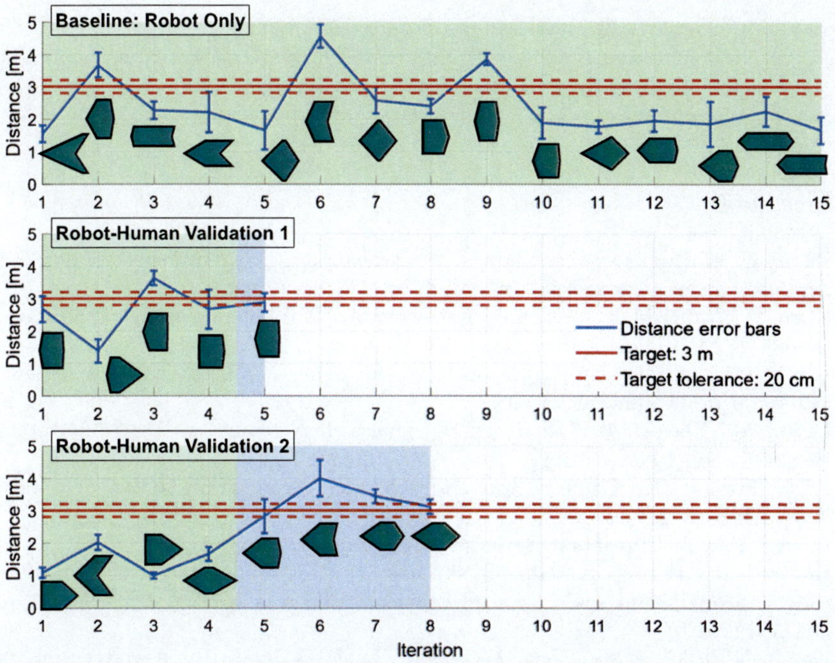

Fig. 10. Validation of the experiment in Fig. 9, where the 'Robot-Human' approach is the best design approach. Two validation experiments are conducted for a target distance of 3 m with a 20 cm tolerance. A 'Robot Only' approach is also shown as a baseline comparison.

have explored a number of iterative collaborative approaches and demonstrated for this specific case-study, that collaborative design offers the most rapid method of maximizing the design objective.

The specific case-study and experiments present only a first exploration of these different methodologies. The focus of paper airplanes is of interest due to the challenges in simulating the designs and also the stochasticity of the process. It is also a design field in which humans have previous history or experience. While the optimal design approach was found to be 'robot followed by human' design, for design problems with different characteristics, it may be that different collaboration methods are optimal. Exploring different design collaboration approaches for a variety of design problems could better explore the coupling between the design problem and the optimal method. The specific implementation of the Bayesian optimization and specifically the exploration-exploitation ratio has an effect on the human-robot collaboration. Further extending the investigation to explore optimization of this ratio in addition to the collaboration method could enable future improvements in collaborative design.

Acknowledgements. Big thanks to our colleagues, Qinghua Guan, Stefan Ilić, Shiv Katiyar, Max Polzin, and Francesco Stella for taking part in the experiments! This project was partially funded by the European Union's Horizon 2020 research and innovation programme under the Marie Skłodowska Curie grant agreement N° 945363.

References

1. Bates, C., Battaglia, P.W., Yildirim, I., Tenenbaum, J.B.: Humans predict liquid dynamics using probabilistic simulation. In: CogSci (2015)
2. Chen, F., Wang, M.Y.: Design optimization of soft robots: a review of the state of the art. IEEE Robot. Autom. Mag. **27**(4), 27–43 (2020)
3. D'Ippolito, B.: The importance of design for firms' competitiveness: a review of the literature. Technovation (2014)
4. Fadini, G., Flayols, T., Del Prete, A., Mansard, N., Souères, P.: Computational design of energy-efficient legged robots: optimizing for size and actuators. In: 2021 IEEE International Conference on Robotics and Automation (ICRA), pp. 9898–9904. IEEE (2021)
5. Frazier, P.I.: A Tutorial on Bayesian Optimization (2018)
6. Howison, T., Hauser, S., Hughes, J., Iida, F.: Reality-assisted evolution of soft robots through large-scale physical experimentation: a review. Artif. Life **26**(4), 484–506 (2020)
7. Ismail, N.I., Ali, Z.M., Ishak, I.S., Noor, R.M., Rabilah, R.: Aerodynamic performances of paper planes. J. Adv. Res. Fluid Mech. Therm. Sci. **77**(1), 124–131 (2021)
8. Junge, K., Hughes, J., Thuruthel, T.G., Iida, F.: Improving robotic cooking using batch Bayesian optimization. IEEE Robot. Autom. Lett. **5**(2), 760–765 (2020)
9. Obayashi, N., Ilić, S., Junge, K., Hughes, J.: Robotic automation and unsupervised cluster assisted modeling for solving the forward and reverse design problem of paper airplanes. Research Square PREPRINT v1 (2022)
10. Puentes, L., Cagan, J., McComb, C.: Automated heuristic induction from human design data. In: International Design Engineering Technical Conferences and Computers and Information in Engineering Conference, vol. 83983, p. V009T09A054. American Society of Mechanical Engineers (2020)
11. Shahriari, B., Swersky, K., Wang, Z., Adams, R.P., De Freitas, N.: Taking the human out of the loop: a review of Bayesian optimization. Proc. IEEE **104**(1), 148–175 (2015)
12. Song, B., Soria Zurita, N.F., Gyory, J.T., Zhang, G., McComb, C., Cagan, J., Stump, G., Martin, J., Miller, S., Balon, C., et al.: Decoding the agility of artificial intelligence-assisted human design teams. Des. Stud. **79**, 101094 (2022)
13. Vujovic, V., Rosendo, A., Brodbeck, L., Iida, F.: Evolutionary developmental robotics: improving morphology and control of physical robots. Artif. life **23**(2), 169–185 (2017)
14. Wang, G., Shan, S.: Review of metamodeling techniques in support of engineering design optimization (2007)
15. Willis, K.D.D., Pu, Y., Luo, J., Chu, H., Du, T., Lambourne, J.G., Solar-Lezama, A., Matusik, W.: Fusion 360 gallery: a dataset and environment for programmatic cad construction from human design sequences. ACM Trans. Graph. (TOG) **40**(4), 1–24 (2021)

16. Wynn, D.C., Eckert, C.M.: Perspectives on iteration in design and development. Res. Eng. Des. **28**(2), 153–184 (2017)
17. Zhao, A., Xu, J., Konaković-Luković, M., Hughes, J., Spielberg, A., Rus, D., Matusik, W.: Robogrammar: graph grammar for terrain-optimized robot design. ACM Trans. Graph. (TOG) **39**(6), 1–16 (2020)

Influence of Music on Enjoyment of Pre-Frail Elderly While Doing Physical Exercises with a Robot Coach

Marlou Kellenaers, Daniel Tetteroo(✉), and Emilia Barakova

Industrial Design, Eindhoven University of Technology, Eindhoven, The Netherlands
{d.tetteroo,e.i.barakova}@tue.nl

Abstract. Pre-frail elderly people have to do physical exercises to reduce the risk of becoming frail but often lack motivation. Socially Assistive Robots could offer a solution because they have shown their effectiveness in achieving measurable progress with the elderly. This paper investigates how the addition of music and autonomy to choose it influences the enjoyment of pre-frail elderly while doing physical exercises under the coaching of a Social Robot. In an experiment with 12 pre-frail elderly coached by a robot to exercise, three different conditions were tested: (1) no music accompanying the exercise, (2) the robot chooses music, and (3) the elderly person chooses music themselves. The results show that the participants enjoyed more the music-enhanced exercises with the robot coach. Also, the music genre needs to be personalized, and the results on the autonomy of choice (robot or themselves), were mixed, however, the majority preferred the robot to choose the music. Further research is needed to make a successful personalized Social Assistive Robot, that utilizes music in its interactions.

Keywords: Pre-frail elderly · Social assistive robot · Robot coach · Enjoyment · Music personalization · Physical exercise

1 Introduction

Regular exercise is seen as one of the most important aspects of general preventive health [1]. However, Schutzer and Graves [2] found that 60% of the adult population in the United States does not exercise regularly and 31% of the adults do not exercise at all. Given the worldwide growth of older adults, and the fact most elderly are not physically active, a public-health problem looms [1].

Another study has found that 87% of the elderly have at least one barrier to participating in physical exercises [2], such as poor health, perceived old age, lack of interest, trouble reaching the sports facility, and others [1–3]. Goal setting [3] enjoyment [1, 4], appropriate music [2], and social support from family [1] could increase the motivation for physical activities. There is no specific research on the barriers and motivators of the pre-frail elderly, but it is assumed that these are the same as for the elderly in general.

A golden standard for the definition of pre-frail does not yet exist, since research on pre-frailty is limited. We adopted the definition of Fried et al. [5] because it is most

P. Borja et al. (Eds.): HFR 2022, 26, pp. 146–157, 2023.
https://doi.org/10.1007/978-3-031-22731-8_11

detailed. According to it, people are pre-frail if they have at least one of the following five characteristics: (1) Weight loss (unintentionally more than 4.5kg in the past year, or 5% of total body weight), (2) Reduced physical activity (daily > 4h sitting, monthly < = 1 occasion of walking and no cycling or jogging), (3) Fatigue (every activity does take effort, it takes effort to get started), (4) Muscle weakness (handheld dynamometer: men 30kg, women 18kg) and (5) Slowness of movement (feeling that everything is going slower, taking more than 6 s for 4 m). A study on the number of frail adults found that 30.4 to 44.9% of all adults in ten European countries show the first signs of frailty and 1.3 to 5.9% of the adults are frail [6]. Pre-frail is the initial stage of frailty. It must be prevented that people become frail, because they have an increased risk of various diseases (hypertension, cardiovascular disease, cancer) and eventually death. Preventing from becoming frail is possible during the pre-frail stage [7]. Research has shown that physical exercises can be effective in preventing, slowing down or reversing the frailty process [8], even in the advanced frail stage [9]. However, a standard prescription with exercises for frail elderly does not exist yet [9].

Socially Assistive Robots (SAR) have been applied as coaches for exercising because they are proactive and may have social facilitating effects [25]. Several successful applications have shown that robots provide help and achieve measurable progress in rehabilitation, learning, and engagement in activities [6, 24]. SARs can also offer a solution for supporting exercising by the elderly if the elderly persons indicate that they would like to receive training support at home [3]. Several factors serve as conditions for acceptance of SARs by older adults [28, 29] such as ease of use, enjoyment [10], factors such as adaptability, and social presence [29]), and even liking and trust [28] as well as the perceived levels of expertise [25]. The behavior of the humanoid robots as Socibot mini was designed to increase trust, promote persuasion [26] and take the roles of an expert [25] so it was chosen for the current study.

According to previous research, listening to music during exercise improves enjoyment and motivation [2, 12, 13]. Moreover, it has been suggested that the positive effect of music on affective states may lead to increased exercise adherence [12]. Mullen et al. [4] claim that elderly people often refer to enjoyment as a motive for exercise participation. Although enjoyment, provided by music can have a significant impact on the motor or cognitive functioning [14], research on the addition of music to the interaction with Socially Assistive Robots is limited, and to our knowledge, has not been explored for pre-frail elderly. Therefore, the research presented in this paper aims at exploring if the addition of music to the SAR exercise coaching interactions influences the enjoyment of pre-frail elderly people and whether the autonomy of choosing the music or leaving this task to the robot has an impact on the enjoyment of exercising.

2 Related Work

There have been several approaches to increasing the motivation of users while doing physical exercises. Showing the performance score [10, 15], reinforcement techniques [15, 16], cognitive challenges [4], and maximizing the potential for user success [17] have all shown to have a positive impact on users' motivation while doing physical exercises. Research on the design of a SAR found that feedback in the form of facial expressions and voice [18], corrective and positive feedback [19] and relational robot behaviors such as praise, politeness, humor, and empathy [20] have also a positive effect on the users' motivation. These are all interesting findings, but they investigated other motivation strategies then the addition of music. Meyns et al. [14] investigated whether children with an oncological disorder (COD) noticed a difference in motivation when performing movement exercises with a therapist or a humanoid robot, and how the addition of music affects their motivation. They concluded that the addition of music and a humanoid robot has a positive effect on motivation while doing physical exercises. However, the motivation and enjoyment of adding music could be completely different for the pre-frail elderly than for children.

3 Methods and Procedures

A within-subject experiment, with three different conditions with respect to incorporating music in the human-robot motivational interaction, was designed. The aim was to find out whether and how these conditions influence the enjoyment of pre-frail elderly while doing physical exercises.

3.1 Participants

The participants had to meet three requirements: (1) they had to be at least 60-years old; (2) they needed to be pre-frail; (3) they had to be able to complete a questionnaire autonomously and answer interview questions. The participants were recruited through De Zorggroep, a care institution in The Netherlands, and both the elderly living in care apartments and those visiting a daycare facility participated in this study. For the second and third requirements, a physiotherapist from the care apartments and a nurse from the daycare selected the pre-frail elderly based on the guidelines of Fried et al. (Fried et al., 2001). In total, 12 participants ($n = 12$) took part in the study, of which three were male and nine were female. The age ranged from 60 to 89 years, (M 77, SD 9.18). No details about the study were hidden from the participants.

3.2 Setting

The study took place at the daycare facility and at the care apartments. These locations were chosen to minimize the effort for the elderly to participate in the study. In both locations, the study was conducted in a controlled setting, in a closed-off indoor room. The two locations showed similarities to the extent that the setting can still be considered as controlled. The first author was present during the study.

3.3 Materials

The SociBot Mini robot[1] was used in the study. Previous research has shown that people prefer a humanoid robot [21] and this robot has been used in music-based interactions

[1] Technical data about Socibot mini through this https://wiki.engineeredarts.co.uk/socibot_at_a_glance.

[25], and acceptance of this robot as a persuader of adults has been established [26]. The robot was controlled via a pre-programmed script enhanced by Wizard of Oz teleoperation by the researcher through a router with an internet connection and a notebook to facilitate a conversation with the elderly. The robot was placed on a table and the participant sat in a chair opposite the robot. To enhance the interaction and social connection between the robot and the participant, it was important to have the participant look at the robot, which ensures that the robot can actively motivate the participant [17]. Another Notebook, next to the robot, displayed a video explaining the physical exercise. A music box, connected to the laptop of the researcher, played the music during the exercises. See Fig. 1 for the study setup.

Fig. 1. Study setup with participant sitting opposite the robot and researcher operating the robot.

3.4 Procedure

Ethical procedure. The ethical review form was submitted and approved by the Ethical Review Board (ERB) of the ID department of Eindhoven University of Technology. Participation was voluntary to all participants, information collected during the research would be confidential at all times, and the approval of participating could be withdrawn at any time. The consent forms, which explains the topic and nature of the study on it, were signed by all the participants.

The study setup. In the study, three different conditions of music while doing physical exercise were tested. Namely: (1) without music, (2) the robot chooses music, and (3) the participant chooses themselves. A physiotherapist advised suitable exercises for the pre-frail elderly from OTAGO thuisoefenprogramma,[2] which are recognized by the central government of the Netherlands. The researcher selected one exercise out of the program, stretching the knee, which would be the least hazardous for the participants. In each condition, the participants had to do the same exercise to exclude the effect of exercise preference on the enjoyment. A script for the robot was written and programmed, to ensure that the study was carried out in the same way for each participant. The robot had no speech recognition. When the participant said something, the researcher operated the robot by choosing from pre-programmed sentences. When the participant asked an unexpected question, the robot would respond with: "Sorry, I don't know the answer. You can ask the researcher for help".

Each study started with a general introduction. Welcoming the participant, asking if they ever talked to a robot (if not, it was made clear that they didn't have to be afraid of the robot) and if they knew what they were going to do. This was done to put the participant at ease and to create a bond with the robot. After the introduction, the robot asked the participant if they were ready or if they had a question. If they were ready, the robot started with one of the three different conditions. Across subjects counterbalancing was applied to control order effects. In every new condition, there was a friendly interaction between the robot and the participant. The robot said for example: "Good to have you back! Do you have a question before we start the next exercise?" Before the participant could start with the exercise, a video of the exercise was shown on the lap top. The robot explained the exercise during the video, to facilitate the interaction between the robot and the participant. The robot said for example: "Sit up straight in the chair. Put your arms relaxed on the handrails and make sure both feet are standing on the ground." After this, the participant could start the exercise and repeat it until he got fatigued. This was done because pre-frail elderly people come in different physical con ditions, 10 repetitions are not possible for one participant and are very easy for the other. While performing the physical exercise, the robot gave positive feedback and acknowledgement to increase the participant's motivation [15, 16, 19]. When the participant started a new condition, the robot kindly welcomed them. And at the end, the robot thanked the participant for the sociability.

In condition 1, the participant performed the exercise without music. In condition 2, the robot chooses music for the participant. The robot says that she has chosen music that most people like while performing the exercise. Before the participants start, the robot says: "I'm curious if you like this music as well". This could have an effect on the participants that they are going to like it, based on response bias theory [22]. The music that the robot plays is senior fitness music found on YouTube and is based on the research of Karageorghis, Jones and Stuart [23], where music with 120 beats per minute (bpm) had the highest enjoyment. In condition 3, the participant could choose music themselves. The researcher searched for the song on YouTube and played it on the music box. If the participant did not know a song, the robot had a suggestion. They could choose between 'De Vlieger' from André Hazes and 'I want to hold your hand' from The Beatles. These songs were chosen for their popularity in the 60s, which make them very likely to be known by the participants.

[2] Het OTAGO thuisoefenprogramma, https://interventies.loketgezondleven.nl/leefstijlinterventies/interventies-zoeken/1500062, last accessed 2019/11/15.

The experimental protocol. The research began with the collection of demographic data such as gender, age, and so on. Following that, the subject was assigned to a condition at random. The subject was required to complete the Physical Activity Enjoyment Scale (PACES) after each condition. PACES is a questionnaire that assesses a participant's level of enjoyment [4]. This questionnaire was adapted from the original one [11] in earlier research to better suit the senior participants [4]. It consists of eight items rated via a 7 point bipolar scale (e.g., It is very pleasant – It is very unpleasant). Participants were given the instruction: "Rate how you feel at the moment about the physical activity you have been doing" [11]. Lower PACES scores reflect higher levels of enjoyment. The participants must fill them in right after the exercise, so they reflect the experience in the condition. After the three conditions, the participant was asked to rank the three conditions from the least fun to most fun in the Funsorter [14] (Fig. 2). This Funsorter method is used to force the participants to make a choice and serve as a basis for the follow-up interview. In the end, a semi-structured interview with open questions was conducted to gain insights about why they liked one condition more than another one.

Fig. 2. Three cards that corresponded to the three conditions (1 - No music; 2 - Robot chooses the music; and 3 - Elderly person chooses the music) were created for the elderly to correctly answer the Funsorter by ordering these from least to most engaging.

The participants performed the three different conditions in succession. In between, they had a short break where they could fill out the PACES questionnaire and have a short conversation with the robot. Testing the three different conditions had a duration of approximately 25 min, depending on the physical condition of the participants. The interview lasted 5–10 min.

3.5 Data Collection and Analysis

Data was collected on (1) the rating of the different items in the PACES questionnaires; (2) keywords from the interviews noted down by the researcher, (3) the results of the

Funsorter, (4) which participants chose their music, and which participants chose from the suggestions in condition 3.

The data gathered through the PACES questionnaire was analyzed by performing a one-way repeated measures ANOVA test in SPSS, to compare the levels of enjoyment in the three different conditions of music. A one-way repeated measures ANOVA was also conducted to test the relationship between the order of conditions and enjoyment. The data from the interview was analyzed through Thematic Analysis. The literal transcriptions of the interviews were not analyzed, only the observational notes because there was a targeted search for answers during the interview.

4 Results

The outcomes from the PACES questionnaire can be seen in Table 1. A one-way repeated measures ANOVA was conducted to compare the level of enjoyment in the three different conditions of music. There was not a significant effect on the level of enjoyment, Wilks' Lambda = 0.68, F (2.10) = 2.31, p = 0.150. This could be the case because of the large standard deviation. The standard deviations were as follows: condition 1 was 13,446, condition 2 was 10,317 and condition 3 was 10,220. Counterbalancing has been applied in the study to control order effects. A one-way repeated measures ANOVA was conducted to compare if the order of conditions had an influence on the level of enjoyment. There was not a significant effect on the order of conditions, Wilks' Lambda = 0.66, F (2.10) = 2.56, p = 0.126.

The outcomes of the Funsorter (Fig. 3), which forces the participants to order the conditions according to how enjoyable they are, showed that 7 out of 12 of the participants favored condition 2, in which the robot selects the music for them. The condition with no music (condition 1) was deemed the least enjoyable by 10 of the 12 participants.

The Funsorter was used as a starting point of the interview. The analysis of the interview, via the Thematic Analysis method, resulted in the findings described below, which are grouped per theme.

Exercising with music. Most of the participants mentioned that it is more pleasant to exercise with music, they find it boring without. They also said that it breaks through the silence and the exercise is less heavy with music. Only one participant (P3) experienced no differences between the conditions and said that music had no influence on her enjoyment or motivation.

The robot chooses the music. It was for most of the participants hard to come up with the name of a song, even if they listen to the radio all day or danced a lot in the past. Only 4 out of 12 participants were able to come up with a song, the others have chosen between the music suggestions. Other reasons why they preferred the robot were because then they hear other music then they are always listening to, it is more convenient than coming up with a song yourself and the music is better suited to the exercise. In conclusion, most of the people preferred the robot to choose music, but all for different reasons.

The user chooses the music. Only one participant (P12) mentioned in the interview that he prefers to choose the music because then he is sure that he likes it. In the Funsorter, 4 participants prefer to choose their own music. However, 3 out of the 4 partic ipants

Table 1. Outcomes PACES questionnaire per condition per participant (n = 12). Lower paces scores reflect higher levels of enjoyment.

Participant number	No music	Robot chooses music	Elderly chooses music
1	36	27	35
2	49	34	36
3	55	48	36
4	11	14	16
5	40	24	19
6	29	27	24
7	13	26	15
8	22	9	11
9	31	33	30
10	20	16	10
11	34	32	37
12	24	27	26

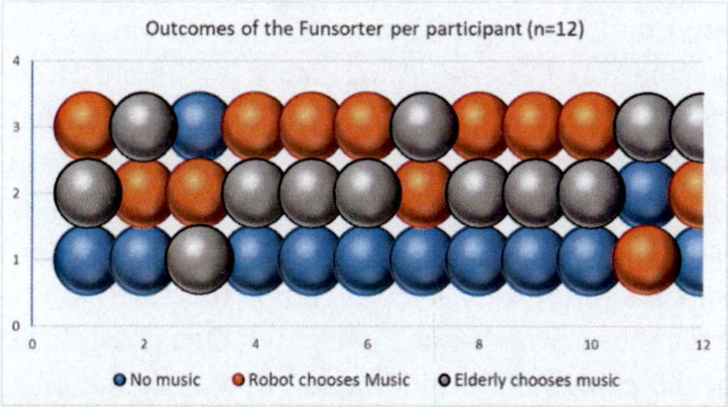

Fig. 3. Outcomes of the Funsorter per participant (n = 12). The participants ranked the three conditions from least fun (score 1) to most fun (score 3). The Blue represents Condition 1 (No music) which is scored as having the lowest level of engagement – mostly 1 point and Orange represents Condition 2 (Robot chooses music) gets rated as the most engaging from the three conditions.

indicated in the interview that they prefer the robot to choose the music. The reported reasons are: because then they hear different music, choosing themselves is difficult because she does not know names or because she has trouble talking due to illness.

Music preference (while exercising). Two participants had no music preference. The choices of the rest of the participants showed that there is no pattern – the preferences varied from country, to music that stimulates, accordion, guitar, calming, and gentle music.

The choice of a robot. One participant (P10) said that the robot was very friendly. She was impressed that the robot followed her with her eyes. Overall it was an interesting experience for the participants. They didn't think they will be helped by a robot in their lives. They also said that they had something to tell their family. In general, the participants were sitting in front of the robot and spoke to the robot with excitement and conveying personal information. This encouraged the interaction and may have had an effect on the participants feeling a social obligation to exercise. None of the participants was frightened by the robot or thought it looked scary.

No correlations were found between the results of the PACES questionnaire and what the participants said in the interview did not result in any findings.

5 Discussion

While the results varied per participant, an overall conclusion can be made that pre-frail elderly prefer to have music while doing physical exercises with a robot, based on the outcomes of the Funsorter. These results comply with the findings of Baur et al. [27] who found that a combination of music and activities promoting creativity in motor training increases enjoyment and thus intrinsic motivation of subjects performing robot-assisted training [27]. In our case, the focus on personal motivation was much higher, since the robot in the case of Baur et al. [27] was a mechanical device that moves the hand of the elderly person, while we use a humanoid robot that attempts to motivate the person through speech interaction. Additionally, personal motivation was stimulated by providing the opportunity for participants to choose music. The majority of the participants preferred the robot choosing the music for them. However, participants' preferences on the type of music were very personal. Participants had different reasons for why they preferred the robot to choose music. This makes it difficult to conclude as to whether they actually prefer the robot to choose music. This finding is consistent with a study by Fasola and Mataric for choosing a game by older adults [20]. They found that the participants did not have a preference for a scenario that offers 'choose yourself' or 'the robot chooses which game you play'. One possible explanation for both studies could be that the enjoyment of not having to choose the music/games does not outweigh the enjoyment of being able to specify a preference. An alternative explanation may relate to the social presence [29] of the robot. The humanoid shape and human-like behavior of Socibot were previously shown to increase the level of trust and compliance of the participants in making choices [26, 28].

In our study, the participants experienced difficulty in choosing a song themselves, because they had to come up with a name of a song. This might have been due to the context in which they were asked to do so: when exercising, one perhaps does not want to hear music one normally listens to in their spare time and vice versa. If this condition would have been implemented differently, for example by choosing a song from a pre-selected menu, the influence on the enjoyment might have been different compared to the condition in which the robot chooses. The presence of the robot, however, can enhance social enjoyment and increase the reliance on the choices and the recommendations of the robot. Especially Socibot with its appearance of a grown-up male with formal appearance and speech was shown to be perceived as a music expert [26].

There is also a need to question the extent to which the participants have understood the PACES questionnaire. The participants were all pre-frail but were all in a different state of mental clarity. For example, one participant was able to fill in the questionnaire independently, while in the case of another participant the researcher had to repeat every statement to the participant to rate the statement on a scale of 1 to 7. This can also lead to a different interpretation of the questionnaire by the participants.

The physical exercise that the participants had to do was very easy for some participants and very difficult for others. Some of them already did this exercise at home, which made them very good at it. Some of the participants' legs weakened during the experiment, making each following condition tougher. This could have an effect on the enjoyment. Also, they had to do the same exercise in every condition, so that the exercise itself would not affect the results. However, this can cause the participant in a second or third condition to have less fun with the exercise. Nonetheless, we did not find an order effect in our study indicating that the effect of fatigue and boredom on the study outcomes was limited.

We note the differences between some participants' responses to the Funsorter and their answers in the interview. This may be because their preferences in the Funsorter did not differ much from each other, and their preferences were rather weak.

This study had several limitations. First, the sample size used in this study was small, resulting in a high standard deviation.

Second, the researcher was present in the experimental room to operate the robot. This may have been uncomfortable for the participant, knowing that the researcher is operating the robot. For example, when the robot compliments them during the execution of the exercise. The participants might have felt more connected to the robot if the researcher had been in another room. Unfortunately, this was not feasible at the locations of this study. Third, it was not possible to have a natural conversation between the robot and the participant. It was only possible to let the robot speak pre-programmed sentences, based on a script. When the participant asked an unexpected question, the robot could not answer, and the participant needed help from the researcher. This could have influenced the experience of the participant with the robot.

Lastly, the SociBot Mini robot doesn't have arms and legs. While Liu et al. [21] showed that the humanoid shape of the robot is beneficial for motivating physical exercises, this robot cannot demonstrate the exercises or cannot practice at the same time as the participant [21]. Had the robot been able to do this, it could have influenced the enjoyment because practicing together can bring additional motivation processes [8].

6 Conclusion

This work investigated whether adding music suggestions or choice possibilities to the behavioral repertoire of a robot coach in physical exercising has a positive effect on the enjoyment and motivation of pre-frail elderly to exercise. We showed that the participants preferred to exercise when the robot coaching was accompanied by music, but a preference for the robot or themselves to choose the music was not found. Despite the small sample size, slightly more participants preferred the robot to make a choice, for different reasons. A general observation was that the participants had different preferences regarding music genre and the preference to choose or let the robot make a choice for them, so further research is needed to clarify how these personal needs can be accommodated for. It is also important to note that pre-frail elderly people face different physical and mental conditions and their condition may deteriorate. Therefore, the music strategy in SAR systems must be continuously adapted to the specific user involved in the interaction, thus further personalizing the robot.

Acknowledgements. We would like to thank Maggie Song and Baisong Liu for their expertise and assistance on the topic of Social Assistive Robots, pre-frail elderly, and physical exercises. We would also like to thank De Zorggroep and the participants of the experiment for their input and cooperation.

References

1. Baert, V., Gorus, E., Mets, T., Geerts, C., Bautmans, I.: Motivators and barriers for physical activity in the oldest old: a systematic review. Ageing Res. Rev. **10**(4), 464–474 (2011). https://doi.org/10.1016/j.arr.2011.04.001
2. Schutzer, K., Sue Graves, B.: Barriers and motivations to exercise in older adults. Prev. Med. **39**(5), 1056–1061 (2004). https://doi.org/10.1016/j.ypmed.2004.04.003
3. Silveira, P., Reve, E.V.H., Daniel, F., Casati, F., Bruin, E.D.D.: Motivating and assisting physical exercise in independently living older adults: a pilot study. Int. J. Med. Inform. **82**(5), 325–334 (2013). https://doi.org/10.1016/j.ijmedinf.2012.11.015
4. Mullen, S.P., et al.: Measuring enjoyment of physical activity in older adults: invariance of the physical activity enjoyment scale (paces) across groups and time. Int. J. Behav. Nutr. Phys. Act. **8**(1), 103 (2011). https://doi.org/10.1186/1479-5868-8-103
5. Fried, L.P., Tangen, C.M., Walston, J., Newman, A.B., Hirsch,C., Gottdiener,J., et al.: Frailty in older adults: Evi-dence for a phenotype. J. Gerontol. Ser. A, Biol. Sci. Med. Sci. **56**(3), M146–M156 (2001). https://doi.org/10.1093/ge-rona/56.3.M146
6. Olde Keizer, R.A.C.M., et al.: Using socially assistive robots for monitoring and preventing frailty among older adults: a study on usability and user experience challenges. Heal. Technol. **9**(4), 595–605 (2019). https://doi.org/10.1007/s12553-019-00320-9
7. Ahmed, N., Mandel, R., Fain, M.J.: Frailty: an emerging geriatric syndrome. Am. J. Med. **120**(9), 748–753 (2007). https://doi.org/10.1016/j.amjmed.2006.10.018
8. Faber, M.J., Bosscher, R.J., China Paw, M.J., van Wieringen, P.C.: Effects of exercise programs on falls and mobility in frail and pre-frail older adults: a multicenter ran- domized controlled trial. Prim. Health Care **17**(3), 35 (2007). https://doi.org/10.7748/phc.17.3.35.s22
9. Espinoza, S., Walston, J.D.: Frailty in older adults: insights and interventions. Clevel. Clin. J. Med. **72**(12), 1105–1112 (2005). https://doi.org/10.3949/ccjm.72.12.1105
10. Novak, D., Nagle, A., Keller, U., Riener, R.: Increasing motivation in robot-aided arm rehabilitation with competitive and cooperative gameplay. J. Neuroeng. Rehabil. **11**(1), 64 (2014). https://doi.org/10.1186/1743-0003-11-64
11. Kendzierski, D., Decarlo, K.J.: Physical activity enjoyment scale: two validation studies. J. Sport Exerc. Psychol. **13**(1), 50–64 (1991). https://doi.org/10.1123/jsep.13.1.50
12. Stork, M.J., Kwan, M.Y.W., Gibala, M.J., Ginis, K.A.M.: Music enhances performance and perceived enjoyment of sprint interval exercise. Med. Sci. Sports Exerc. **47**(5), 1052–1060 (2015). https://doi.org/10.1249/mss.0000000000000494
13. Vuilleumier, P., Trost, W.: Music and emotions: from enchantment to entrainment. Ann. N. Y. Acad. Sci. **1337**(1), 212–222 (2015). https://doi.org/10.1111/nyas.12676

14. Meyns, P., et al.: Do a humanoid robot and music increase the motivation to perform physical activity? A quasi-experimental cohort in typical developing children and preliminary findings in hospitalized children in neutropenia. Int. J. Hum. Comput. Stud. **122**, 90–102 (2019). https://doi.org/10.1016/j.ijhcs.2018.07.010
15. Colombo, R., Pisano, F., Mazzone, A., Delconte, C., Micera, S., Carrozza, M. C., … Minuco, G.: Design strategies to improve patient motivation during robot-aided rehabilitation. J. Neuro Eng. Rehabil. **4**(1) (2007). https://doi.org/10.1186/1743-0003-4-3
16. Schneider, S., Goerlich, M., Kummert, F.: A framework for designing socially assistive robot interactions. Cogn. Syst. Res. **43**, 301–312 (2017). https://doi.org/10.1016/j.cogsys.2016.09.008
17. Fasola, J., Mataric, M.J.: Robot motivator: Increasing user enjoyment and perfor- mance on a physical/cognitive task. In: 2010 IEEE 9th International Conference on Development and Learning (2010). https://doi.org/10.1109/devlrn.2010.5578830
18. Lotfi, A., Langensiepen, C., Yahaya, S.: Socially assistive robotics: robot exercise trainer for older adults. Technologies **6**(1), 32 (2018). https://doi.org/10.3390/technologies6010032
19. Görer, B., Salah, A.A., Akın, H.L.: An autonomous robotic exercise tutor for elderly people. Auton. Robot. **41**(3), 657–678 (2016). https://doi.org/10.1007/s10514-016-9598-5
20. Fasola, J., Mataric, M.J.: Using socially assistive human-robot interaction to motivate physical exercise for older adults. Proc. IEEE **100**(8), 2512–2526 (2012). https://doi.org/10.1109/jproc.2012.2200539
21. Liu, B., Markopoulos, P., Tetteroo, D.: How anthropomorphism affects user ac ceptance of a robot trainer in physical rehabilitation. In: Proceedings of the 12th International Joint Conference on Biomedical Engineering Systems and Technologies (2019). https://doi.org/10.5220/0007343600300040
22. Haselton, M.G., Nettle, D., Murray, D.R.: The Evolution of Cognitive Bias. Handb. Evol. Psychol. 1–20 (2015). https://doi.org/10.1002/9781119125563.evpsych241
23. Karageorghis, C., Jones, L., Stuart, D.: Psychological effects of music tempi during exercise. Int. J. Sports Med. **29**(7), 613–619 (2008). https://doi.org/10.1055/s-2007-989266
24. Feng, Y., Yu, S., van de Mortel, D., Barakova, E., Hu, J., Rauterberg, M.: LiveNa ture: ambient display and social robot-facilitated multi-sensory engagement for people with dementia. In: Proceedings of the 2019 on Designing Interactive Systems Conference, pp. 1321–1333 (2019)
25. Song, H., Zhang, Z., Barakova, E. I., Ham, J., Markopoulos, P.: Robot role design for implementing social facilitation theory in musical instruments practicing. In: Proceed ings of the 2020 ACM/IEEE International Conference on Human-Robot Interaction, pp. 253–260 (2020)
26. Ghazali, A.S., Ham, J., Barakova, E., Markopoulos, P.: Assessing the effect of persuasive robots interactive social cues on users' psychological reactance, liking, trusting beliefs and compliance. Adv. Robot. **33**(7–8), 325–337 (2019)
27. Baur, K., Speth, F., Nagle, A., Riener, R., Klamroth-Marganska, V.: Music meets robotics: a prospective randomized study on motivation during robot aided therapy. J. Neuroeng. Rehabil. **15**(1), 79 (2018)
28. Ghazali, A.S., Ham, J., Barakova, E., Markopoulos, P.: Persuasive robots acceptance model (PRAM): roles of social responses within the acceptance model of persuasive robots. Int. J. Soc. Robot. **12**, 1075–1092 (2020)
29. Heerink, M., Kröse, B., Evers, V., Wielinga, B.: Assessing acceptance of assistive social agent technology by older adults: the almere model. Int. J. Soc. Robot. **2**(4), 361–375 (2010)

Online vs. Offline Adaptive Domain Randomization Benchmark

Gabriele Tiboni[1(✉)], Karol Arndt[2], Giuseppe Averta[1], Ville Kyrki[2], and Tatiana Tommasi[1]

[1] Politecnico di Torino, Torino 10129, Italy
{gabriele.tiboni,giuseppe.averta,tatiana.tommasi}@polito.it
[2] Aalto University, Espoo 02150, Finland
{karol.arndt,ville.kyrki}@aalto.fi

Abstract. Physics simulators have shown great promise for conveniently learning reinforcement learning policies in safe, unconstrained environments. However, transferring the acquired knowledge to the real world can be challenging due to the reality gap. To this end, several methods have been recently proposed to automatically tune simulator parameters with posterior distributions given real data, for use with domain randomization at training time. These approaches have been shown to work for various robotic tasks under different settings and assumptions. Nevertheless, existing literature lacks a thorough comparison of existing adaptive domain randomization methods with respect to transfer performance and real-data efficiency. This work presents an open benchmark for both offline and online methods (SimOpt, BayRn, DROID, DROPO), to investigate current limitations on multiple settings and tasks. We found that online methods are limited by the quality of the currently learned policy for the next iteration, while offline methods may sometimes fail when replaying trajectories in simulation with open-loop commands. The code used is publicly available at https://github.com/gabrieletiboni/adr-benchmark.

Keywords: Robot learning · Sim-to-real · Domain randomization · Benchmark

1 Introduction

Recent advancements in the field of deep Reinforcement Learning (RL) have shown promising results for many robotic applications, by allowing to solve tasks through simple trial-and-error. On one side, this has opened a new path for collaborative robots that autonomously learn complex skills. On the other, learning through interactions with the environment still requires a substantially large number of training data and poses important safety risks [9]. To this end, researchers have adopted strategies to conveniently train robots for safe optimal control, such as learning from human demonstrations or by interacting with unconstrained simulated environments. There has been a growing interest in the

P. Borja et al. (Eds.): HFR 2022, 26, pp. 158–173, 2023.
https://doi.org/10.1007/978-3-031-22731-8_12

latter approach in recent years, driven by several success stories of robotic tasks being learned exclusively in simulation, from robotic in-hand manipulation [16], to drone flight [15] and locomotion for quadruped robots [22]. Albeit promising, however, the effectiveness of a sim-to-real transfer strongly depends on the quality and the accuracy of the physics engines that simulate the underlying scenario. Indeed, directly transferring robot policies learned in a poorly calibrated simulator—a.k.a. the *source domain*—usually leads to low performance on real hardware [11]—the *target domain*. Such discrepancy is commonly known as the *reality gap* and represents the main limitation for RL in robotics. Generally speaking, the reality gap may be attributed to a combination of factors, such as mismatched dynamics models and unmodeled phenomena. Additionally, humidity and temperature changes, together with wear and tear of robotic setups, often lead to dynamically changing real-world environments.

To bridge the reality gap and learn more generalizable RL agents, several works have proposed to train robots over varying simulated environments, i.e. by randomizing dynamics parameters according to a predefined probability distribution. This technique, known as *Domain Randomization* (DR), has demonstrated to be a promising method to learn robust robot policies that can be directly transferred to the target domain [28]. More specifically, DR requires designing a source domain distribution over physical parameters that allow for both training of a single well-performing policy on varying dynamics, and for efficient transfer to the real system. This problem attracted researchers to move from point-estimate tuning techniques to full posterior distribution inference for use with Domain Randomization. A recent survey [15] refers to such methods as *Adaptive Domain Randomization* (ADR), as the primary goal is to automatically infer source domain distributions, as opposed to using manually engineered static distributions. Despite the plethora of recently published ADR methods

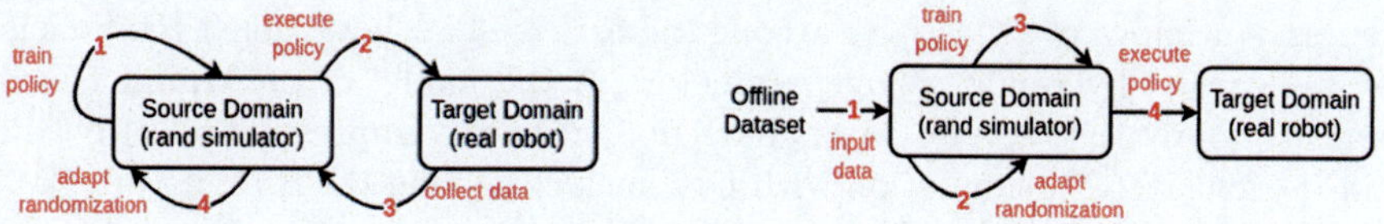

Fig. 1. Conceptual illustration of online (left) and offline (right) adaptive domain randomization.

[3,12,16,20,23,25], their problem setting, robotic tasks, and imposed assumptions often differ. Furthermore, each method presents superior performance w.r.t. their proposed baselines, but these have been hardly ever compared to each other. In accordance with [15], we claim that while these works have critically different assumptions, the existing literature lacks a thorough comparison of real-world transfer performance of existing ADR methods. A similar comparison may shed light on which methods are more suitable for certain problem settings than others, besides investigating current limitations for future studies.

This work presents an open benchmark of four ADR methods (SimOpt [3], BayRn [12], DROID [25], DROPO [23]) tasked with sim-to-sim transfers under noisy conditions and unmodeled phenomena. We focus on target domain performance as the metric of interest, rather than inspecting the accuracy of the inferred parameter posteriors. To further investigate each method's limitations, we explore how performances vary according to the amount of data collected from the target domain, which is often critically limited and expensive to obtain in robotics scenarios. In particular, we distinguish between *online* approaches [3,12], which iteratively gather on-policy data from the target domain, and *offline* approaches [23,25], which rely on fixed offline datasets (see Fig. 1). For example, offline methods can be fed with human demonstrations which feel more natural to non-expert users when interacting with the real setup, as opposed to rolling out RL policies iteratively with careful supervision. However, online approaches might reach higher performance in the long run by adapting to the target domain through multiple iterations. To the best of our knowledge, this is the first work aiming to provide insights into how algorithms from the two aforementioned ADR settings perform along a variety of axes and tasks.

The contributions of this work are: (1) shedding light on the critically different assumptions and limitations of online vs. offline ADR methods; (2) providing a thorough performance analysis over four ADR methods w.r.t. target domain performance and data efficiency; (3) investigating how different data collection strategies affect the performances of offline methods.

2 Related Works

Domain Randomization (DR) has been gaining wider adoption in recent years as a method for sim-to-real transfer, driven by its ease of implementation and effectiveness. Antonova et al. [1] were among the first to show how robust RL policies may be learned by randomizing parameters in the model of the training environment. Soon after, several works have demonstrated impressive robotic skills trained exclusively in simulation with DR, such as solving the Rubik's cube [16], learning locomotion tasks [22], autonomously opening a door [5], or pushing objects with arbitrary friction properties [17]. Valassakis et al. [26] have further investigated how DR compares to other sim-to-real transfer techniques for simple robotic manipulation tasks. Moreover, a recent study [4] has attempted to provide a theoretical framework around DR for dynamics to explain its impressive empirical results. While not explored in this work, the same concept has also been applied in the field of deep RL for vision-based policies, by randomizing appearance properties of the simulated environment [8,21,24].

As the complexity of the task increases, the main challenge lies in how to design source domain distributions to both encompass real-world physical parameters and make up for mismatched dynamics models [27]. When fixed uniform distributions are picked, the approach is generally referred to as Uniform Domain Randomization (UDR). UDR may be the result of prior task-specific knowledge or tedious back-and-forth tuning until the desired target performance is achieved.

To relax the manual engineering efforts associated with this approach, Adaptive Domain Randomization (ADR) methods attempt to automatically infer the source domain distribution. Early attempts by Rajeswaran et al. [19] in this direction demonstrated that physical parameters could be inferred by collecting data from the target domain, despite showing experiments in simulation only. More recent works gained popularity by applying a similar concept to actual sim-to-real transfer tasks, and using the inferred source domain distribution for domain randomization at training time: SimOpt [3] infers dynamics parameters by minimizing a discrepancy metric between real and simulated trajectories, while BayRn [12] directly optimizes for real-world returns by framing inference as an end-to-end Bayesian Optimization problem. While training always happens in simulation, these methods require iterative target domain data collection—as they roll out the currently trained policy to ask for real-world feedback—potentially posing data efficiency concerns and assuming the real hardware to be available at training time. We refer to these as *online* ADR methods.

An opposite line of work has been recently proposed by exploiting offline fixed datasets to infer source domain distributions: DROID [25] minimizes joint torques when replaying real-world trajectories in simulations, while DROPO [23] infers dynamics parameters given real-world transitions in a maximum-likelihood framework. These methods make no assumptions on how the given dataset has been collected, making them suitable for use with human demonstrations—which are often conveniently available in many robotic setups—or data collected by previously trained policies for other tasks. By nature, offline ADR approaches thrive on safety-critical tasks where rolling out multiple intermediate policies may be too expensive. On the other hand, these may fall short to readily adapt to the target domain given limited data, as opposed to online methods.

Our work aims to provide the first empirical insights to support or re-evaluate the above premises, by comparing the four previously described methods under noisy conditions and unmodeled phenomena in simulation, by means of target domain performance and data efficiency. We chose these methods as they all leverage feed-forward network architectures, gradient-free optimization for non-differentiable simulators, and parametric source domain distributions. We acknowledge some similarities with [11], a recent open benchmark on system identification for physics simulator tuning. However, this study purely focused on point-estimate parameter inference—i.e. system identification—and did not discuss the data efficiency requirements. In addition, each algorithm is tested in direct parameter estimation under noiseless and perfectly modeled conditions.

As the lack of a common benchmark for DR has been supported by a recent survey [15], we encourage future works to extend the comparison to other ADR methods and settings. For instance, recent approaches have studied the underlying problem from a more general perspective, allowing to infer free-form source domain distributions given a starting data collection policy [14,20].

Finally, other works have approached the task in an unsupervised problem setting, i.e. *without* making use of target domain data: [10] drives policy training on progressively harder dynamics parameters to improve generalization, [13] estimates the transfer performance with reference simulated environments, [16] encourages wider source domain distributions as long as the training performance is sufficiently high.

3 Methodology

This section introduces the formal problem formulation, a detailed overview over online vs. offline ADR methods, and our experimental setup.

3.1 Problem Setup

Problem Formulation. Consider the source domain environment to be modeled as a discrete-time dynamical system, defined by a continuous state space $\mathcal{S} \in \mathbb{R}^{n_s}$, a continuous action space $\mathcal{A} \in \mathbb{R}^{n_a}$, and an initial state distribution $\mu(s_0)$. In simulation, the environment can be further parameterized by its dynamics parameters $\xi \in \mathbb{R}^{n_\xi}$, e.g. masses, friction coefficients, robot link sizes. The system dynamics are therefore described by the transition probability density function $\mathcal{P}_\xi(s_{t+1}|s_t, a_t)$, given the current state s_t and action taken a_t, at time step t. At each step, the agent is given a scalar reward feedback r_t according to the function $R(s_t, a_t, s_{t+1})$, assumed to be deterministic for the sake of simplicity. Overall, the source environment forms a Markov Decision Process (MDP) described by the tuple $\mathcal{M}_\xi = \langle \mathcal{S}, \mathcal{A}, \mathcal{P}_\xi, \mu, R \rangle$. We assume ξ to be random variables that obey a parametric distribution $p_\phi(\xi)$, parameterized by ϕ—e.g. mean and variance for Gaussian distributions.

Under this formulation, the goal of an RL agent is to maximize the expected cumulative rewards (i.e. the return), by acting with a stochastic policy $\pi(a_t|s_t)$. In particular for the domain randomization setting, the goal is to train a robust policy over the source domain distribution $p_\phi(\xi)$,

$$\pi_\phi^* = \arg\max_\pi \mathbb{E}_{\xi \sim p_\phi(\xi)} \left[E_{\pi, \mathcal{P}_\xi, \mu_0} \left[\sum_{t=0}^{T} \gamma^t r_t \right] \right] \tag{1}$$

with discount factor $\gamma \in (0, 1]$. In practice, DR can be easily implemented into existing RL algorithms by sampling new dynamics parameters $\xi \sim p_\phi(\xi)$ at the start of each training episode, resulting in the outer expectation in Eq. 1. ADR methods further include an inference phase in the overall pipeline to estimate the parameters ϕ of the source domain distribution $p_\phi(\xi)$. In this work, we consider Supervised ADR methods, i.e. approaches that use a collection of state-transitions $\mathcal{D} = \{(s_0, a_0, s_1, r_1), \ldots, (s_{T-1}, a_{T-1}, s_T, r_T)\}$ from the target domain environment $\mathcal{M}_{real}$. In particular, we assume that the target MDP $\mathcal{M}_{real}$ shares the same state space, action space and reward function of $\mathcal{M}_\xi$, but generally differs by dynamics.

Online ADR methods, such as SimOpt [3] and BayRn [12], rely on iterative data collection from the target domain at optimization time (see Fig. 1). They generally involve a bi-level optimization problem where RL policy training with DR in Eq. 1 is interleaved by policy rollouts in the real setup. At the i-th iteration, target transitions $\mathcal{D}_i$ are collected with the currently learned policy $\pi^*_{\phi_i}$ and used to adapt the source domain distribution $p_{\phi_{i+1}}(\xi)$, repeating the process until convergence.

While policies are still learned exclusively in simulation, these methods assume the real setup to be available for rolling out intermediate policies. Note how this assumption may be particularly expensive in robotic tasks, as careful expert supervision is needed when dealing with real hardware—e.g. for resetting the episodes, running safety checks, or interpreting and anticipating undesired behavior. Furthermore, intermediate policies may be the main point of failure themselves, as things can hardly be recovered if a policy happens to learn unwanted behavior in the middle of the process—e.g. it is unsafe to be run on the real setup, or it is not able to collect informative data, such as in the case of a robotic arm which moves through air without actually interacting with the environment. Finally, these approaches may not be fully parallelized and run in a cluster in an end-to-end fashion, as previously noted by a related work [12].

Offline ADR methods make use of previously collected target domain data $\mathcal{D}$ to infer an optimal source domain distribution $p_{\phi^*}(\xi)$, and later train a single policy $\pi^*_{\phi^*}$ that can be directly transferred to the real world. An illustration of the pipeline of such methods is depicted in Fig. 1, on the right.

In contrast to online methods, these make no assumptions on how real data has been collected. For this reason, offline ADR methods can be fed with human demonstrations, or data collected by previously trained policies for similar tasks. As such, previous studies in this line of work [23,25] claim to adopt a safer and less restrictive pipeline. Note, indeed, how offline methods can potentially be run in online fashion—by collecting novel data with the newly learned policy and repeating the process—while the opposite is not always possible. Moreover, this approach allows for end-to-end parallelization—without expert personnel in the middle—and easier benchmarking of different algorithms by keeping a fixed shared dataset. Despite the promising assumptions, these methods are by definition limited to the information provided by the previously collected trajectories, and may fail to readily adapt to the real domain.

3.2 Benchmark Methods

Uniform Domain Randomization. We implement static domain randomization as a baseline, also known as Uniform Domain Randomization (UDR). Instead of manually tuning uniform distributions—which would be hard to fairly and systematically benchmark—we followed the same approach proposed by previous works [12,23]: we average the results of 10 policies trained on randomly sampled uniform bounds, within a predefined search space over the dynamics parameters. This would essentially reflect the performances of a random search, allowing up to 10 evaluations.

Bayesian Domain Randomization (BayRn) [12] was introduced as an affordable extension to UDR for adapting source domain distributions. BayRn optimizes the parameters ϕ through Bayesian Optimization (BO), using the final target domain return as optimization metric. By definition, BayRn falls under the area of online ADR methods. We parameterize source distributions as uniform when implementing BayRn, and we initialize the Gaussian Process with the results of the 10 UDR policies, as suggested by the authors. This makes BayRn slightly less comparable to the other ADR methods tested, which do not receive initial information. However, as we only allow a maximum of 5 real-world iterations in our benchmark protocol, we believe this step to be crucial for obtaining meaningful results through BO. Note that BayRn does not explicitly use sensor measurements from the target domain, but critically requires reward computation on real hardware.

SimOpt [3] uses a discrepancy metric between real and simulated trajectories to progressively optimize $p_\phi(\xi)$, with a trust region in the dynamics parameter space to stabilize the optimization process. Data is iteratively collected on the real setup with the currently trained policy, and later replayed in simulation with closed-loop actions from the same policy. We implement SimOpt using the Relative Entropy Policy Search (REPS) implementation by [6]. In order to maximize data efficiency, we only collect a single target trajectory per policy and perform multiple REPS updates at each iteration. A hyperparameter search is carried out over the number of intermediate REPS updates and number of sampled parameters per update, and we end up using 5 and 1000, respectively.

Finally, we implement a variant of SimOpt to investigate its performances when only a single real-world iteration is allowed, similarly to offline methods. Throughout the experiments, we refer to this variant as *SimOpt-1*, which is implemented by executing the total number of REPS updates (25) during the first and only iteration.

DROID [25] proposes to use expert human demonstrations for adapting source domain distributions $p_\phi(\xi)$. Demonstrations are replayed in simulation by executing the same actions as in the real setup, allowing to regress dynamics parameters in simulation through CMA-ES [7]. As we only show experiments on torque-control tasks, we adapt the original algorithm to minimize the L2-norm between observations rather than joint torques. Moreover, we don't assume to provide optimal offline data for the underlying task, therefore we don't consider the penalty factor in the original objective function.

DROPO [23] proposes to optimize $p_\phi(\xi)$ with a maximum-likelihood approach, based on offline-collected data. Under this formulation, CMA-ES is used to regress both means and variances of the source domain distribution, by replaying offline data in simulation similarly to DROID. Nevertheless, DROPO further assumes that sensor measurements from the real setup give access to the full state configuration, in order to reset the simulated environment arbitrarily along the trajectory. We test DROPO using the open-source implementation [23], and tune the hyperparameter ϵ as suggested by the authors. Moreover, we set the sample size to be 10 times larger the dimensionality of tested dynamics parameter spaces, to allow for meaningful likelihood inference.

3.3 Benchmark Tasks

We test the aforementioned ADR methods in four OpenAI gym [2] environments, depicted in Fig. 2. These robotic tasks are commonly used to benchmark sim-to-real transfer methods in the absence of real hardware [11]. To further simulate the reality gap, we consider each environment in three different versions: *vanilla*, *noisy*, and *unmodeled*. The first case corresponds to the basic condition with no mismatch between source and target. The other two allow us to compare performances under noisy conditions—by injecting noise during target domain data collection—and in presence of unmodeled phenomena—by misidentifying a subset of dynamics parameters. More specifically, we simulate unmodeled phenomena as in [19,23]: a selected number of parameters is taken out of the optimization problem and underestimated by 20% w.r.t. their original ground truth values. Note that the environments are designed by increasing difficulty, by progressively expanding the state space, dynamics parameter space, and number of unmodeled parameters. As the Humanoid environment contains a richer state space which includes masses and constraint forces, we only feed inference methods with the main bodies kinematics properties—i.e. the first 45 dimensions—while leaving full information to policies at training time. An overall description of the benchmark tasks is reported in Table 1.

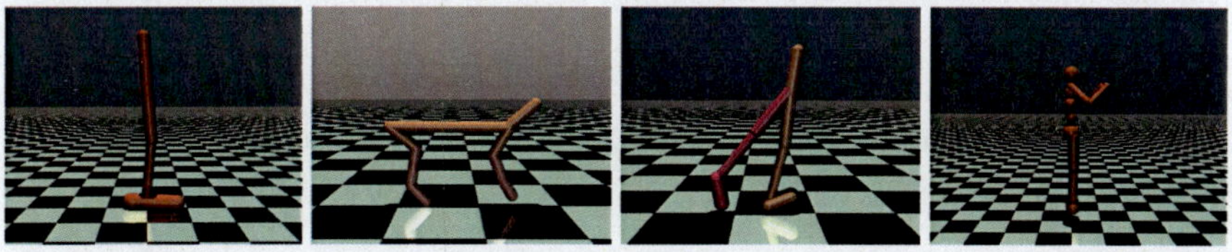

(a) (b) (c) (d)

Fig. 2. Illustration of the four OpenAI gym tasks used in this benchmark: (a) Hopper, (b) HalfCheetah, (c) Walker2D, and (d) Humanoid.

3.4 Benchmark Protocol

As ADR methods share the same final goal, we propose a systematic framework to compare online vs. offline approaches by means of target domain policy performance and data efficiency. However, given the different assumptions and requirements highlighted in Sect. 3.2, these metrics should by no means be considered as the sole axes of evaluation. Nevertheless, this work may help identifying which method is most suitable for a given problem setting, and how these algorithms perform when all requirements are fulfilled.

Table 1. List of OpenAI gym [2] environments used in this benchmark.

Environment	$\mathcal{S}$	$\mathcal{A}$	ξ	Unmod. ξ	Noise	Parameters
Hopper	11	3	4	1	10^{-4}	Link masses
HalfCheetah	17	6	8	3	10^{-4}	Link masses, friction
Walker2d	17	6	13	4	10^{-3}	Link masses and lengths, friction
Humanoid	376	17	30	7	10^{-3}	Link masses, joint damping

The dimensionality of the state space, action space, dynamics parameter space, and unmodeled parameter space is reported, together with the noise level.

We benchmark online methods by allowing a maximum of 5 iterations. At each iteration, we limit data collection to a single trajectory clipped to a length of 200 transitions. This way, we keep the size of target datasets limited to a maximum of 1000 transitions, amounting to about 10 seconds-worth of locomotion data. We feed offline methods with the cumulative trajectories collected by online methods up until the current iteration. In particular, we use data from SimOpt, as, similarly to DROID and DROPO, it makes use of sensory measurements from the target domain. By doing so, we ensure that SimOpt, DROID and DROPO have access to the same type and amount of information at all times.

Dynamics parameters are normalized to the interval $[0, 4]$, according to a predefined bounded search space. All methods start from a conservative prior source domain distribution centered in 2 with identity covariance matrix. An exception to this is BayRn, which deals with uniform distributions and gets initialized with the results of the random search by UDR. In the case of the SimOpt single-iteration variant (SimOpt-1), we use the initial distribution to train a data collection policy, which is later used to collect the same amount of data used by offline methods, respectively for each iteration. We report the performances at the zero-th iteration as the target return of a policy trained on the initial conservative guess, marked as the starting point for all methods.

For policy training with Reinforcement Learning, we use the Soft Actor-Critic (SAC) implementation by *stable-baselines3* [18]. We parametrized both actor and critic policies as 3-layer MLP neural networks, with 128 neurons in the hidden layers. We train each policy on 12 parallel environments, for a maximum of 5M timesteps overall. All SAC hyperparameters are kept at default values, except for the learning rate which is tuned separately for each environment by training on ground truth dynamics. To ensure fair comparison among different policies, we stop the training process when a task-specific reward threshold is reached.

Finally, we report returns as the average performance over 3 random seeds—by repeating both the inference phase and policy training—normalized by the training reward threshold for the sake of clarity.

4 Results

4.1 Vanilla Parameter Estimation

As a starting point, we benchmark the outlined methods in a direct parameter estimation setting. Under this formulation, the two domains are solely mismatched by discrepancies in dynamics parameter values. ADR methods thus attempt to estimate source domain distributions that resemble the ground truth parameters, by collecting data from the target environment. While previous works mostly investigated the inference results under this setting [11], we directly focus on the policy performance in the ground truth environment, with respect to the amount of data seen by each algorithm.

The results are depicted in Fig. 3, in terms of returns normalized by the training reward threshold. We observe that all methods are able to successfully transfer in the low-dimensional Hopper task, while some discrepancy is noted for more complex environments. In particular, we found that open-loop replaying of target trajectories by DROID generally leads to less stable results, with respect to SimOpt's approach of collecting sim trajectories by rolling out the currently converged policy. This effect is particularly visible in the Walker2D task where multiple trajectories lead to lower performances for DROID, e.g. if the Walker-agent happens to fall down while executing the target actions, the trajectory discrepancy metric may get uninformative. SimOpt authors have stated similar observations when experimenting with open-loop replaying in their experiments (see Appendix A in [3]). We suspect that the trust region imposed by SimOpt may further prevent diverging trajectories, as the next-iteration parameter search likely falls closer to the current policy training dynamics w.r.t. CMA-ES unconstrained search. Moreover, we noticed that, while executing actions in

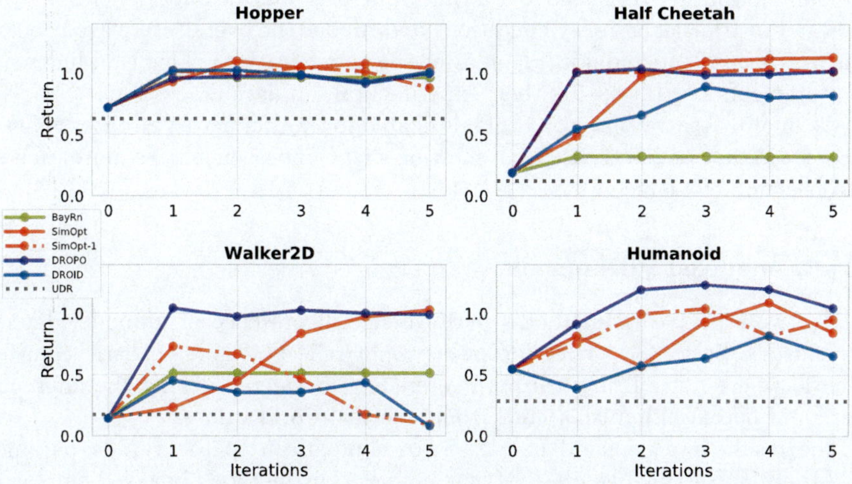

Fig. 3. Policy returns in the target domain (normalized by training reward threshold) in the vanilla parameter estimation sim-to-sim setting.

the same way as DROID, DROPO does not suffer from this. To this end, we found that intermediately resetting the simulator state is crucial to prevent sim trajectories to diverge from real ones, and provide informative distance measures.

Overall, we observe that SimOpt is able to progressively get better results and solve the task in all cases within 5 iterations, i.e. with 5 or less trajectories collected from the target domain. Interestingly enough, the quality of the policy also matters for SimOpt, as SimOpt-1 fails in the Walker2D task when collecting all trajectories with the poorly performant initial policy. On the other hand, BayRn fell short on higher-dimensional tasks when only 5 iterations are allowed, as Gaussian Processes are notoriously problematic to scale to high parameter dimensions (≥ 8 in our case). Due to this, we do not report results for Bayesian Optimization on the Humanoid environment.

Finally, we found that DROPO is able to solve all tasks and match the long-term results of SimOpt in this setting, even with a single target trajectory.

4.2 Observation Noise

We test the transfer performance of all benchmark methods when the target domain is assumed to provide noisy measurements, as in the case of real hardware. For this purpose, we feed each algorithm with target observations containing injected noise according to the predefined values in Table 1.

We report the results for this setting in Fig. 4. In general, we draw similar conclusion as in the direct parameter estimation setting: (i) all methods succeed on the Hopper task, (ii) DROID and SimOpt-1 may fail unexpectedly when sim trajectories diverge during replay (see Walker2D in Fig. 4), (iii) BayRn fails to transfer on problems with higher-dimensional dynamics parameter spaces. Moreover, in this setting we experienced generally lower SimOpt performances, especially for harder tasks: we found that SimOpt's intermediate policies may at times fail to transfer in noisy conditions and hinder the overall iterative process (see SimOpt on Humanoid in Fig. 4 after the first iteration). This is in line with the statements by [12] during their experimental evaluation.

We finally observe that DROPO is still able to transfer across all tasks given a single trajectory of 200 transitions, showing slight improvements as more (noisy) data becomes available.

4.3 Unmodeled Phenomena

Finally, we report each method's performance in presence of unmodeled phenomena (see Sect. 3.3). This approach would reflect a more realistic transfer scenario where not all physical parameters are considered during the inference phase and potentially mismatched from true real-world values.

The results are depicted in Fig. 5. In comparison to the vanilla parameter estimation setting, we observe that offline ADR methods behaved similarly, while SimOpt and its single-iteration variant struggled to solve the task in

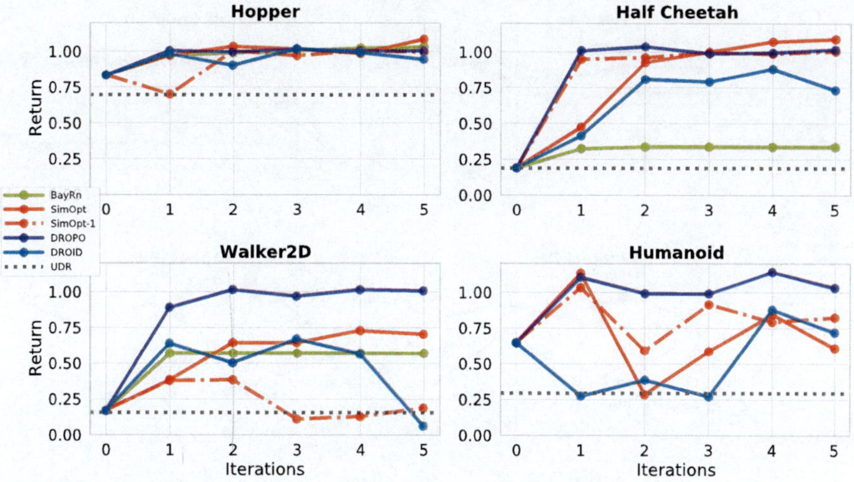

Fig. 4. Normalized returns in the target domain under noisy observations.

the Humanoid environment. We believe this to be caused by intermediate ill-performing policies which prevented a successful parameter estimation and hindered the overall process. Besides this occurrence, SimOpt demonstrated stable and successful transfers within 5 trajectories in the remaining tasks. Surprisingly, BayRn was shown to obtain superior performance in the Hopper task than other methods, likely by avoiding explicit parameter estimation and directly optimizing for target domain returns. Overall, we found DROPO to be the best performing method under this setting, despite requiring significantly more resources for hyperparameter tuning: DROPO inherently requires multiple runs to decide on the best regularization value ϵ, which was found to significantly impact on final performances. Nevertheless, we stuck to the original suggested procedure for tuning this parameter [23]. In addition, we experienced noticeably higher inference times by DROPO, as likelihood estimation requires a considerable amount of Monte-Carlo sampling. On the other hand, DROID can be run with minimal hyperparameter tuning and faster objective function evaluations, but still generally suffered from divergent trajectories during parameter search on the harder tasks. Furthermore, we observed that DROID consistently converged to nearly zero-variance distributions, as a result of the CMA-ES covariance matrix decaying to infinitesimal values after finding a local optimum. This behavior limits the potential of domain randomization, effectively reducing it to system identification.

4.4 Off-Policy Data Collection

We designed the main experimental evaluation by feeding offline methods with the cumulative data collected by SimOpt, in order to compare the methods with the same target domain trajectories. As this would hardly be the case in a real

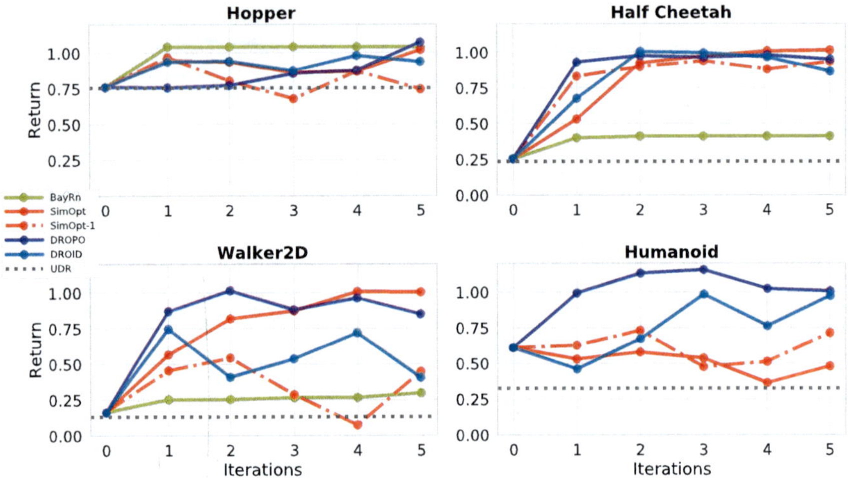

Fig. 5. Normalized returns in the target domain in presence of unmodeled phenomena.

application, in this section we investigate how different data collection strategies impact on offline ADR performances. In particular, we test how target returns with data from SimOpt compare to (i) randomly collected data, and (ii) data collected with a policy trained on the prior source distribution $\mathcal{N}(\mathbf{2}, \mathbb{1})$—as in the first iteration of the main results. We test the performances in the Hopper and Walker2D tasks, under noisy conditions, and report the results in Fig. 6.

We observe that different strategies did not significantly affect the performances in the Hopper environment when sufficient transitions are provided. However, we noticed that DROID is significantly more sensitive to the quality of collected data in the Walker2D task, with respect to DROPO. As previously argued, we attribute this phenomenon to DROID's open-loop replaying of target actions in simulation, which may lead to uninformative distance measures—e.g. when the Walker2D-agent falls down, or when the sim trajectories diverge from the target one due to highly mismatched dynamics parameters. Nevertheless, we claim that tasks different from locomotion may pose further requirements to data collection procedures, e.g. robotic manipulation would at least require the robot to interact with the objects in the scene.

Finally note that, while we could not test this strategy in sim-to-sim settings, offline methods may still be fed with data collected by kinesthetic guidance—i.e. human demonstrations—as opposed to online methods.

5 Conclusions

In this paper we provide an extensive experimental evaluation of four Supervised Adaptive Domain Randomization methods in terms of target policy performance and data efficiency, when tested on challenging sim-to-sim transfer scenarios.

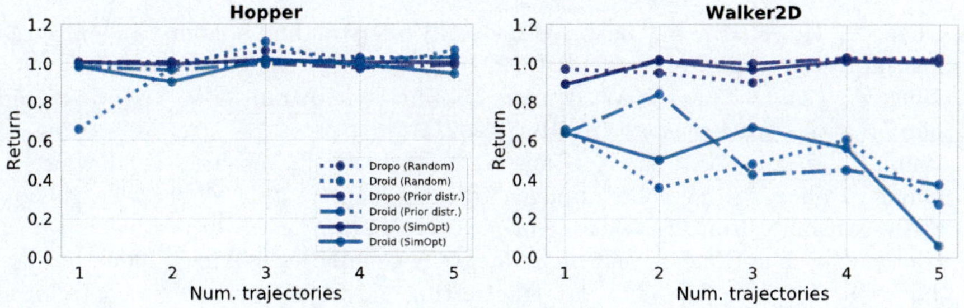

Fig. 6. Target domain performance with different data collection strategies under noisy conditions.

Additionally, we highlight the different assumptions and requirements of each benchmark method, while splitting them into online vs. offline settings w.r.t. to data collection procedures and optimization pipelines. We found that online methods are able to solve most target tasks within 5 data collection iterations—i.e. 10 seconds-worth of data—with least hyperparameter tuning and inference time. Interestingly, we observed that their performance is affected by the quality of the currently learned policy, as they may sometimes fail unexpectedly due to intermediate ill-behaving policies, posing it as the main limitation of online methods. On the other hand, offline methods generally led to better jump-start performance with fewer target transitions available, even when compared to a single iteration of SimOpt. In particular, DROPO achieved the highest average return among the tested tasks, given stricter assumptions—i.e. full knowledge of real-world state configurations is assumed. In contrast, we observed that open-loop replaying of trajectories by DROID may lead to divergent sim trajectories and, in turn, less informative trajectory discrepancy measures.

This work provides the first empirical insights into the performances of current ADR algorithms with respect to the type and amount of data required. These considerations can help researchers in the field to continue on improving sim-to-real transfer methods with Domain Randomization, and make informed decisions when applying these in future works.

Acknowledgments. We acknowledge the computational resources generously provided by HPC@POLITO and by the Aalto Science-IT project.

References

1. Antonova, R., Cruciani, S., Smith, C., Kragic, D.: Reinforcement learning for pivoting task (2017). arXiv Preprint: arXiv:1703.00472v1
2. Brockman, G., Cheung, V., Pettersson, L., Schneider, J., Schulman, J., Tang, J., Zaremba, W.: Openai gym (2016). arXiv:1606.01540v1
3. Chebotar, Y., Handa, A., Makoviychuk, V., Macklin, M., Issac, J., Ratliff, N., Fox, D.: Closing the sim-to-real loop: adapting simulation randomization with real world experience. In: ICRA (2019)

4. Chen, X., Hu, J., Jin, C., Li, L., Wang, L.: Understanding domain randomization for sim-to-real transfer. In: ICLR (2022)
5. Ding, Z., Tsai, Y., Lee, W.W., Huang, B.: Sim-to-real transfer for robotic manipulation with tactile sensory. In: IROS (2021)
6. Finn, C., Zhang, M., Fu, J., Tan, X., McCarthy, Z., Scharff, E., Levine, S.: Guided policy search code implementation (2016). http://rll.berkeley.edu/gps. softwareavailablefromrll.berkeley.edu/gps
7. Hansen, N.: The CMA Evolution Strategy: A Comparing Review (2006)
8. James, S., Davison, A., Johns, E.: Transferring end-to-end visuomotor control from simulation to real world for a multi-stage task. In: PMLR, pp. 334–343 (2017)
9. Kober, J., Bagnell, J.A., Peters, J.: Reinforcement learning in robotics: a survey. Int. J. Robot. Res. **32**(11), 1238–1274 (2013)
10. Mehta, B., Diaz, M., Golemo, F., Pal, C.J., Paull, L.: Active domain randomization. In: CoRL (2020)
11. Mehta, B., Handa, A., Fox, D., Ramos, F.: A user's guide to calibrating robotics simulators. In: CoRL (2020)
12. Muratore, F., Eilers, C., Gienger, M., Peters, J.: Data-efficient domain randomization with Bayesian optimization. IEEE ICRA Robot. Autom. Lett. **6**(2), 911–918 (2021)
13. Muratore, F., Gienger, M., Peters, J.: Assessing transferability from simulation to reality for reinforcement learning. IEEE TPAMI **43**(4), 1172–1183 (2021)
14. Muratore, F., Gruner, T., Wiese, F., Belousov, B., Gienger, M., Peters, J.: Neural posterior domain randomization. In: Faust, A., Hsu, D., Neumann, G. (eds.) Proceedings of the 5th Conference on Robot Learning and Machine Learning Research, Vol. 164, pp. 1532–1542. PMLR (2022)
15. Muratore, F., Ramos, F., Turk, G., Yu, W., Gienger, M., Peters, J.: Robot learning from randomized simulations: a review. Front. Robot. AI **9**, 799–893 (2022)
16. OpenAI, Akkaya, I., Andrychowicz, M., Chociej, M., Litwin, M., McGrew, B., Petron, A., Paino, A., Plappert, M., Powell, G., Ribas, R., Schneider, J., Tezak, N., Tworek, J., Welinder, P., Weng, L., Yuan, Q., Zaremba, W., Zhang, L.: Solving Rubik's cube with a robot hand (2019). arXiv Preprint: arXiv:1910.07113v1
17. Peng, X.B., Andrychowicz, M., Zaremba, W., Abbeel, P.: Sim-to-real transfer of robotic control with dynamics randomization. In: ICRA (2018)
18. Raffin, A., Hill, A., Gleave, A., Kanervisto, A., Ernestus, M., Dormann, N.: Stable-baselines3: Reliable reinforcement learning implementations. J. Mach. Learn. Res. **22**(268), 1–8 (2021). http://jmlr.org/papers/v22/20-1364.html
19. Rajeswaran, A., Ghotra, S., Ravindran, B., Levine, S.: Epopt: learning robust neural network policies using model ensembles. In: ICLR (2017)
20. Ramos, F., Possas, R.C., Fox, D.: BayesSim: adaptive domain randomization via probabilistic inference for robotics simulators. In: RSS (2019)
21. Sadeghi, F., Levine, S.: CAD2RL: real single-image flight without a single real image. In: RSS (2017)
22. Tan, J., Zhang, T., Coumans, E., Iscen, A., Bai, Y., Hafner, D., Bohez, S., Vanhoucke, V.: Sim-to-real: learning agile locomotion for quadruped robots. In: RSS (2018)
23. Tiboni, G., Arndt, K., Kyrki, V.: DROPO: sim-to-real transfer with offline domain randomization (2022). arXiv Preprint: arXiv:2201.08434v1
24. Tobin, J., Fong, R., Ray, A., Schneider, J., Zaremba, W., Abbeel, P.: Domain randomization for transferring deep neural networks from simulation to the real world. In: IROS (2017)

25. Tsai, Y., Xu, H., Ding, Z., Zhang, C., Johns, E., Huang, B.: DROID: minimizing the reality gap using single-shot human demonstration. IEEE Robot. Autom. Lett. **6**(2), 3168–3175 (2021)
26. Valassakis, E., Di Palo, N., Johns, E.: Coarse-to-fine for sim-to-real: sub-millimetre precision across wide task spaces. In: IROS (2021)
27. Vuong, Q., Vikram, S., Su, H., Gao, S., Christensen, H.: How to pick the domain randomization parameters for sim-to-real transfer of reinforcement learning policies? In: ICRA (2019)
28. Zhao, W., Queralta, J.P., Westerlund, T.: Sim-to-real transfer in deep reinforcement learning for robotics: a survey. In: 2020 IEEE Symposium Series on Computational Intelligence (SSCI), pp. 737–744. IEEE (2020)

On the Use of Energy Tanks for Robotic Systems

Federico Califano[1(✉)], Ramy Rashad[1], Cristian Secchi[2], and Stefano Stramigioli[1]

[1] Robotics and Mechatronics department, University of Twente, Enschede, The Netherlands
{f.califano,r.a.m.rashadhashem,s.stramigioli}@utwente.nl
[2] Department of Sciences and Methods for Engineering, University of Modena and Reggio Emilia, Modena, Italy
cristian.secchi@unimore.it

Abstract. In this document we describe and discuss *energy tanks*, a control algorithm which has gained popularity inside the robotics and control community over the last years. This article has the threefold scope of i) introducing to the reader the topic in a simple yet precise way, starting with a throughout description of the energy-aware framework, where energy tanks find their genesis; ii) summarising the range of applications of energy tanks, including an original reflection about different formulations of those; iii) discussing limits and future challenges involving energy tanks and energy-aware control in general.

1 Introduction and Background

In order to understand in an insightful way the idea behind energy tanks, we have to introduce its originating system theoretic fields, i.e., *port-based modelling* and *passivity-based control*. We first present a motivational introduction collecting the relevant conceptual features and then introduce the basic mathematical notation needed to fully understand the ideas and the techniques behind energy tanks.

The port-based modeling framework [10] relies on *energy*, one of the fundamental concepts in engineering practice and physics. Consistently with a thermodynamic approach, dynamical systems are understood and modelled as devices able to transform and route physical energy across the network they create. They do so by means of *power ports*, representing the entities interconnecting separate systems, and by means of which physical systems exchange energy. Since energy is the physical quantity which is shared among all physical domains, this paradigm yields a great success in modeling multi-physical systems and has one of its most fortunate outcomes in the *port-Hamiltonian* (pH) system theory [6]. In such theory, the classic Hamiltonian modeling framework is extended exploiting bond-graph concepts and port-based thinking in order to lead to a system representation where energy flows are evident and power ports are exploited for representing the power flowing through the system.

P. Borja et al. (Eds.): HFR 2022, 26, pp. 174–188, 2023.
https://doi.org/10.1007/978-3-031-22731-8_13

Furthermore, pH system are closed with respect to power preserving interconnections. This means that when multiple pH systems are interconnected by their power ports and when no energy is either produced nor dissipated in the interconnection (i.e. energy is just exchanged), the interconnected system is still pH and its properties can be easily inferred from its generating parts, e.g. the total energy of the interconnected system is given by the sum of its constituent ones.

It becomes thus possible to model increasingly large networks in which energy transfers between subsystem are systematically monitored and possibly used for control purposes. This framework differs from most control theoretic approaches in which dynamical systems are seen as signal-processing devices. In fact, in the so-called *control by interconnection* method, peculiar of the pH framework, the controller itself is treated as a virtual physical system instead of a signal processor, with its own power ports and pH structure. In this way the closed-loop system, generated by the power preserving interconnection of the pH plant and the pH controller, benefits of maintaining a pH structure as well, that can be used to develop novel control techniques.

These techniques make an important contact with the branch of nonlinear control theory of *passivity-based control* [35]. Passivity is a system theoretic property which, roughly speaking, guarantees that a dynamical system does not generate an unbounded amount of energy autonomously. This property, which is stronger than stability, is a desirable one for the closed-loop system for different and sometimes controversial reasons, which we will discuss in detail later, and which constitute the real essence of the motivation behind energy tanks.

1.1 Passivity and Port-Hamiltonian Systems

We now introduce the mathematical preliminaries needed to understand the main properties of pH systems and energy tanks. We keep the discussion as simple as possible to convey the main ideas needed for introducing energy tanks, and refer to [6,35] for a throughout description of these preliminaries.

A nonlinear affine control system with state $x \in \mathbb{R}^n$ and input-output pair $(u, y) \in \mathbb{R}^m \times \mathbb{R}^m$ in the form

$$\dot{x} = f(x) + g(x)u \tag{1}$$

$$y = h(x) \tag{2}$$

is said to be passive with respect to a non negative state-dependent storage function $V(x)$ if the following inequality is true for any possible input signal $u(t)$ and at any positive time t:

$$V(x(t)) - V(x(0)) \leq \int_0^t y^T(s)u(s)ds. \tag{3}$$

Passivity implies stability, which is a property of the autonomous system (i.e., $u = 0$), under the weak conditions that qualify the storage function as a Lyapunov function [35]. Passivity lends itself to a physical analogy: the storage function can be interpreted as a (generalized) energy function and the inner product of the input and of the output as a (generalized) power flow. For these reason, input and output can be called power conjugated variables.

The input-state-output formulation of a port-Hamiltonian system is given by the following instance of (1–2):

$$\dot{x} = [J(x) - R(x)]\partial_x H(x) + g(x)u, \tag{4}$$

$$y = g^T(x)\partial_x H(x). \tag{5}$$

This structure exhibits explicitly some important geometrical properties of the system, like the *Hamiltonian* $H(x)$, mapping the state variables to the total energy of the system; the symmetric positive semi-definite matrix $R(x)$, representing the energy dissipating elements; and the skew-symmetric matrix $J(x)$, representing the internal interconnection structure along which energy is distributed and generalising the Symplectic/Poisson structures of analytic mechanics. We indicate with $\partial_x H$ the differential of the Hamiltonian, represented as a column vector. A graphical representation of the system is depicted in the block diagram of Fig. 1a.

An immediate property of pH systems (4–5) is that they are passive with respect to the Hamiltonian as storage function:

$$\dot{H} = (\partial_x H(x))^T \dot{x} = -(\partial_x H(x))^T R(x)\partial_x H(x) + y^T u \le y^T u, \tag{6}$$

where we used the skew-symmetry of $J(x)$ and symmetry and positive semi-definiteness of $R(x)$. Integrating the latter equation in time yields the passivity condition (3), which we rewrite to enhance its clear energetic interpretation, indicating with $E_d(t) := \int_0^t (\partial_x H(x))^T R(x)\partial_x H(x) \ge 0$ the dissipated energy:

$$H(x(t)) - H(x(0)) = \int_0^t y^T(s)u(s)ds - E_d(t) \le \int_0^t y^T(s)u(s)ds. \tag{7}$$

The left hand side of the equation represents the stored energy in the system, which is always less or equal than the supplied energy through the I/O port, represented by the right term. The positive dissipated energy $E_d(t)$ determines the convergence rate to the equilibrium in case of an uncontrolled system, and can be interpreted as a *passivity margin*, i.e., the bigger the dissipated energy is, the bigger is the margin for which the passivity inequality results satisfied.

The port-Hamiltonian system of equations (4–5) can be restructured to highlight more its constituting energetic subsystems. The system components are now classified into energy storage, energy dissipating and energy supply elements, as depicted in the bond graph of Fig. 1b and its block-diagram equivalent Fig. 1c. Central to this formulation of the pH system is the so called *Dirac structure* $\mathcal{D}(x)$, which represents the relation between all the power conjugated variables encoded in the pH system. This relation can be represented by the following skew-symmetric map

$$\begin{pmatrix} \dot{x} \\ y_r \\ -y \end{pmatrix} = \begin{pmatrix} J(x) & -I & g \\ I & 0 & 0 \\ -g^T & 0 & 0 \end{pmatrix} \begin{pmatrix} \partial_x H(x) \\ u_r \\ u \end{pmatrix}, \tag{8}$$

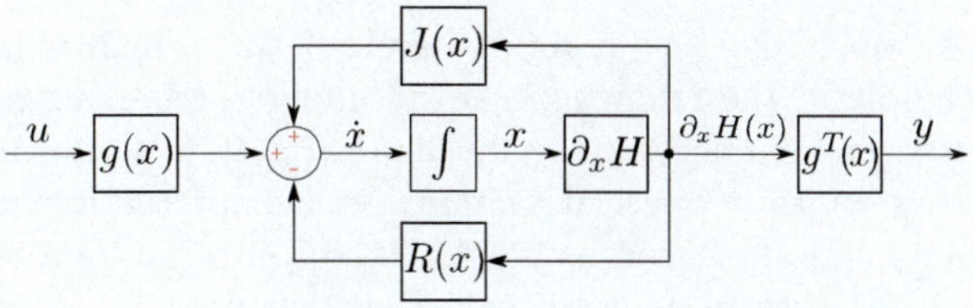

(a) signal-based block diagram

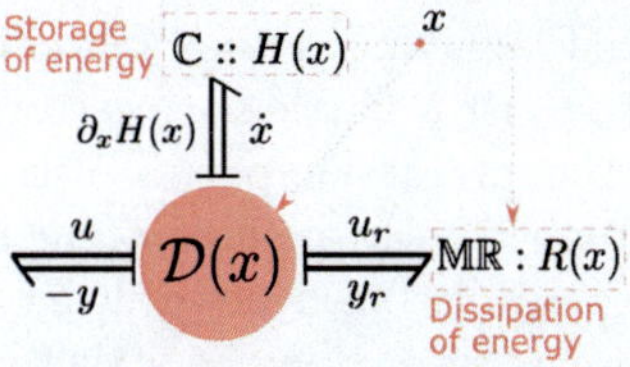

(b) energy-based bond graph

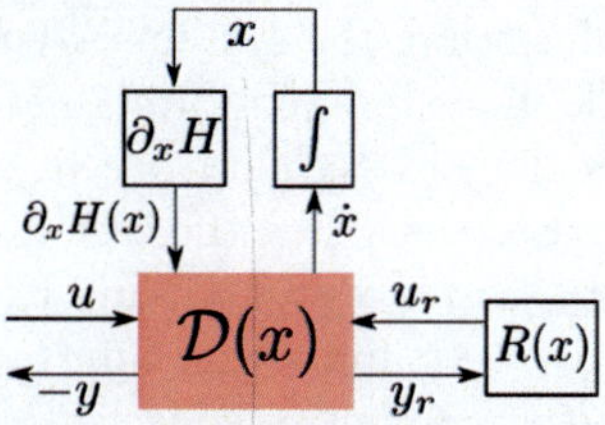

(c) energy-based block diagram

Fig. 1. Graphical representation of a generic port-Hamiltonian system

where I is the n-dimensional identity matrix. This representation unravels the power conjugated variables corresponding to the dissipative port (u_r, y_r), related by means of the algebraic equation $u_r = R(x)y_r$. The latter, together with the second equation in (8), can be combined to form (4). The sum of all the power terms represented by the pairing of the conjugated variables gives zero because of the skew-symmetry of the map, i.e., $(\partial_x H(x))^T \dot{x} + y_r^T u_r - y^T u = 0$, which encodes exactly the power balance (6) once the previously described algebraic relation on the dissipative port is used.

1.2 Passivity-Based and Energy-Aware Control

When it comes to control, port-Hamiltonian theory makes close contact with passivity-based control, a set of techniques aiming at designing a controller in such a way that the closed-loop system is passive, directly achieving stability [35]. This objective is automatically encoded in the pH structure, since the form (4–5) is conserved under power-preserving interconnection of pH systems. As a consequence, the passive system (4–5) can represent both the open-loop system (i.e., the uncontrolled plant) and a *target* closed-loop system, with desired and physically interpretable properties. The control objective becomes then to design a pH controller such that the desired target pH closed-loop system is achieved. Instances of these techniques are the *energy balancing* (EB-PBC) method, in which the target Hamiltonian function is assigned; and the *interconnection damping assignment* method (IDA-PBC), in which all the system matrices of the target system are assigned [6]. In what follows we elaborate on the motivation for achieving a closed-loop passive system in this framework.

A physical system, like e.g., a robot manipulator, which will be taken as archetypical example in the following, can be understood as a pH system with two distinct ports: one connecting the mechanical system to an unknown environment (which is another physical system), called interaction port; and one connecting it to its actuators, commanded by the control system to be designed (represented as a pH system as well), called control port, as shown in Fig. 2a. The described passivity-based control techniques in the pH setting have the objective of shaping the systems' dynamics by means of the controller such that the closed-loop system exhibits a new, desired behavior at the interaction port. The achievement of stability in this context is just a particular case of possible closed-loop dynamics, i.e., the case in which the interaction port is disconnected. Along the task it executes, the robot will interact with the environment, whose dynamics is often unknown. Contrarily to what is often done in e.g., robust control theory, making specific simplifying assumptions on the structure of the environment as a dynamical system (like linearity) is considered misleading and conceptually wrong. What matters for safety and performance purposes is a stable interconnection between the system and its environment along the task execution. The interaction will follow physical laws, dependent on the unknown parameters of the environment. The following claim, referred to as *passivity as must*, is taken as a primary objective in this framework: *The controlled robot should result in a passive behaviour seen from the environment port.* The motivation behind this claim lies on the fact that if the controlled robot is passive, then the interconnection with any passive environment would indeed result in a stable interconnection. Furthermore, as formally proven in [32], it is always possible to construct a passive environment such that its interconnection with a non passive controlled system results in an unstable behavior. This set of considerations made passivity of the controlled system a necessary condition in the field.

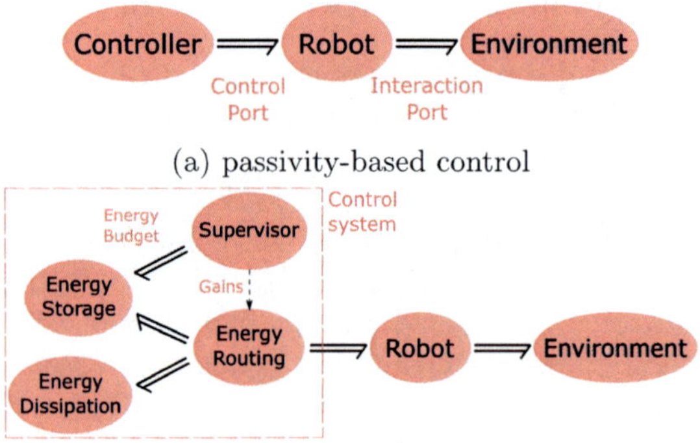

(b) energy-aware control

Fig. 2. Comparison between the standard passivity-based control and energy-aware control frameworks

Designing control systems with passivity as the sole goal is often considered over-conservative and non optimal in terms of performance. In fact, it is no surprise that a design which is completely agnostic with respect to an environment that the system has to interact with, is questionable in terms of performance of a task, which potentially comprises an interaction with the environment itself. Furthermore some tasks (e.g., periodic locomotion) require a continuous injection of energy in the system to be executed, and cannot be achieved with a pure passive design, which we stress serves at the purpose of stability of interconnection with a passive environment. On the other hand, the *energy-aware control framework* [32] extends the standard passivity-based framework with a set of modules such that the goal is not primarily achieving passivity, but being able to take decisions observing the energy flows happening insight the whole control network. With reference to Fig. 2b, a fundamental constituent of the energy-aware framework is *energy routing*, first introduced in [7]. This technique introduces the possibility to design nontrivial power-preserving interconnections between two or more systems such that the energy could be transferred from one system to another without violating passivity [32]. These tools, which can be used together as abstractly represented in Fig. 2b to form an energy aware control system, serve in linking a low-level passivity based control framework to one more aware of performance and integrated at a decision level.

The energy routing technique is fundamental for directly controlling the energy flow and, consequently for implementing a desired dynamic behavior for the controlled system. Starting from the operating principles outlined in [7], the *energy tank* has been firstly proposed in [26]. Using energy tank it is possible to passively implement any desired control action according to the available energy budget [32]. The energy budget is provided by a high-level control module referred to as the *supervisor* in Fig. 2b. This supervisor module is interconnected to the control system and can provide to it new energy if needed. In this perspective, if the supervisor does not provide new energy to the controlled system, then the passivity requirement allows flows of energy coming only from the environment. The supervisor implements a task-dependent *energy budget protocol*, which serves to optimally perform the task.

2 Energy Tanks

Mathematically an energy tank is a dynamical system which constitutes an atomic energy storing element. Using a pH formulation (4–5), and indicating the variables with a subscript t, it can be represented as

$$\dot{x}_t = u_t \tag{9}$$

$$y_t = \partial_{x_t} T(x_t), \tag{10}$$

where $T(x_t)$ is the non negative energy function for the tank. This is just a simple integrator with possibly nonlinear output, and is used to store information about the current energy budget in the system. In particular the goal is to interconnect the tank (9–10) to the plant (4–5), in a way that allows the implementation of some control action which fulfills the execution of some task, and at the same time meeting the desired passivity constraint. This is possible thanks to a suitable power-preserving interconnection between the two systems, exploiting the previously discussed idea of energy routing, i.e., the lossless energy transfer between the tank and the plant. As described in the last section, let us consider the pH plant (4–5) distinguishing between the interaction port (with input u_e) and the control port (with input u_c). Mathematically this is achieved by setting $u = u_c + u_e$, and as a consequence both the control and the interaction input are conjugated to the same original output $y_c = y_e = y$. The power preserving interconnection between the plant and the tank reads formally

$$\begin{pmatrix} u_c \\ u_t \end{pmatrix} = \begin{bmatrix} 0 & \frac{w}{\partial_{x_t} T(x_t)} \\ -\frac{w^T}{\partial_{x_t} T(x_t)} & 0 \end{bmatrix} \begin{pmatrix} y_c \\ y_t \end{pmatrix}, \tag{11}$$

where w is the desired task-dependent control action to be passively implemented. We represent the interconnection of the systems using both block diagrams and bond graphs, respectively in Fig. 3b and 3a, where we abstractly represent the interconnection as a state-modulated structure $\mathcal{D}_t$. Notice that $\mathcal{D}_t$, due to its skew-symmetry, is a power preserving interconnection. In fact, it can be easily seen that $u_c^T y_c = -u_t y_t$. Nevertheless, $\mathcal{D}_t$ is not a Dirac structure since it is modulated by the co-energy variable $\partial_{x_t} T(x_t)$. This has profound implications on the geometric structure of the controlled system, that is no more port-Hamiltonian but that can be modeled as a port contact system, as discussed in [25].

This interconnection produces two key effects, whose combination embodies the role of energy tanks. First, it correctly implements the desired action, i.e., from the side of the plant (4), one obtains

$$\dot{x} = [J(x) - R(x)]\partial_x H(x) + g(x)w + g(x)u_e. \tag{12}$$

Secondly, the interconnection is power-preserving, which can be inferred from the skew-symmetry of the matrix in (11). To show this explicitly let us consider the closed-loop system, with closed-loop energy $\mathcal{H} := H + T$:

$$\begin{pmatrix} \dot{x} \\ \dot{x}_t \end{pmatrix} = \begin{pmatrix} J(x) - R(x) & g(x)\frac{w}{\partial_{x_t} T(x_t)} \\ -\frac{w^T}{\partial_{x_t} T(x_t)} g^T(x) & 0 \end{pmatrix} \begin{pmatrix} \partial_x \mathcal{H} \\ \partial_{x_t} \mathcal{H} \end{pmatrix} + \begin{pmatrix} g(x) \\ 0 \end{pmatrix} u_e \tag{13}$$

$$y_e = \begin{pmatrix} g^T(x) & 0 \end{pmatrix} \begin{pmatrix} \partial_x \mathcal{H} \\ \partial_{x_t} \mathcal{H} \end{pmatrix} \tag{14}$$

Evaluating the variation of the closed-loop Hamiltonian along the system trajectories one obtains $\dot{\mathcal{H}} = (\partial_x \mathcal{H})^T R(x) \partial_x \mathcal{H} + y_e^T u_e \leq y_e^T u_e$, which proofs passivity of the closed-loop system with respect to the interaction port (u_e, y_e), with $\mathcal{H}$ as storage function, and with the same dissipation rate of the original system. This means that the (undefined in sign) power $y^T w$ flowing on the port of the original system is at any time exchanged with the tank without being

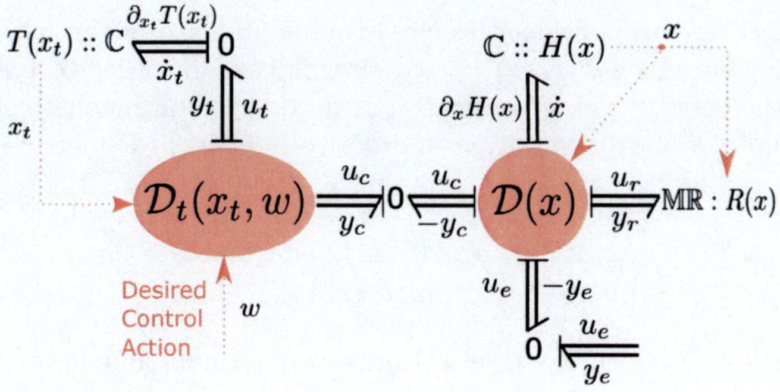

(a) bond graph

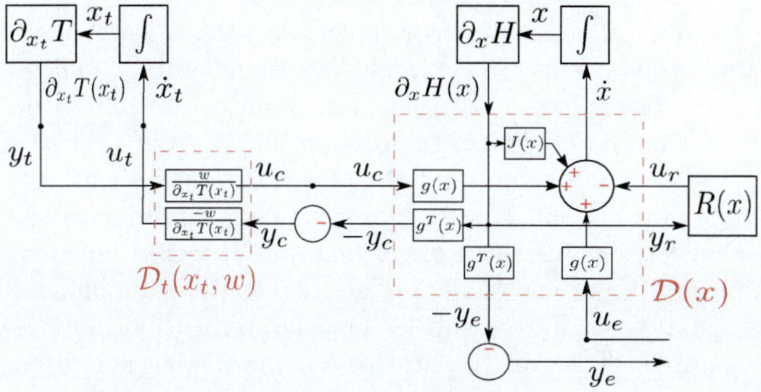

(b) block diagram

Fig. 3. Basic energy tank control implementation graphically represented by bond graphs (a) and block diagrams (b).

dissipated or generated. In other words, and considering for simplicity a case in which no natural dissipation ($R(x) = 0$) and no external interaction ($u_e = 0$) is present for the original system, one obtains $\dot{\mathcal{H}} = 0$, or equivalently $\dot{T} = -\dot{H}$, which means that the power flowing in/out the tank is equal to the power flowing out/in the original system, i.e., the energy routes from one system to another without being created or destroyed.

These properties introduce an important advantage in the context of energy-aware control, since the input w is free to be chosen arbitrarily, without the need of being generated by a pH controller representing a passive physical system (e.g., a spring and a damper, resembling the traditional PD controller). The energy in the tank $T(x_t)$ plays the role of the amount energy that is still at disposal of the control mechanism implementing the action w before loosing the described formal passivity. In fact, it is important to notice that the interconnection (11), as well as the closed-loop system (13) become singular when $\partial_{x_t} T(x_t) = 0$, representing the moment at which it becomes impossible to passively perform the desired action w. This passivity property manifests with respect to the closed-loop storage function $\mathcal{H}$, which in turns depends on the choice of the energy tank function $T(x_t)$. Let us consider the most common choice of the latter, namely $T(x_t) = \frac{1}{2}x_t^2$, for which $\partial_{x_t}T(x_t) = x_t$. In this case, in order to meaningfully implement the energy tank algorithm, it is needed to continuously observe the energy in the tank and its variation in order to implement a control action αw, rather than w, where

$$\alpha = \begin{cases} 1, \text{ if } (T \geq \epsilon \geq 0) \text{ or } (T < \epsilon \text{ and } \dot{T} > 0) \\ 0, \text{ if } (T < \epsilon \text{ and } \dot{T} \leq 0), \end{cases} \tag{15}$$

for some small positive energy level ϵ. In this way formal passivity of the closed-loop system is recovered completely since the system is just detached from the tank and the controller in the moment in which no energy budget is left, possibly having to wait till the supervisor replenishes the tank. A less understood fact is that the interconnection (11) might fail to generate a well-posed closed-loop system even for choices of $T(x_t)$ which analitically exclude the condition $\partial_{x_t}T(x_t) = 0$. It is in fact tempting, as proposed in [37], to choose as energy tank function $T(x_t) = e^{x_t}$, and thus avoid the aforementioned singularity condition and the construction in (15). Nevertheless this choice produces new problems, like the fact that the closed-loop storage function $\mathcal{H}$ would not qualify as Lyapunov function for the origin of the closed-loop system, renouncing to formal stability. An instance of an even subtler problem with this choice is illustrated in the next example, and is to the best of our knowledge not discussed in the literature.

Example 1 *Consider a simple mass subject to a force F, with kinetic energy $H(p) = \frac{1}{2m}p^2$, momentum $p := mv$, being v the velocity of the mass. The pH equations of motion are $\dot{p} = F = u_c; y_c = \partial_p H(p) = v$. Let us consider the interconnection (11) with the tank (9–10) using $T(x_t) = e^{x_t}$. The initial conditions are set as $p(0) = 0$ (i.e., $H(0) = 0$), and $x_t = 0$ (i.e., $T(0) = 1$). Let us perform a task in which we constantly accelerate the mass, applying a constant force $\bar{F}$, i.e., we set $w = \bar{F}$ in (11). As usual $-\dot{H} = \dot{T}$ holds, and thus the time evolution of the energy in the tank satisfies $T(t) = T(0) - \int_0^t \dot{H}\,dt = 1 - \bar{F}\int_0^t v\,dt = 1 - \frac{\bar{F}^2t^2}{2m}$ where we used the explicit solution for the velocity $v(t) = \frac{\bar{F}t}{m}$. We notice that since $T \geq 0$ by definition, the interconnection (11) fails to work at $\bar{t} = \frac{\sqrt{2m}}{\bar{F}}$. To see this let us consider the tank dynamics (9), where the input is u_t is determined by the second equation in (11). The equation reads $\dot{x}_t = -\frac{\bar{F}v}{e^{x_t}} = -\frac{\bar{F}^2t}{me^{x_t}}$, where we substituted again the time dependent solution for the velocity. The latter differential equation in x_t possesses a solution which escapes to $-\infty$ at the previously mentioned finite time, making the whole tank algorithm and any closed-loop analysis invalid for a wider time interval.*

This simple example is very instructive to show that special care must be taken when implementing the interconnection (11), and that the exponential choice for the energy of the tank does not prevent the singularity since $\partial_{x_t}T(x_t) = 0$ could happen anyhow due to a finite-time escape of the state in (9). Furthermore, since $-\dot{H} = \dot{T}$ holds for any form of $T(x_t)$, and since $\dot{H}$ does not depend on

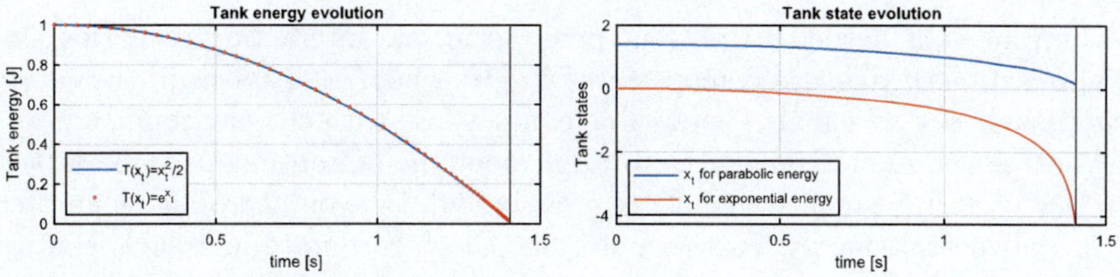

Fig. 4. Simulations of tank energy and tank states (with initial conditions chosen to have the same initial energy) for parabolic and exponential choice of $T(x_t)$ for Ex. 1, with $m = 1$, $\bar{F} = 1$.

the form of $T(x_t)$, the evolution of the tank energy in time does not depend on the specific form of $T(x_t)$. As a consequence, the specific choice of energy $T(x_t)$ corresponds to a specific choice of coordinate for the energy variable x_t, and does not influence any objective energy behaviour. In other words, using $T(x_t) = e^{x_t}$ (resp. $T(x_t) = x_t^2/2$) means just using a coordinate that "observes" a zero energy at $x_t = -\infty$ (resp. at $x_t = 0$). To conclude, stating that a specific choice of $T(x_t)$ avoids the singularity in (11) is only a coordinate-based illusion: the singularity is only "pushed" at $-\infty$ in the chart where one reads the x_t state in the exponential version. This is shown in Fig. 4, where the simulations corresponding to the task described in Ex. 1 are reported for both choices of the tank energy. We conclude that, unless one finds numerical advantages in the digital implementation of one specific version of $T(x_t)$, the quadratic choice is one that does not cause a loss of theoretical generality, and the conditional check in (15) must be performed in any case to preserve passivity.

2.1 Extensions and Applications

In the aforementioned discussion, the basic energy tank (9–10) has been introduced as a simple integrator with one power port (u_t, y_t) connected to the plant via (11) with the initial energy budget $T(x_t(0))$ being the only control parameter to be specified by the user. Many extensions of this basic energy tank implementation have been introduced in the literature and is an active area of research. In what follows we present some of these extensions.

One common extension is to exploit (some of) the dissipated energy of the controlled system (i.e., E_d in (7)) to refill the energy tank in a controlled manner [5,20,21]. Another extension is to put an upper bound on the allowable stored energy in the tank which prevents storage of very high virtual energy that could potentially lead to unsafe behavior between the robot and its environment. Such extension, implemented e.g. in [5,21], was realized by the addition of a virtual overflow valve that would start dissipating the power flowing to the tank once the energy level inside reaches a user-specific maximum.

The idea of using virtual valves to implement high level control objectives using energy tanks was further extended in [30] by considering not only the energy level of the tank but also by observing the amount of extracted power and limiting it if needed to prevent possible unsafe interaction scenarios [34]. An application of this was demonstrated in [21] where high spikes in the energy tank's power act as signals for possible contact-loss with the environment.

Another common extension is to implement the discontinuous tank detachment law (15) in a smooth manner, e.g. as in [30], to avoid introducing chattering in the control signals. Furthermore, one can incorporate high level control-objectives by detaching the tank not only when the energy level depletes but in other scenarios such as when a contact-loss is detected with the environment as in [24]. In the recent work of [31], an adaptive extension to energy tanks was proposed to deal with dynamic passive environments relying on varying the power limits of the energy tanks online through an iterative learning procedure.

In conclusion, the above non-exhaustive list of extensions demonstrates that energy-aware control using energy-tanks is an active research topic. Furthermore, it is being utilized in a wide variety of application domains in robotics as summarized in Table .

Table 1. The use of energy-tanks in different robotic application domains.

Domain	**References**	**Domain**	**References**
Aerial physical interaction	[2,20,21]	Robotic surgery	[8,9,33]
Collaborative robotics	[19,34]	Shared control	[27,28]
Generic manipulation	[5,15,24]	Teleoperation architectures	[12,13,18]
Immitation learning	[14]	Teleoperation of aerial robots	[16,36]
Multi-robot systems	[11,22,29]	Teleoperation of legged robots	[23]
Rehabilitation robotics	[17]		

3 Discussion and Future Challenges

The mechanism of energy tanks has distinguished itself for elegance and simplicity, providing a very convenient way to formally establish passivity of a controlled system. Although the success of the approach, it also raised some critiques and perplexities, which range from the entire energy-aware paradigm to the more specific implementation of energy tanks. In what follows we collect some of these issues with the hope to act as catalysts for fruitful discussions and reflections.

Passivity as must The passivity as must requirement is a claim sitting at the very foundation of the energy-aware framework, and it is worth reflecting on its real significance since it is absolutely not a commonly shared necessary condition in the control community. The requirement is in no way connected to a performance metric for the execution of a task. Instead it is based on the general commonsense-based rule that it is wrong to assume any model for the environment to study the stability of the closed-loop system when interfacing the unknown environment along the task execution. Only with a passively controlled system we can indeed infer stability of the interconnection with any passive environment. Although the argument is logically valid, it is worth questioning if, in the context of the execution of a complex task, the distinction between a passive and an active environment is significant in the design process.

The framework tends to focus on the environment as a system rather that on optimising performance over a task. It is therefore legitimate to question under which conditions it makes sense to start with the passivity as must requirement, reflecting on the relation between classes of tasks and classes of environments.

Safety There is often a confusion between the concepts of passivity and safety. Indeed often it is naively said that passivity is necessary for safety, but this is not the case. First of all, when considering the interaction with the environment, safety is not a mathematically defined system theoretic property, and depends on the class of tasks, of systems and on the specific safety hazards that are being considered. The formal passivity condition for the closed-loop system interconnected with energy tanks does not avoid practically unsafe behaviours, especially for high energy in the tank $T(x_t)$. In fact there is no guarantee that the high (yet bounded) amount of energy in the tank is extracted in a short time interval, leading to dangerous power ejections from the system. This aspect is problematic with the use of energy tanks in their basic form, since both high initialisation of energy and a possible refilling of the tank along the task execution can enhance the problem. The safety aspect has been treated with the use of energy tanks in e.g., [19,34], without removing some perplexity. Indeed, in these works the initially passive controller is integrated with a safety protocol, which happens to destroy passivity. Motivated by the passivity as must claim, this situation is then recovered by means on energy tanks, used to passivise the control action commanded by the safety protocol. In this case the passivity as must assumption seems to create circular patterns in the design of the control strategy, and the underlying unanswered question remains: why should one chose to use energy tanks to passivise a control action which is already designed to be safe?

Need of energy budgeting Once a complex task has to be executed in an energy-aware control framework, the supervisor has the role of injecting, following a task-dependent protocol, the energy to achieve the task. This is where passivity breaks down, and the more abstract concept of energy-awareness comes into the game. From an energy tank perspective the energy budgeting protocol should respond to the questions: What is the initial energy budget in the tank? Under what circumstances does the tank need to be refilled? In general, in order to assure a good performance along the task, complex energy budgeting protocols need to be implemented in a case by case fashion. This procedure becomes more involved for complex tasks, and may again raise the question under which class of tasks this is a framework worth to stick with from the beginning.

Towards Energy-Aware Optimisation To recap, the energy-aware framework is founded on a claim imposing the necessity of a controlled system which is passive at the environmental port, independently on the task to be performed. This can be always achieved using the basic energy tank architecture. There is no standard way to address the problem of performance optimisation and energy budgeting in this framework, which become particularly critical in situations in which the task cannot be accomplished passively (e.g., periodic locomotion). The safety aspect appears transversally as a third, separated issue to be solved, additionally to passivity of the internal controller and optimisation of the performance along the task. These issues are deeply intertwined, and need to be addressed together.

A task-depended solution to this passivity-safety-performance problem represents the main open challenge in this energy-aware framework. Energy tanks are perfect candidates to be used is this context since they: i) can passivise any control action; ii) encode information about energy content and power flows undergoing the system, making them extremely useful to deal with safety and energy efficiency metrics; iii) allow for the implementation of arbitrary control actions, for which optimisation over a performance (and/or safety) metric is very tempting. As possible modification of the mentioned objective, we stress that energy tanks should act as energy and power observers, and not be blindly used to recover passivity of the controlled system, which is ultimately the main reason why perplexities raised in the framework. This could lead to a relaxation of the passivity as must claim, in favour of an energy-aware solution achieving safety and performance optimisation along a task execution. Preliminary works that addressed the usability of energy tanks in this direction include [1,3,4,30,31].

4 Conclusions

In this paper we discussed the genesis of energy tanks in the context of its containing system theoretic frameworks, that are port-based modeling and energy-aware control. In addition to listing extensions of the basic energy tank implementation and its range of applications, we raised a series of discussion topics involving passivity, safety, and performance optimisation in relation with the use of energy tanks, with the goal of clarifying the future challenges in establishing this control technique for robotic applications.

References

1. Benzi, F., Ferraguti, F., Riggio, G., Secchi, C.: An energy-based control architecture for shared autonomy. IEEE Trans. Robot. 1–19 (2022)
2. Brunner, M., Giacomini, L., Siegwart, R., Tognon, M.: Energy tank-based policies for robust aerial physical interaction with moving objects, pp. 2054–2060 (2022)
3. Califano, F., van Dijk, D., Roozing, W.: A task-based post-impact safety protocol based on energy tanks. IEEE Robot. Autom. Lett. **7**(4), 8791–8798 (2022)
4. Capelli, B., Secchi, C., Sabattini, L.: Passivity and control barrier functions: optimizing the use of energy. IEEE Robot. Autom. Lett. **7**(2), 1356–1363 (2022)

5. Dietrich, A., Xuwei, W., Bussmann, K., Ott, C., Albu-Schffer, A., Stramigioli, S.: Passive hierarchical impedance control via energy tanks. IEEE Robot. Autom. Lett. **2**(2), 522–529 (2017)
6. Duindam, V., Macchelli, A., Stramigioli, S., Bruyninckx, H.: Modeling and Control of Complex Physical Systems (2009)
7. Duindam, V., Stramigioli, S.: Port-based asymptotic curve tracking for mechanical systems. Eur. J. Control **10**(5), 411–420 (2004)
8. Ferraguti, F., Preda, N., Manurung, A., Bonfe, M., Lambercy, O., Gassert, R., Muradore, R., Fiorini, P., Secchi, C.: An energy tank-based interactive control architecture for autonomous and teleoperated robotic surgery. IEEE Trans. Robot. **31**(5), 1073–1088 (2015)
9. Ferraguti, F., Secchi, C., Fantuzzi, C.: A tank-based approach to impedance control with variable stiffness. In: Proceedings—IEEE International Conference on Robotics and Automation, pp. 4948–4953 (2013)
10. Montbrun-Di Filippo, J., Delgado, M., Brie, C., Paynter, H.M.: A survey of bond graphs : theory, applications and programs. J. Franklin Inst. **328**(5):565–606 (1991)
11. Franchi, A., Secchi, C., Son, H.I., Bülthoff, H.H., Giordano, P.R.: Bilateral teleoperation of groups of mobile robots with time-varying topology. IEEE Trans. Robot. **28**(5), 1019–1033 (2012)
12. Franken, M., Stramigioli, S., Misra, S., Secchi, C., MacChelli, A.: Bilateral telemanipulation with time delays: a two-layer approach combining passivity and transparency. IEEE Trans. Robot. **27**(4), 741–756 (2011)
13. Heck, D., Saccon, A., Beerens, R., Nijmeijer, H.: Direct force-reflecting two-layer approach for passive bilateral teleoperation with time delays. IEEE Trans. Robot. **34**(1), 194–206 (2018)
14. Kastritsi, T., Dimeas, F., Doulgeri, Z.: Progressive automation with DMP synchronization and variable stiffness control. IEEE Robot. Autom. Lett. **3**(4), 3789–3796 (2018)
15. Kronander, K., Billard, A.: Passive interaction control with dynamical systems. IEEE Robot. Autom. Lett. **1**(1), 106–113 (2016)
16. Mersha, A.Y., Stramigioli, S., Carloni, R.: On bilateral teleoperation of aerial robots. IEEE Trans. Robot. **30**(1), 258–274 (2014)
17. Najafi, M., Rossa, C., Adams, K., Tavakoli, M.: Using Potential field function with a velocity field controller to learn and reproduce the therapist's assistance in robot-assisted rehabilitation. IEEE/ASME Trans. Mechatron. **25**(3), 1622–1633 (2020)
18. Pacchierotti, C., Tirmizi, A., Bianchini, G., Prattichizzo, D.: Enhancing the performance of passive teleoperation systems via cutaneous feedback. IEEE Trans. Haptics **8**(4), 397–409 (2015)
19. Raiola, G., Cardenas, C.A., Tadele, T.S., De Vries, T., Stramigioli, S.: Development of a safety- and energy-aware impedance controller for collaborative robots. IEEE Robot. Autom. Lett. **3**(2), 1237–1244 (2018)
20. Rashad, R., Engelen, J.B.C., Stramigioli, S.: Energy tank-based wrench/impedance control of a fully-actuated hexarotor: a geometric port-hamiltonian approach. In: Proceedings—IEEE International Conference on Robotics and Automation, vol. 2019 (2019)
21. Rashad, R., Bicego, D., Zult, J., Sanchez-Escalonilla, S., Jiao, R., Franchi, A., Member, S., Stramigioli, S.: Energy aware impedance control of a flying end-effector in the Port-Hamiltonian framework. IEEE Trans. Robot. (In press) (2022)
22. Riggio, G., Fantuzzi, C., Secchi, C.: On the use of energy tanks for multi-robot interconnection. In: IEEE International Conference on Intelligent Robots and Systems, pp. 3738–3743 (2018)

23. Risiglione, M., Sleiman, J.P., Minniti, M.V., Cizmeci, B., Dresscher, D., Hutter, M.: Passivity-based control for haptic teleoperation of a legged manipulator in presence of time-delays. In: IEEE International Conference on Intelligent Robots and Systems, pp 5276–5281 (2021)
24. Schindlbeck, C., Haddadin, S.: Unified passivity-based Cartesian force/impedance control for rigid and flexible joint robots via task-energy tanks. In: Proc. Int. Conf. Robot. Autom. **2015**, 440–447 (2015)
25. Secchi, C., Stramigioli, S., Fantuzzi, C.: Port-contact systems in bilateral telemanipulation. IFAC Proc. Volumes **40**(12), 474–479 (2007)
26. Secchi, C., Stramigioli, S., Fantuzzi, C.: Position drift compensation in port-Hamiltonian based telemanipulation. In: IEEE International Conference on Intelligent Robots and Systems, pp. 4211–4216 (2006)
27. Selvaggio, M., Robuffo Giordano, P., Ficuciellol, F., Siciliano, B.: Passive task-prioritized shared-control teleoperation with haptic guidance. In: Proceedings—IEEE International Conference on Robotics and Automation, pp. 430–436 (2019)
28. Selvaggio, M., Notomista, G., Chen, F., Gao, B., Trapani, F., Caldwell, D.: Enhancing bilateral teleoperation using camera-based online virtual fixtures generation. In: IEEE International Conference on Intelligent Robots and Systems, pp. 1483–1488 (2016)
29. Shahriari, E., Birjandi, S.A.B., Haddadin, S.: Passivity-based adaptive force-impedance control for modular multi-manual object manipulation. IEEE Robot. Autom. Lett. **7**(2), 2194–2201 (2022)
30. Shahriari, E., Johannsmeier, L., Haddadin, S.: Valve-based virtual energy tanks: a framework to simultaneously passify controls and embed control objectives. In: Proceedings of the American Control Conference, pp. 3634–3641 (2018)
31. Shahriari, E., Johannsmeier, L., Jensen, E., Haddadin, S.: Power flow regulation, adaptation, and learning for intrinsically robust virtual energy tanks. IEEE Robot. Autom. Lett. **5**(1), 211–218 (2020)
32. Stramigioli, S.: Energy-aware robotics. In: Lecture Notes in Control and Information Sciences **461**, 37–50 (2015)
33. Su, H., Schmirander, Y., Li, Z., Zhou, X., Ferrigno, G., De Momi, E.: Bilateral teleoperation control of a redundant manipulator with an RCM Kinematic constraint. In: Proceedings—IEEE International Conference on Robotics and Automation, pp. 4477–4482 (2020)
34. Tadele, T.S., De Vries, T.J.A., Stramigioli, S.: Combining energy and power based safety metrics in controller design for domestic robots. In: Proceedings—IEEE International Conference on Robotics and Automation, pp. 1209–1214 (2014)
35. van der Schaft, A.: L2-Gain and Passivity Techniques in Nonlinear Control, 3rd edn. Springer Publishing Company, Incorporated (2016)
36. Zhang, D., Tron, R.: Stable haptic teleoperation of UAVs via small L2 gain and control barrier functions. In: IEEE International Conference on Intelligent Robots and Systems, pp. 8352–8357 (2021)
37. Zheng, M., Yuan, T., Huang, T.: Time-varying impedance control of Port Hamiltonian system with a new energy-storing tank. Complexity (2018)

Robotic Strawberry Flower Treatment Based on Deep-Learning Vision

Jelena Vuletić(✉), Marsela Polić, and Matko Orsag

Faculty of Electrical Engineering and Computing, University of Zagreb, 10000 Zagreb, Croatia
jelena.vuletic@fer.hr, marsela.polic@fer.hr, matko.orsag@fer.hr

Abstract. In this paper we propose a robotic system for flower removal and pollination with the deep learning-based 3D perception module. The robotic system consists of a collaborative robot equipped with an RGB-D camera used in the detection pipeline. A semi-automated 3D based image annotation method (A3IA) is developed for the purpose of automatically generating the training data. The efficiency of the detection model trained on automatically annotated data is compared to that of the model trained on synthetic data. Experimental validation of the proposed methods is conducted and robustness and precision of both detection and positioning are reported.

Keywords: Robotic system for flower pollination · Semi-automated image annotation · sim2real learning · Synthetic data

1 Introduction

Various agrotechnical procedures of a repetitive nature have already been identified as perfect automation candidates in both indoor and outdoor farming. The SpECULARIA [17] project explores the possibility of deploying a heterogeneous team of robots in small indoor organic farms with the goal of reducing the human labor in labor-intensive activities, such as harvesting. Though numerous solutions have already been proposed for robotic harvesting of different crops and cultures, including strawberries [24], apples [11] and peppers [18], harvesting robots are not yet commercially successful due to the limitations of existing systems.

Along with harvesting, pruning and thinning are also labour-intensive activities critical for yield improvement. The major challenges in robotic pruning include targeted branch detection and navigation to the cutting point [7]. Robotic flower thinning, performed to increase average fruit size and quality, faces similar problems. Furthermore, the small target size poses an additional challenge on both detection and manipulation task. However, there have been some successful efforts in this area, reported in [25]. While early approaches

P. Borja et al. (Eds.): HFR 2022, 26, pp. 189–204, 2023.
https://doi.org/10.1007/978-3-031-22731-8_14

focused on mass flower removal, more recent work explore the possibility of precise flower detection using either multispectral [9] or RGB cameras [5]. In this work, we rely on the capabilities of a consumer RGB-D camera, Intel RealSense D435, in a robotic flower thinning application.

Fig. 1. End effector mounted on the collaborative robot Franka Panda in a position ready for the strawberry flower manipulation.

Similar to flower thinning, robotic pollination is an example of another important agrotechnical activity relying on flower detection and manipulation. Closed greenhouse environments, such as the one in the focus of the SpECULARIA project, usually lack natural pollinators, which could be solved by the introduction of commercial robotic pollinators. Authors in [21] suggest using a ground vehicle equipped with a robotic arm and an Intel RealSense D435 camera for automated pollination. They propose a two-stage framework for flower detection consisting of a naive Bayes' classifier based segmentation that produces a set of patches denoting flowers and non-flowers, and a convolutional neural network (CNN) based classifier that performs a binary classification on the previously extracted patches. Finally, the position of the flower is estimated using depth image, and a visual servoing procedure is applied when reaching the flower. The same sensory setup is deployed in the work presented in this paper, but instead of two-stage flower detection framework, we propose a single-stage detection based on the transfer learning paradigm for a complex CNN detection model. Furthermore, according to the results presented in [18], since the servoing increases the task execution time without significantly improving the estimation of the detected object position, we propose skipping the servoing stage. Instead, in this work we present a method for minimizing the probability of collisions between the robot and the plant. Our proposed solution for precise flower detection and plant manipulation can be used for both robotic flower thinning and pollination.

Though CNN based models are the state-of-the-art for object detection, their demand for large training datasets still presents an obstacle for researchers. In order to avoid the tedious task of manual data labeling, the possibility of developing automated and semi-automated robotic systems for data annotation is being widely researched [3,10,13,20,22]. While some authors use visual markers for automatic annotation [13], it has been shown that precision decreases significantly when using visual markers compared to the precision that can be achieved using an articulated robot and an RGB-D camera [10]. Similar to the work presented in [10], we used a Franka Emika Panda robot and an Intel RealSense D435 camera for our semi-automated annotation system. However, while the authors in [10] present the system for labeling 2D objects, we aim at automating 3D objects labeling. A similar problem of automatic 3D objects labeling has been tackled in [3], where an RGB camera was mounted on the robot, and the initial target object pose was extracted either by marking the object with the augmented reality pen or by manually drawing a mask around the target object in at least two frames. We propose simplifying the extraction of the initial pose of the target object by using an RGB-D camera and a custom GUI.

An alternative approach for automated data generation is to use synthetic data [2,4]. Although using synthetic datasets in combination with real datasets improves model accuracy, models trained only on synthetic datasets often generalize poorly to real data, and bridging the gap between synthetic and real data remains an open research problem [6,16]. Two common approaches to overcome the existing domain gap are domain adaptation [12] and domain randomization [23]. In this work, the possibility of using only synthetic data with the domain randomization method for real-world flower detection is investigated. Finally, the robustness and accuracy of the models trained on synthetic and automatically annotated data are compared on the real-world robotic application.

1.1 Contributions

In this paper, we claim the following contributions:

1. *A3IA*. Semi-automated 3D based image annotation method (A3IA) presented in this work is experimentally validated using a robotic system for flower detection, removal and pollination. Method is semi-automated as it requires human-in-the-loop to annotate the first sample, which is further explained in Sect. 3.1.
2. *Publicly released datasets.* Both synthetic and automatically annotated strawberry flowers datasets are publicly released and can be downloaded from [19].
3. *Quantitative comparison of models trained on different automatically generated datasets.* Accuracy and recall of deep learning detection models trained on either synthetic or automatically annotated strawberry flower dataset are reported.

Fig. 2. Visualization of different tools for flower removal and pollination: stick (left), knife (in the middle) and gripper (right).

2 Method

The system consists of a Franka Panda collaborative robot, controlled via ROS using MoveIt motion planning package. Real-time patched Linux kernel was used to control the cobot. Due to incompatibility of the real-time kernel and NVidia drivers, a remote master ROS setup was used and the detection module was run on the other computer using NVidia GeForce MX330 GPU and NVidia's CUDA Toolkit. The consumer RGB-D camera, Intel RealSense D435, was mounted on Panda hand in a so-called *eye-in-hand* configuration, which allows the scene to be captured from multiple perspectives using only a single camera.

2.1 Calibration Procedure

In order to properly use camera information for positioning the end effector, the robot-camera system has to be calibrated. The calibration is conducted in an autonomous manner, as described in [15]. The result of the autonomous calibration process is a transformation $\mathbf{T}_c^e \in \mathbb{R}^{4\times4}$ between the camera optical frame L_c and the robot end effector frame L_e.

A simple contrasting blob (L_B) is used as a target in the camera calibration procedure. Its position $\mathbf{p}_W^B$ in the world frame L_W does not need to be a priori known. Once the target is recorded and detected in the image from up to n poses $i = 1, ..., n$, its 3D position in the camera reference frame is extracted from the organized point cloud. The resulting set of detections contains the positions $\mathbf{p}_{c\ i}^B \in \mathbb{R}^3$ of the target in the local camera frame along with the joint configuration $\mathbf{q} \in \mathbb{R}^7$ in which the frame was recorded. Joint values enable us to calculate the transform $\mathbf{T}_0^f(\mathbf{q}_i)$ between the robot base L_0 and the flange L_f, where the camera is mounted. Finally, treating the transformation between the flange and the camera as an unknown $\mathbf{T}_f^c$, the optimization procedure can be employed. The optimization procedure minimizes the target position dissipation when the target is transformed into the global coordinate frame:

$$\underset{\mathbf{T}_f^c \in \mathbb{R}^{4\times4}}{\arg\min} \sum_{i\neq j} \left\| \mathbf{T}_0^f(q_i)\mathbf{T}_f^c\mathbf{p}_{c\ i}^B - \mathbf{T}_0^f(q_j)\mathbf{T}_f^c\mathbf{p}_{c\ j}^B \right\|, \tag{1}$$

Additionally, a pointer end effector is mounted on the Panda flange (Fig. 1). The pointer serves as a good approximation model for different tools in flower manipulation procedures, pollinating, cutting and picking, which are shown in Fig. 2. As with the camera frame, in order to be able to reach the flower with the tip of the end effector L_e, the robot-end effector transformation had to be obtained. Similar calibration procedure was utilized, but instead of extracting the target from the point cloud, the tip of the end effector was repeatedly positioned on the same target, each time with the different joint configurations. A minimization problem identical to Eq. 1 yielded the transform from the mounted end effector tip to the Panda flange frame $\mathbf{T}_f^e$. From this point, it is straightforward to calculate the transformation $\mathbf{T}_c^e$ between the camera frame and the end effector tip.

2.2 Flower Approach Procedure

In SpECULARIA, we use mobile robots to bring plants grown in the greenhouse to the manipulator. Using the information from this structured farming concept enables us to roughly estimate the position $\mathbf{p}_0^\sigma$ of the strawberry container L_σ in the manipulator base frame. This knowledge is used to build a bounding cylinder of the best view points around the plant:

$$\mathbb{BC} = \left\{ \mathbf{x}, \mathbf{v}\colon \begin{array}{c} \mathbf{x} = \mathbf{p}_0^\sigma + r \cdot \mathbf{Rot}(\hat{\mathbf{z}_0}, \phi)\mathbf{x}_0 + h \cdot \hat{\mathbf{z}_0} \\ \mathbf{v} = -\frac{r \cdot \mathbf{Rot}(\hat{\mathbf{z}_0}, \phi)\mathbf{x}_0 + h \cdot \hat{\mathbf{z}_0}}{\| r \cdot \mathbf{Rot}(\hat{\mathbf{z}_0}, \phi)\mathbf{x}_0 + h \cdot \hat{\mathbf{z}_0} \|} \end{array} \right.$$

$$\phi \in \left\langle -\frac{\pi}{2}, \frac{\pi}{2} \right\rangle, h \in \langle h_{min}, h_{max} \rangle \tag{2}$$

where each point $\mathbf{x}$ is placed around the $\mathbf{p}_0^\sigma$ at a radius r, as shown in Fig. 3. $\mathbf{Rot}(\hat{\mathbf{z}_0}, \phi)$ denotes the elementary rotation matrix around the base axis $\hat{\mathbf{z}_0}$, for an angle ϕ. The approach angle $\mathbf{v}$ for every point is calculated so that it always points towards the $\mathbf{p}_0^\sigma$. Finally, the height of the cylinder (i.e. h_{min}, h_{max}) is defined to fit the robots workspace.

From here it is straightforward to plan a lawnmower path, and a $C2$ continuous trajectory that will enable smooth image capturing along this path as shown in Fig. 3. For clarity, in this work we concentrate on the predefined number of positions on the cylinder. In our experiments, robot recorded the plant from 14 positions $c2_i$ lying on the $C2$ continuous path. In each of the positions, an RGB-D camera image and a point cloud are used for flower detection. If the flower is detected in the image, the centroid of the flower $\mathbf{cf}_i$ is extracted from the organized point cloud by filtering the point cloud with the bounding box of the detected object. The centroid is transformed to the global reference frame, and compared to an existing database of detected flowers. If $\mathbf{cf}_i$ is within a bounding box of an existing flower, the detection rate for that database element is increased. If the detected centroid does not fit within the bounding box of a known flower, it is added to the database. Every detection is stored along with the corresponding image, point cloud and current camera pose.

Fig. 3. The robot captures the plant from a predefined number of positions on the plant's bounding cylinder ($\mathbb{BC}$). Continuous trajectory of the robot ($C2$) is shown in red.

Once the entire cylinder is traversed, the stored centroids are compared and the flower is chosen as the target if it is visible from at least two frames. If multiple flowers are present, we first opt for the one closest to the end effector. As during the first growing season all flowers are being removed, and all flowers have to be pollinated in the subsequent season, we do not discriminate among detected flowers. The global position of the target flower centroid is denoted with **cf**.

Once the target flower is chosen, a motion planning method is deployed that minimizes the probability of collision with other plant parts during the approach. The approach optimization searches for the approach pose with the minimum set of obstacle points, where the set of obstacle points $\mathbb{O}$ can be defined as in Eq. 3:

$$\mathbb{O}_{c2_i} = \{\mathbf{p} \in \{\mathbb{PC} \setminus \eta_{cf}\} | z^{p}_{c2_i} < z^{cf}_{c2_i}\}, \tag{3}$$

where a point $\mathbf{p}$ from the point cloud $\mathbb{PC}$ belongs to the set of obstacles if it is not in the neighborhood of the centroid η_{cf}, and if its z coordinate z^{p}_{c} in the local camera frame is smaller than the flower centroids z coordinate z^{cf}_{c}. The set of obstacle points is calculated for each search pose $\mathbf{c2}_i$. The approach position is chosen as the one with least obstacles, i.e. according to Eq. 4:

$$\mathbf{c2}^* = \underset{\mathbf{c2}_i}{\operatorname{argmin}} |\mathbb{O}_{c2_i}| \tag{4}$$

Finally, to obtain the approach pose, the orientation of the robot end-effector has to be defined in the chosen position $\mathbf{c2}^*$. Instead of reusing the $\mathbf{c2}^*$ orientation, we calculate the orientation according to the set of Eqs. 5.

$$\begin{aligned} \hat{\mathbf{z}}_{c2} &= \frac{\mathbf{c2}^* - \mathbf{cf}}{\|\mathbf{c2}^* - \mathbf{cf}\|} \\ \hat{\mathbf{y}}_{c2} &= \hat{\mathbf{z}}_0 \times \hat{\mathbf{z}}_{c2} \\ \hat{\mathbf{x}}_{c2} &= \hat{\mathbf{z}}_{c2} \times \hat{\mathbf{y}}_{c2}, \end{aligned} \tag{5}$$

Fig. 4. Examples of detected strawberry flowers on the real images. Detections from the network trained on the synthetic dataset are shown in blue, while detections of the network trained on the automatically annotated dataset are shown in red. The same threshold of 0.7 was used for both networks (detections with lower score are not shown).

where $\mathbf{x}^{c2^*}$ and $\mathbf{x}^{cf}$ are the positions of the chosen position on the cylinder, and of the target flower centroid, expressed in the robot reference frame, and end-effector orientation is defined such that the local $\hat{\mathbf{x}}$ axis roughly corresponds to the base frame $\hat{\mathbf{z}}_0$ axis. This choice of end effector approach pose reduces the possibility of harming the plant, as the flower is approached from the pose with the minimum share of plant parts between the flower and the end-effector. Finally, we plan to the flower from the chosen approach pose. Once the end effector reaches the flower, it is able to execute the desired procedure with the appropriate tool, such as pollination with a stick, flower removal with a knife, or similar. Several possible solutions for different agrotechnical procedures are shown in Fig. 2.

3 Detection and Generation of Training Data

A single stage object detector, SSD [14], with the MobileNet-V2 used as the base network, is trained for detection of strawberry flowers. Two approaches are tested and compared, one relying on synthetic data, and the other on automatically annotated data obtained with the A3IA method proposed within this work. These approaches are designed and tested since a suitable open sourced dataset of flowers was not available, and present a less labour intensive alternative to manually generating a custom dataset. Both datasets are publicly released and can be downloaded from [19]. Transfer learning is deployed in both cases for the base network, with initial weights pretrained on the COCO dataset. During training, weights of all the layers were tuned. Both models are trained using TensorFlow Object Detection API. Models were trained for 2000 steps with a batch size of 8. Random horizontal flip and random image crop were used for data augmentation. Momentum optimizer and cosine decay learning rate schedule were utilized. Examples of flower detection results on real images for both networks are shown in Fig. 4.

Fig. 5. Examples of the automatically annotated images. Bounding boxes are automatically generated by using the known position of the object in the global reference frame and the transformation from the camera to the robot in each frame.

3.1 Robotic System for Semi-automated Dataset Collection

The collaborative robot with an eye-in-hand RGB-D camera executes a trajectory similar to that in the search stage of the flower manipulation procedure, but with a significantly higher number of positions on the plant's bounding cylinder, shown in Fig. 3. After recording an image and a point cloud from the initial position on the cylinder, user is required to annotate the targeted object, in our case the flower, on the RGB image, by drawing a minimum bounding rectangle around the targeted object using a custom GUI. Once the object is annotated, 3D position of the object is obtained by applying bounding rectangle to the organized point cloud. The targeted object position is transformed from the local camera frame to the global reference frame.

Then, multiple images and point clouds are recorded from the remaining positions on the cylinder, and the corresponding annotations are generated automatically using the known position of the object in the global reference frame. For each frame, global position of the target object is transformed into the current local camera frame, and a sphere with a radius of 2 cm is generated about this point. This geometric primitive roughly approximates strawberry flowers geometry, and all the point cloud points inside this sphere are considered as flower points. These points are projected from the organized point cloud to the corresponding RGB image. The minimum bounding rectangle around the projected points is generated and its coordinates are stored together with the corresponding image and point cloud as an annotated sample. This procedure is conducted for all of the frames (positions on the cylinder), resulting in a CSV file containing ground truth bounding boxes. Examples of automatically generated bounding boxes are shown in Fig. 5. By repeating the same procedure for 5 different strawberry plants, we generated a dataset of 375 annotated images of strawberry flowers. One series of images, consisting of 65 images of a single strawberry plant, is additionally manually annotated and used as a test set.

3.2 Synthetic Dataset Generation

The synthetic strawberry plant dataset is generated using 3D modeling software, Blender [8]. Leaf models are based on photographs of strawberry leaves which

Fig. 6. Examples of synthetic images of strawberry plants rendered in Blender using custom dataset generation pipeline.

are imported in Blender as planes and morphed for a more realistic shape. Pot, stems and flowers are modeled from scratch, but real images of strawberry flowers were used as texture maps for the modeled flowers. Some strawberry fruit models were modeled from scratch, while others were generated using open source photogrammetric tool Meshroom [1]. Leaves, flowers and strawberries were placed in separate collections and each of them had an instance of stem as a parent.

The custom dataset generation pipeline was implemented in Blender through its embedded Python interpreter as a Python script. In the proposed pipeline, user defines the desired number of passes. Each pass consists of 10 frames. Camera position and lightning settings vary across the frames. Strawberry plant is randomly generated at the beginning of each pass by spawning pseudo-random number of leaves, flowers and fruit. The positions, sizes and orientations of the generated instances are randomly set, but kept within a predefined range.

Original images are rendered using Cycles render engine, while Eevees render engine was used when rendering each object separately, which was necessary for generation of bounding boxes. Renders of separate objects were placed in separate maps, based on the predefined object ID, which denoted the class of an object. Finally, OpenCV Python library was used to extract the bounding boxes, which were defined as a minimum rectangle around the rendered object.

Using the described pipeline, we generated synthetic dataset of 1980 images of strawberry plants. The background was uniform in 5% of the images, while the background for the remaining 95% of the images was randomly chosen among 100 images of different textures and environments. Examples of rendered images are shown in Fig. 6.

4 Experimental Validation

A series of experiments was conducted to evaluate precision of the detection, robustness of the collision avoidance method and accuracy of positioning. Precision of both CNN models is statistically evaluated and the method is validated for both the model trained with automatically annotated data and the model trained with synthetic data. A performance comparison of the two approaches is also given.

4.1 Detection Precision

The proposed 3D object detection module consists of an object detection CNN with an RGB image as an input, and a filtering module that uses the CNN detection bounding boxes with an organized point cloud. Both image and point cloud are captured with the same off-the-shelf RGB-D camera, Intel RealSense D435. Detection module outputs 3D position of an object in the global reference frame.

The first set of experiments was conducted to evaluate the accuracy of 2D object detection modules, the CNN models. Both CNN models were tested on a subset of images collected autonomously with the collaborative robot, as described in the previous section. Test set consisted of 65 manually annotated images of a single strawberry plant. Accuracy and recall, reported in Table 1, were calculated for intersection over union (IoU) thresholds of 0.3 and 0.5, where detection is considered as true positive if IoU of predicted and ground truth bounding box is higher than the threshold and as false positive if not. If the ground truth bounding box is not detected in an image, it is considered as false negative.

Table 1. Accuracy and recall of CNN models.

	IoU = 0.3		IoU = 0.5	
	Accuracy	Recall	Accuracy	Recall
Synthetic	0.92	0.24	0.92	0.24
Auto-annotated	0.88	0.7	0.4	0.33

While the model trained on the synthetic dataset consistently achieves high accuracy, its recall is quite low, as it often fails at detection of highly occluded flowers. However, as we are recording plant from 14 positions when planning flower approach, while requiring that the flower is detected from at least 2 positions, we are more interested in the accuracy of the detection and low recall is balanced with multiple detection trials. On the other hand, recall of the model trained with A3IA approach is higher, as the test data is recorded in similar conditions as the training data and the model saw similar examples of occlusion in the training phase. Interestingly, accuracy of the model trained on automatically annotated data significantly drops with the higher IoU threshold. As visible in Fig. 5, area of automatically annotated bounding box does not fit the area of the actual flower perfectly as the flower is approximated with the sphere of fixed radius of 2 cm, resulting in a lower IoU score. An improvement could be introduced by more accurate approximation of the flower geometry in the annotation stage.

The second set of experiments was conducted to evaluate the 3D detection pipeline precision, by 10 experiment repetitions of detecting a single strawberry flower based on a set of search positions on the plant's bounding cylinder. A single experiment consisted of building a bounding cylinder, recording the plant from the predefined positions, flower detection, and approach planning and execution. The procedure was repeated for both CNN models on the same plant. Ideally, the obtained position estimations would be identical in all experiment repetitions for both models. However, due to occlusions, RGB-D camera imprecision and possible inaccuracy in the camera-robot transform calibration, small deviation in the estimated flower position is detected among the estimations obtained with the same model. The results presented in Table 2 show that dissipation of the estimates is negligible, i.e. at micrometer level for each axis for both models. The mean estimates, also shown in Table 2, differ among the models at a millimeter level, which is due to differences in annotation process. While annotations for the synthetic data are pixel-wise perfect, precision of the automatically generated annotations depends on the precision of the both human annotator in the loop, and of the depth module of the camera. Additionally, due to possible occlusions in some frames, the constructed sphere might enclose points which do not belong to the flower, yielding an inaccurate bounding box.

Table 2. Flower detection robustness analysis.

	Synthetic		Auto-annotated	
	Mean	Std.	Mean	Std.
x [mm]	34.81	0.14	37.03	0.18
y [mm]	520.63	0.18	514.79	0.27
z [mm]	549.68	0.26	548.71	0.47

4.2 Collision Avoidance Robustness

The approach position chosen so as to minimize possible collisions between the robot and the plant has also been recorded in 10 experiment repetitions on the single plant for both detection models. As shown in Table 3, the choice of the approach position is model-agnostic, meaning that approximately the same position was chosen when detection is performed with either one of the trained models. The maximum difference between the chosen approach positions is under millimeter level in all of the spatial axes. The maximum Euclidean distance among the chosen positions, as well as the maximum angle between the approach vectors is sufficiently small in both intra-model and inter-models scenario, as shown in Table 4. An angle between the two approach vectors, $\vec{u}$ and $\vec{v}$, is defined as in Eq. 6:

$$\alpha = a\cos\frac{\vec{u}\cdot\vec{v}}{|\vec{u}|\cdot|\vec{v}|} \tag{6}$$

Though the difference between mean flower positions estimated with different models exists, as reported in Table 2, the maximum distance between approach positions chosen with different models is comparable to the intra-models maximum distances, which suggests a high robustness of collision avoidance method to the inaccuracies of the detection module.

Table 3. Approach positioning robustness analysis.

	Synthetic		Auto-annotated		Inter-models	
	Δ_{max}	Std.	Δ_{max}	Std	Δ_{max}	Std.
x [mm]	0.1	0.03	0.16	0.05	0.16	0.04
y [mm]	0.07	0.02	0.1	0.03	0.12	0.03
z [mm]	0.1	0.04	0.09	0.03	0.13	0.04

Table 4. Maximum Euclidean distance (d) between the approach positions and maximum angle (α) between the approach vectors.

	Synthetic	Auto-annotated	Inter-models
d_{max} [mm]	0.13	0.18	0.2
α_{max}	0.15°	0.29°	1.21°

For the A3IA annotation method, we also investigated the dissipation of the number of obstacle points recorded from different poses across experiment repetitions, as shown in Fig. 7. The results are reported for poses in which the flower was detected in all experiment repetition. The variation of the number of obstacle points recorded from a single pose across experiment repetitions is a consequence of both minor positioning deviations of the robot and of RGB-D camera imprecision. However, there is a significant difference between the number of recorded points across the poses, with the number of points recorded from pose number 5 being notably the smallest in all experiment repetitions. For reference, on average 7.1×10^3 points have been detected as obstacle points from the pose number 5, which is repeatedly chosen as an approach pose, while the total average number of points recorded from the same pose was 178.2×10^3.

4.3 Positioning Accuracy

The proposed method, suitable for both flower thinning and pollination, has been validated by positioning the end effector on 20 different strawberry flowers and manually inspecting the positioning accuracy. In order to compare their performance, experiments have been conducted with both CNN models in the detection stage. The positioning was considered to be successful if the robot equipped with an appropriate tool would be capable of performing flower thinning and flower pollination task from the reached position. As shown in Table 5, 90% (18/20) of the attempts were successful when the detection model was trained on the synthetic dataset. Somewhat lower positioning success of 85% (17/20) was reached when the model was trained on the automatically annotated data.

For both models, 10% (2/20) of the attempts failed due to inability of the robot to reach the goal position. Even though the flower was correctly detected in those attempts, planning failed either because the flower was out of the robot reach or IK failed to find the feasible solution. All detections were successful

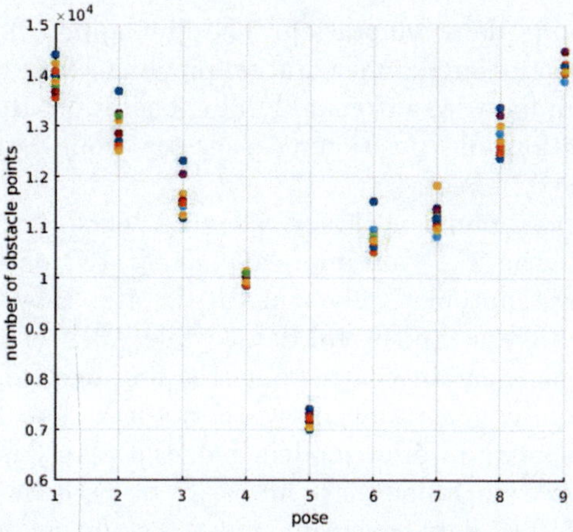

Fig. 7. Number of obstacle points ($|\mathbb{O}|$) for the approach poses in which the flower has been detected in all experiment repetitions. Detection was performed with the model trained on the automatically annotated data. Pose number 5 was repeatedly chosen as the pose with the minimum probability of collision.

when the model trained on the synthetic data was employed in the detection stage. With the model trained on the automatically annotated data one attempt failed due to erroneous detection. The photographs of the positioning can be found at [19], along with the accompanying video.

Table 5. Flower positioning results. Success (S), Failed detection (FD), Failed planning (FP).

	Synthetic		Auto-annotated	
	abs	%	abs	%
S	18	90	17	85
FD	0	0	1	5
FP	2	10	2	10

5 Conclusion

Precise object detection and positioning is essential in various agrotechnical procedures including but not limited to flower thinning and pollination. This work presents the method for precise flower positioning with deep learning based perception module evaluated under experimental conditions. As the need for large datasets is one of the biggest obstacles in the development of deep learning based perception methods, we deploy two data generation methods which alleviate the cost of time consuming data annotation. The first approach utilizes synthetic data rendered using a custom data generation pipeline, while the second is based on the robotic system for semi-automated data generation. Robustness and precision of both detection and positioning accuracy using both approaches are reported and discussed in Sect. 4.

Comparison of positioning accuracy achieved when employing the model trained on either synthetic or automatically annotated data suggests superiority of synthetic over automatically annotated data. However, generation of realistic models can take up to several days, while A3IA enables generation of tens of annotated images by annotating only the first one, which takes a couple of seconds. Additionally, generation of synthetic data does require some level of expert knowledge in the initial modeling of the plant, while our proposed automatic data generation framework can easily be used by non-experts. This simple, yet highly effective data generation method could enable farmers to easily apply the proposed flower manipulation method to other cultures by generating training datasets on their own.

Though we have demonstrated the efficiency of the automatic data annotation on the example of the strawberry flowers, we believe the proposed framework can serve as a good starting point for development of more complex automatic annotation systems. As a part of the future work, we plan to improve the annotation method by employing clustering when selecting the voxels which belong to the target object. This way, the method would not depend on the object shape and size and could be applied on other plant parts and objects in general. Under assumption that the minor differences between the mean position estimates among the models are the consequence of imprecision of the human annotator in the loop, smarter voxel selection method could also compensate the imprecision of the annotator and result in minimizing the difference between the mean position estimates between the two models.

Acknowledgements. This work has been supported by Croatian Science Foundation under the project Specularia UIP-2017-05-4042 [17]. The work of doctoral student Jelena Vuletić has been supported in part by the "Young researchers' career development project-training of doctoral students" of the Croatian Science Foundation funded by the European Union from the European Social Fund.

References

1. AliceVision: Meshroom: a 3D reconstruction software (2018). https://github.com/alicevision/meshroom
2. Barth, R., IJsselmuiden, J., Hemming, J., Van Henten, E.J.: Data synthesis methods for semantic segmentation in agriculture: a capsicum annuum dataset. Comput. Electron. Agric. **144**, 284–296 (2018)
3. De Gregorio, D., Tonioni, A., Palli, G., Di Stefano, L.: Semiautomatic labeling for deep learning in robotics. IEEE Trans. Autom. Sci. Eng. **17**(2), 611–620 (2020)
4. Di Cicco, M., Potena, C., Grisetti, G., Pretto, A.: Automatic model based dataset generation for fast and accurate crop and weeds detection. In: 2017 IEEE/RSJ

International Conference on Intelligent Robots and Systems (IROS), pp. 5188–5195. IEEE (2017)
5. Dias, P.A., Tabb, A., Medeiros, H.: Apple flower detection using deep convolutional networks. Comput. Industry **99**, 17–28 (2018)
6. Doersch, C., Zisserman, A.: Sim2real transfer learning for 3d human pose estimation: motion to the rescue. In: Wallach, H., Larochelle, H., Beygelzimer, A., d'Alché-Buc, F., Fox, E., Garnett, R. (eds.) Advances in Neural Information Processing Systems, vol. 32. Curran Associates, Inc. (2019)
7. He, L., Schupp, J.: Sensing and automation in pruning of apple trees: a review. Agronomy **8**, 211 (2018)
8. Hess, R.: Blender Foundations: The Essential Guide to Learning Blender 2.6. Focal Press (2010)
9. Horton, R., Cano, E., Bulanon, D., Fallahi, E.: Peach flower monitoring using aerial multispectral imaging. J. Imaging **3**(1) (2017)
10. Ilin, V., Kalinov, I., Karpyshev, P.A., Tsetserukou, D.: Deepscanner: a robotic system for automated 2d object dataset collection with annotations. arXiv:2108.02555 (2021)
11. Jia, W., Zhang, Y., Lian, J., Zheng, Y., Zhao, D., Li, C.: Apple harvesting robot under information technology: a review. Int. J. Adv. Robot. Syst. **17**(3) (2020)
12. Kar, A., Prakash, A., Liu, M.Y., Cameracci, E., Yuan, J., Rusiniak, M., Acuna, D., Torralba, A., Fidler, S.: Meta-sim: learning to generate synthetic datasets. In: ICCV (2019)
13. Kiyokawa, T., Tomochika, K., Takamatsu, J., Ogasawara, T.: Fully automated annotation with noise-masked visual markers for deep-learning-based object detection. IEEE Robot. Autom. Lett. **4**(2), 1972–1977 (2019)
14. Liu, W., Anguelov, D., Erhan, D., Szegedy, C., Reed, S., Fu, C.Y., Berg, A.: Ssd: Single shot multibox detector. In: Computer Vision—ECCV 2016, vol. 9905, pp. 21–37 (2016)
15. Maric, B., Polic, M., Tabak, T., Orsag, M.: Unsupervised optimization approach to in situ calibration of collaborative human-robot interaction tools. In: 2020 IEEE International Conference on Multisensor Fusion and Integration for Intelligent Systems (MFI), pp. 255–262. IEEE (2020)
16. Pashevich, A., Strudel, R., Kalevatykh, I., Laptev, I., Schmid, C.: Learning to augment synthetic images for sim2real policy transfer. In: 2019 IEEE/RSJ International Conference on Intelligent Robots and Systems (IROS), pp. 2651–2657 (2019)
17. Polic, M., Ivanovic, A., Maric, B., Arbanas, B., Tabak, J., Orsag, M.: Structured ecological cultivation with autonomous robots in indoor agriculture. In: 2021 16th International Conference on Telecommunications (ConTEL), pp. 189–195 (2021)
18. Polic, M., Tabak, J., Orsag, M.: Pepper to fall: a perception method for sweet pepper robotic harvesting. Intelligent Service Robotics (2021)
19. Polic, M., et al.: Strawberry flowers (2021). https://sites.google.com/view/specularia-pepper-picking/home/flower-picking. Accessed 2021-09-09
20. Ruiz-Sarmiento, J., Galindo, C., Gonzalez-Jimenez, J.: Olt: a toolkit for object labeling applied to robotic RGB-D datasets. In: 2015 European Conference on Mobile Robots (ECMR), pp. 1–6 (2015)
21. Strader, J., Yang, C., Gu, Y., Nguyen, J., Tatsch, C., Du, Y., Lassak, K., Buzzo, B., Watson, R., Cerbone, H., Ohi, N.: Flower interaction subsystem for a precision pollination robot. In: 2019 IEEE/RSJ International Conference on Intelligent Robots and Systems (IROS), pp. 5534–5541 (2019)

22. Stumpf, D., Krauß, S., Reis, G., Wasenmüller, O., Stricker, D.: Salt: a semi-automatic labeling tool for RGB-D video sequences. In: 16th International Conference on Computer Vision Theory and Applications, pp. 595–603 (2021)
23. Tobin, J., Fong, R., Ray, A., Schneider, J., Zaremba, W., Abbeel, P.: Domain randomization for transferring deep neural networks from simulation to the real world. In: 2017 IEEE/RSJ International Conference on Intelligent Robots and Systems (IROS), pp. 23–30. IEEE (2017)
24. Xiong, Y., Ge, Y., Grimstad, L., From, P.J.: An autonomous strawberry-harvesting robot: design, development, integration, and field evaluation. J. Field Robot. **37**(2), 202–224 (2020)
25. Zhang, Q., Karkee, M., Tabb, A.: The use of agricultural robots in orchard management. In: Billingsley, J. (ed.) Robotics and Automation for Improving Agriculture. Burleigh DODDS Science Publishing (2019)

Simulative and Experimental Evaluation of a Soft-DTW Neural Network for sEMG-Based Robotic Grasping

Alessandra Bernardini(✉), Roberto Meattini, Gianluca Palli, and Claudio Melchiorri

DEI—Department of Electrical, Electronic, Information Engineering, University of Bologna, Viale del Risorgimento 2, Bologna 40136, Italy
alessandr.bernardin5@unibo.it, roberto.meattini2@unibo.it, gianluca.palli@unibo.it, claudio.melchiorri@unibo.it

Abstract. In this paper, we present a neural network architecture for minimally supervised regression of surface electromyographic (sEMG) signals into control commands to drive robotic grasping devices. The proposed architecture overcomes one of the limitations of state-of-the-art supervised regression approaches, which require an instant by instant labelling of the training dataset. This is achieved by deploying a differentiable version of the Dynamic Time Warping (DTW) similarity measure as loss function of a feed-forward neural network. The effectiveness of this approach was assessed both with simulation and experimental studies. We first used a model of the sEMG generation process to test the feasibility of the method. Then, we evaluated the proposed approach in a two-step experimental session involving a group of 10 subjects: an offline experiment was conducted to investigate neural network performance with desynchronized labelling, whereas an online experiment was carried out to control both a simulated and a real robotic hand. The obtained results demonstrate that the presented method allows minimally supervised regression of sEMG signals, reporting performances comparable with standard supervised approaches. In this relation, we show that the proposed soft-DTW neural network enables successful myocontrol of robotic hands even in presence of substantial temporal misalignments between sEMG trainset and related labelling, while supervised regression totally loses its capabilities. This means that the presented approach allows a greatly simplified training procedure that can pave the way to an innovative myocontrol framework characterized by highly simplified training procedures for the user without performance degradation.

Keywords: Myoelectric control · Multifingered hands · Telerobotic and teleoperation

P. Borja et al. (Eds.): HFR 2022, 26, pp. 205–217, 2023.
https://doi.org/10.1007/978-3-031-22731-8_15

1 Introduction

The problem of finely and reliably controlling robotic grasping devices is still an open challenge nowadays. Although many efforts have been made in this direction, modern devices present a significant unbalance between technological advancements and controlling capabilities.

Looking at the related works in the domain of robotic hands control, researchers have extensively developed Human-Robot interfaces (HRi) using surface electromyography (sEMG) signals recorded from surface electrodes. The reason can be found in the relative ease of use and low invasiveness of myoelectric equipment, whose only request is to place electrodes on the operator's forearm. Specifically, among the many techniques used to extract meaningful control commands from sEMG signals, supervised and unsupervised machine learning are the main paradigms developed so far. Supervised classification and regression approaches require labelled datasets, from which the algorithm is trained to learn the correct input-output function. On the one hand, supervised classification is used to extract discrete commands from sEMG signals, generating in this way a sequential control of the robotic device but suffering from performance degradation when simultaneous or complex motions have to be predicted [1]. On the other hand, supervised regression techniques allow the user to continuously regulate grasping motions [2], in a way similar to real human control. Anyway, the need for point-by-point labelling of the sEMG training dataset negatively affects the applicability of regression approaches, due to the fact that complicated and long training sessions are required to generate a correct input-label matching [3]. Oppositely, unsupervised learning algorithms such as Non-negative Matrix Factorization [4], Principal Component Analysis [5] and Autoencoders [6] do not require point-by-point labelling of the training dataset. However, data fitting capabilities rapidly degradate when complex hand gestures are considered with unsupervised approaches.

In this paper, we present a neural network architecture to perform regression of sEMG signals without the need for instant-by-instant training data labelling. Specifically, we use a differentiable version of the Dynamic Time Warping (DTW) similarity measure (called soft-DTW) as loss function of a simple feed-forward neural network. DTW is a technique used to evaluate the distance between time series, independently from length, phase misalignments and nonlinear temporal distortions [7]. These positive aspects of DTW can be exploited to overcome the limitations of state-of-the-art supervised regression approaches, namely to construct a network that is robust to training target misalignments.

Results are shown considering both simulation and experimental studies. As first step, a model of the myoelectric signal generation process is used to assess the feasibility of our proposed architecture. Then, offline and online experiments are carried out to evaluate neural network performances, and to control both a real and a simulated robotic hand during grasping and motion following tasks, respectively.

The article is structured as follows. Section 2 describes the neural network architecture, as long as the soft-DTW based loss function used for minimally supervised regression. In Sect. 3, the simulation study is reported: the model of sEMG signals generation process is presented, along with the simulation results. Section 4 shows experimental protocols and results, while Sect. 5 draws the conclusions.

2 Soft-DTW Neural Network

In this section, we present the neural network architecture and the soft-DTW based loss function which allow us to perform training without the need of the recorded dataset to be temporally aligned with the reference grasping signal.

2.1 Neural Network Architecture

A feed-forward neural network is an artificial network that does not present loops in its architecture. It consists of an input layer, a certain number of hidden layers and an output layer. The fundamental building block, namely the generic i-th layer of the network, is made of N_i neurons whose value is obtained through an affine transformation followed by a desired activation function. Specifically, the input vector $y^{(i-1)}(t) \in \mathbb{R}^{N_{i-1}}$ of the i-th layer is weighted by the matrix $W^{(i)} \in \mathbb{R}^{N_i \times N_{i-1}}$, translated with the bias vector $b^{(i)} \in \mathbb{R}^{N_i}$, and finally passed through the activation function $\mathcal{F}(\bullet)$. Thus, the output vector of the i-th layer $y^{(i)}(t) \in \mathbb{R}^{N_i}$ can be expressed as

$$y^{(i)}(t) = \mathcal{F}\big(W^{(i)} y^{(i-1)}(t) + b^{(i)}\big) \, . \tag{1}$$

In our architecture, the input is given by the Root Mean Square (RMS) value of the sEMG signals recorded by 8 bipolar surface electrodes, represented by the vector $E(t) = [e_1(t)\ e_2(t) \ldots e_8(t)]^T$ (see Sect. 4.1 for more details about the acquisition process), while the output layer is made of only one neuron, representing the grasp closure level $c(t) \in \mathbb{R}$, used to regulate the opening/closing motion of the hand (see Sect. 4.2). Hyperbolic tangent sigmoid activation functions were chosen for all the hidden layers of the network, whereas a linear transfer function was used to generate the desired output signal. Note that the number of hidden layers of the soft-DTW neural network, as well as the number of neurons in each layer, were not a priori fixed, but they were selected as the ones maximizing network performance. Thus, as can be guessed, the objective of our network was to extract from sEMG signals the control command $c(t)$ allowing a continuous regulation of the grasping motion.

2.2 Soft-DTW Based Loss Function

All the parameters describing the neural network architecture have to be carefully selected in order to fit the desired objective output. To achieve this, we trained our network with a scaled conjugate gradient descent algorithm combined with the proposed DTW-based loss function.

As already stated, DTW is a similarity measure used to compute the best alignment between time series [8]. The main problem with this technique relies on the fact that DTW is not differentiable (the minimum operator cannot be differentiated), thus it cannot be used as fitting loss of a neural network. To solve this issue we considered the soft-DTW divergence [9] (which is differentiable, nonnegative and minimized when the two time series are identical) between the network output and the reference target. Denoting as $C \in \mathbb{R}^{1\times d}$ the one-dimensional network output (the grasp closure level) and as $T \in \mathbb{R}^{1\times d}$ the target signal (with d the number of recorded samples), the soft-DTW divergence measure [9] is given by

$$D_{\lambda}^{\Delta}(C,T) = SDTW_{\lambda}(\Delta(C,T)) - \frac{1}{2}SDTW_{\lambda}(\Delta(C,C)) - \frac{1}{2}SDTW_{\lambda}(\Delta(T,T)) \,. \quad (2)$$

$SDTW_{\lambda}(\Delta(C,T))$ computes the soft-DTW distance between the vectors C and T [9] using the cost of Δ (in our case the square Euclidean one)

$$SDTW_{\lambda}(\Delta(C,T)) = \min_{A\in\mathcal{A}(C,T)}{}^{\lambda} \; \langle A, \Delta(C,T)\rangle \,, \quad (3)$$

where $\langle A, \Delta\rangle$ is the Frobenius inner product between matrices A and Δ, which are the alignment matrix and the cost matrix between vectors C and T, respectively. Remember that the alignment matrix $A \in \{0,1\}^{d\times d}$ between C and T (whose generic elements are $c_i \in \mathbb{R}$ and $t_i \in \mathbb{R}$) is a matrix made of zeros and ones, where 1 means that element c_i is aligned with t_i, 0 otherwise. Instead, the cost matrix $\Delta \in \mathbb{R}^{d\times d}$ stores the squared euclidean distances between each element of C and T. Moreover $\mathcal{A}(C,T)$ represents the set of all monotonic alignment matrices between C and T. Finally, $\min^{\lambda}$ denotes the smooth min operator, whose expression is

$$\min_{x}{}^{\lambda} f(x) = -\lambda \, \log \sum_{x} \exp\left(\frac{-f(x)}{\lambda}\right) , \quad (4)$$

where λ is a parameter used to regulate the tradeoff between smoothness and approximation, and such that for $\lambda \to 0^{+}$, we recover the traditional min operator. In our case $\lambda = 0.1$.

As we said, the soft-DTW divergence is used as loss function of our neural network, thus it must be explicitly differentiated in order to be used in the gradient back-propagation phase

$$\nabla_{C} D_{\lambda}^{\Delta}(C,T) = \left(\frac{\partial\Delta(C,T)}{\partial C}\right)^{\top} \mathbb{E}[A;\Delta(C,T)] - \frac{1}{2}\left(\frac{\partial\Delta(C,C)}{\partial C}\right)^{\top} \mathbb{E}[A;\Delta(C,C)] \,, \quad (5)$$

where $\partial\Delta(C,T)/\partial C$ is the Jacobian of the cost matrix (between C and T) with respect to C, while $\mathbb{E}[A;\Delta(C,T)]$ is the expected alignment matrix induced by $\Delta(C,T)$. Since the soft-min operator in Eq. (4) introduces a Gibbs distribution over the alignments in $\mathcal{A}(C,T)$, we can express the expected alignment matrix $\mathbb{E}[A;\Delta(C,T)]$ as:

$$\mathbb{E}[A;\Delta(C,T)] = \sum_{A\in\mathcal{A}(C,T)} \frac{\exp(-\langle A,\Delta(C,T)\rangle/\lambda)}{\sum_{A'\in\mathcal{A}(C,T)}(-\langle A',\Delta(C,T)\rangle/\lambda)} A \in (0,1]^{d\times d} \,. \quad (6)$$

As we can see, the expected alignment matrix can be seen as a weighted average of all the paths in $\mathcal{A}(C,T)$ (refer to [9] for further details).

Lastly, we need to specify that the target T used to train the network is built a posteriori, once the training dataset has been recorded. More in detail, the user is first instructed to perform a certain number of power grasp opening/closing motions, without any visual reference and starting from the fully open hand. Then the target reference is built as a trapezoidal profile between 1 (corresponding to the hand fully open) and -1 (corresponding to the closed hand), uniformly distributed along the d recorded samples.

3 Simulation Studies

The first phase of our study consisted in analyzing the functioning of the soft-DTW neural network by using a simulation of the sEMG signals generation process.

3.1 Model of sEMG Signals

To study the behaviour of the network on a simulated myoelectric signal, we leveraged on the generative model presented in [10], according to which the sEMG signal can be expressed in terms of control information coming from the higher levels of the nervous system. Following this model, we developed a Matlab simulation of the sEMG signals recorded by 8 bipolar surface electrodes placed on the forearm, during the execution of hand motions. Specifically, we considered the opening/closing motion of the hand as a single DoF ($N = 1$) governed by the function $D(t)$ (which represents the neural command associated with that motion). Moreover, we assumed $M = 4$ forearm muscles responsible for the opening/closing motion. As a consequence, we can get the activation functions of each muscle $X(t) = [x_1(t), x_2(t), x_3(t), x_4(t)]^T$ as [10]:

$$X(t) = S \times D(t)\ , \tag{7}$$

where $S \in \mathbb{R}^{M \times N}$ is the so-called synergy matrix, used to specify the contribution of each muscle to the opening/closing motion of the hand. In our case, muscles 1 and 2, which are the extensor and flexor muscles, respectively, are assumed to contribute more to the considered motion with respect to 3 and 4. Thus the considered synergy matrix is reasonably chosen to be:

$$S = \begin{bmatrix} 0.8 & -0.8 & 0.2 & -0.2 \end{bmatrix}^T . \tag{8}$$

Once the activation functions of all muscles are given, we can model the sEMG signal recorded by a generic electrode as [10]

$$e(t) = \sum_{m=1}^{M} \sum_{j=1}^{N_m} v_{mj}(t) * f_{mj}[x_m(t)]\ , \tag{9}$$

where $N_m = 50$ is the number of motor units in the m-th muscle, while v_{mj} and f_{mj} model the effect of the volume conductor and the firing sequence of motor units, respectively. The filtering action of human tissues, namely the volume conductor effect, was implemented following [11], in which the authors model the motor unit potential detected by a rectangular plate electrode of any size. Moreover, to represent the firing sequence of motor units, we combined the exponential threshold model of [12], to specify which motoneurons are active, and the Integral Pulse Frequency Modulation (IPFM) strategy explained in [13], to describe the impulse generation process. Finally, as explained in Eq. (9), we obtained the sEMG signal recorded by a surface electrode by convolving, for all muscles and for all motor units, the potential detected by the electrode, with the corresponding motor unit's firing sequence. Additionally, to further include the effect of crosstalk between neighbour surface electrodes, we incorporated in the simulation a scaling operation in order to vary the sensitivity of electrodes to adjacent muscles.

3.2 Soft-DTW Neural Network Applied to Simulated sEMG

In the simulation phase, we were interested in studying the effectiveness of our soft-DTW neural network in extracting control commands from sEMG, while having the target T not synchronized with the input sEMG signal. Thus, as first step, a simulated myoelectric signal was generated considering three repetitions of the power grasp motion. In Fig. 1a the simulated sEMG signal recorded by 8 surface electrodes is shown. Then, to investigate the potentialities of our network we constructed three different targets for training:

(i) The *synchronized target* obtained from the neural command used to simulate the sEMG signal (according to Eq. (7) and Eq. (9))
(ii) The *target shrunk by 1/3* (obtained shortening the synchronized signal by 1/3 of its length)
(iii) The *target shrunk by 2/3* (obtained shortening the synchronized signal by 2/3 of its length)

Note that, for training reasons (i.e. to have the output and target sequences of our network of the same length), in the shrunk targets we held the last value until matching the length of the original synchronized target.

In the simulation study, we trained our network on the RMS value of the simulated signal considering each time a different target and conducting a grid search to select the best number of layers and neurons per layer. The outputs of the soft-DTW neural network can be seen in Fig. 1b, compared with the outcomes of a classical Mean Square Error (MSE)-based neural network trained with the same target signals, but using the mean square error as loss function. As we can see, both the soft-DTW and the MSE-based neural networks perform well with the synchronized target. Oppositely, in case of the target shrunk by 1/3 and the target shrunk by 2/3, the neural network trained with the MSE-based loss function is not able to catch the desynchronization between the input signal and the target reference, thus it cannot reconstruct the output correctly. Indeed, only the red signal (the output of the soft-DTW neural network) is able to follow, in all the cases, the correct synchronized motion represented by the green dotted line in the figure. This shows the robustness of the network with respect to target misalignments.

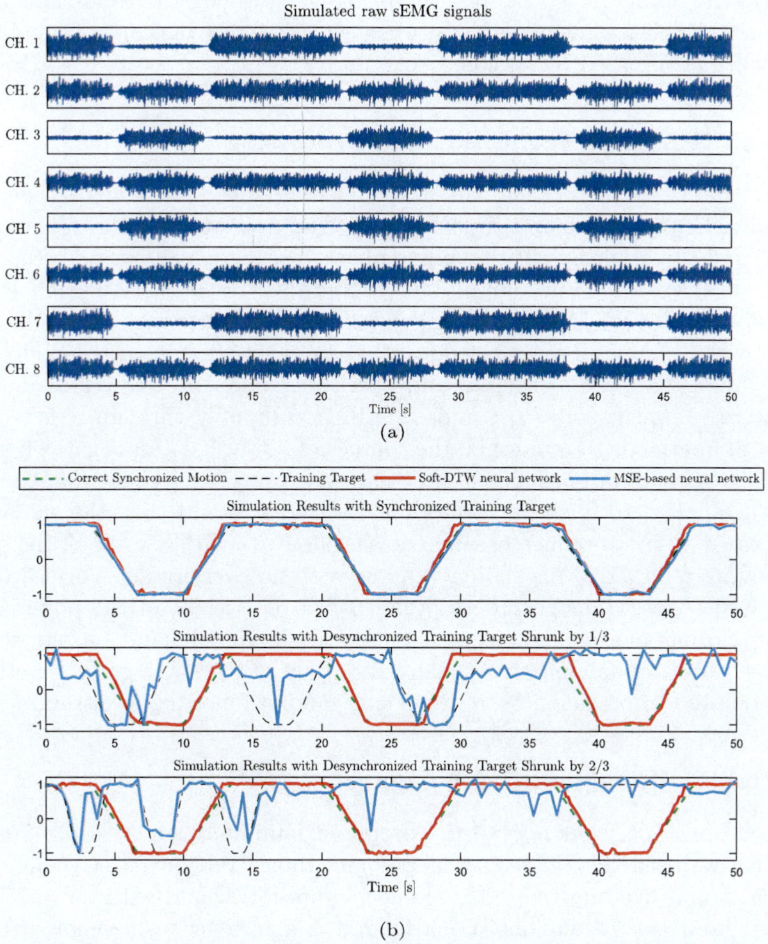

Fig. 1. Results of the simulation session. **a** Simulated raw sEMG signals obtained for three power grasp motions. **b** Example of soft-DTW and MSE-based neural networks training outputs for the three different targets: synchronized target, target shrunk by 1/3 and target shrunk by 2/3

4 Experimental Studies

In the following, we first explain the experimental setup used in the second stage of our studies, namely the devices used for signal acquisition/processing and the simulated and real robot hands used to test the proposed soft-DTW neural network. Then, the experimental session is articulated into an offline and online phase: in the offline phase we were interested in studying the robustness of the network, while the online experiment was devoted to test the control capabilities of the soft-DTW based regression.

4.1 sEMG Signals Acquisition and Processing

A total number of 10 healthy subjects (1 female, age: 25, 9 males, age: 30.5 $\pm$ 4) participated in the experiments: half of them were involved in the offline session, and the other five in the online phase. During experimental recordings, participants were seated in front of a screen and their forearm sEMG signals were recorded using the OYMotion gForcePro armband, a device with 8 built-in EMG sensors with differential dry electrodes, that we positioned in correspondence of the *Flexor Digitorum Superficialis* and *Extensor Digitorum Communis* muscles. Data were sampled at 1 kHz and sent to a computer through a low-power Bluetooth interface. To remove all the sources of noise that can negatively affect the sEMG signal, we considered the following processing chain: 50 Hz notch filter was applied to filter out powerline interference, and then the signal was band passed 20 Hz to reduce baseline noise. Lastly, the RMS value of the signal was computed on a 200 ms sliding window with no overlapping. For the offline study, subjects were instructed to perform five repetitions of the power grasp opening/closing motion, following a reference simulated hand on the screen. Differently, during online dataset acquisition, only two power grasp repetitions were requested to participants, without any motion following constraint.

4.2 Robotic Hand and Grasp Control

To evaluate the performances of the proposed minimally supervised regression approach, we used the University of Bologna Hand, version IV (UB Hand IV) [14], which is a five-fingered anthropomorphic robotic hand with $n_J = 15$ DoF actuated by $n_T = 25$ tendons. Each tendon is driven by a servomotor that is controlled in position mode, thus the vector of motor torques $\tau(t) \in \mathbb{R}^{n_T}$ is given by

$$\tau(t) = K_m(\theta_m^{ref}(t) - \theta_m(t)), \tag{10}$$

where K_m is the diagonal matrix of motors stiffnesses, while $\theta_m^{ref}(t) \in \mathbb{R}^{n_T}$ and $\theta_m(t) \in \mathbb{R}^{n_T}$ are the desired and actual positions of the motors. Note that $\theta_m^{ref}(t)$ is obtained from the joint space reference angles $\theta_j^{ref}(t)$ using the mapping matrix H. Lastly, the vector $\theta_j^{ref}(t)$ of joint reference angles, is obtained as

$$\theta_j^{ref}(t) = S_G\alpha(t), \tag{11}$$

where S_G is the grasp synergy matrix, that is a matrix used to map the closure hand reference $\alpha(t)$ into the motion of the hand's joints. Since in our application the hand reference can vary in the range [0, 1], where 0 corresponds to the open configuration and 1 to the closed one, we selected the weights of the synergy matrix to perform the correct robotic hand motion in that range. Moreover, we must highlight that the hand closure reference $\alpha(t)$ is simply a properly scaled version of the soft-DTW neural network output, the grasp closure level $c(t)$, modified in order to match the interval $[0, 1]$.

4.3 Offline Experimental Session

Driven by the positive outcomes of simulation studies, the objective of the offline experimental session was to investigate the performances of the proposed soft-DTW neural network on a real sEMG signal. Indeed, similarly to want we have done in the simulation phase, we studied the behaviour of the network when the target T was not synchronized with the actual hand motion.

Offline Experimental Protocol For the offline study, the five participants were requested to execute five repetitions of the power grasp motion, while their forearm sEMG signals were acquired as illustrated in Sect. 4.1. The recorded datasets were then used to train the soft-DTW neural network with three different target outputs: as in the simulation, we considered the synchronized target (i.e. the motion reference signal of the simulated hand), the target shrunk by 1/3 and the target shrunk by 2/3. During the training phase, a nested cross-validation (nested CV) was conducted for each target type, in order to estimate the performance of the network. Specifically, the nested CV was articulated into an inner loop, devoted to the tuning of the hyperparameters of the network (namely the number of hidden layers and neurons per layer), and an outer loop, used to evaluate the performance of the network tuned in the inner loop. Since the training dataset was split into 5 folds (one for each power grasp motion), the results of the five outer loops were averaged to obtain an estimate of neural network performance. Moreover, also in the offline experiments, we compared the outcomes of the soft-DTW neural network with the results of a MSE-based neural network trained in the same way. In both cases, we used the standard DTW as performance metric.

Offline Experimental Results In the boxplots of Fig. 2, results of the nested CV are reported, comparing the behaviour of the soft-DTW neural network and MSE-based neural network with the three different targets used for training. The Analysis of Variance (ANOVA) test was conducted to analyse the significance of these results. As first step, a two-way ANOVA was performed, considering the effects of neural network type and output type on our results. The test showed a statistically significant interaction between the two factors (the normality assumption was not violated, while the Greenhouse-Geisser correction was applied due to sphericity violation). As a consequence, more interestingly, we conducted a one-way ANOVA test, which revealed a statistically significant difference, $F(1.13, 7.34) = 18.57$, $p < 0.001$. Hence, the Games-Howell test was performed to compute a pairwise comparison between groups. According to this test, no statistically significant difference was present between the results obtained with the soft-DTW neural network trained with the three targets and the MSE-based neural network trained with the synchronized target. These results support the use of the soft-DTW neural network for minimally supervised regression of sEMG signals since they show its robustness with respect to desynchronizations of the target output (i.e. labelling desynchronizations). Additionally, Fig. 3 shows the outcomes of a single CV outer loop for subject 2. As we can see, both neural networks trained with the synchronized target show positive results, while the MSE-based neural network trained with the target shrunk by 1/3 and 2/3 is not able to handle labelling desynchronizations. Indeed the output cannot follow the correct synchronized motion (green dotted line).

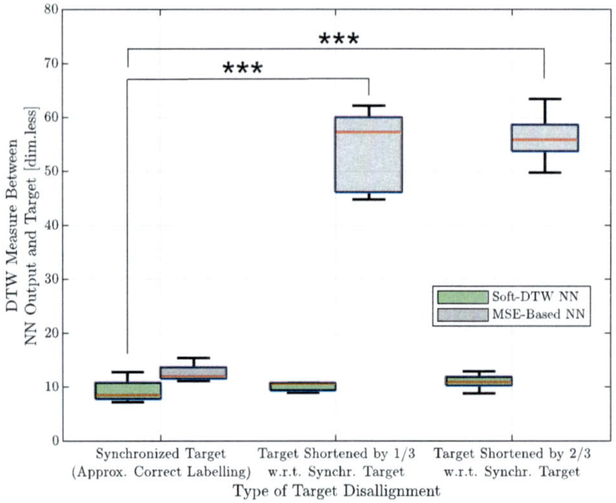

Fig. 2. Results of the offline experiment: boxplot of the CV results over the five subjects (statistically significant difference is represented with the symbol "***")

4.4 Online Experimental Session

In the online experimental session, we were interested in analyzing the online control capabilities of our soft-DTW based regression approach. To this aim, we articulated the online session in two experiments in which the five participants were asked to control the simulated and the real UB Hand. In the following we denote these experiments as *Task 1* and *Task 2*, respectively. Note that in both online experiments, subjects were requested to first perform two power grasp

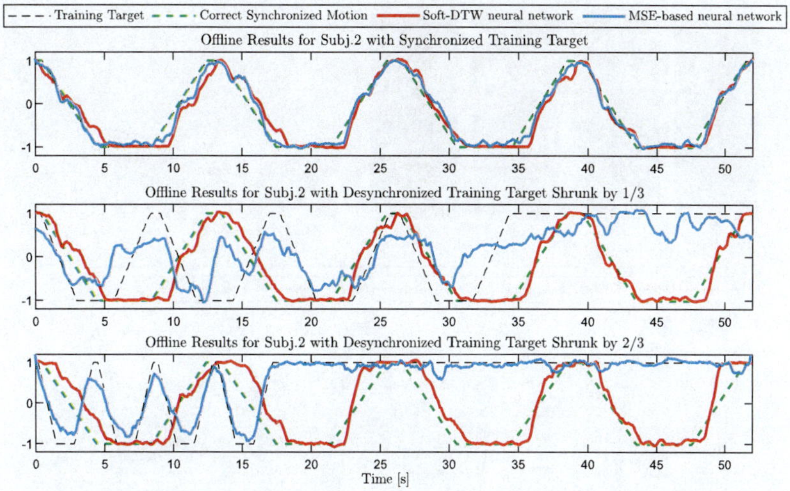

Fig. 3. Results of the offline experiment: example of soft-DTW and MSE-based neural networks training outputs for the three different targets

repetitions without any motion reference, as presented in Sect. 4.1, to train the soft-DTW neural network.

Task 1 In the first experiment, the subjects were instructed to control the simulated UB Hand in such a way to follow the movements of a reference hand as closely as possible. Specifically, the reference hand was programmed to perform 4 repetitions of the power grasp opening/closing motion with two different final closure levels. In Fig. 4, we present the results of the online *Task 1*, averaged among the five participants. As we can see, subjects were able to follow the reference hand quite well, with an average absolute error that remained below 0.7 for the whole task execution.

Task 2 The second task of the online experimental session dealt with the control of the real UB Hand IV. We asked the five participants to regulate the grasp closure level of the UB Hand (see Sect. 4.2) in such a way to perform a stable grasp of a cylindrical object five times. More in detail, the subject had to grasp with the robotic hand a cylindrical object given by the experimenter, hold it firmly for 19 s and lastly release it to the experimenter when he requested the object. On the basis of this protocol, the subjects involved in the online experiment performed a total of 25 "grasp-hold-release" tasks. To have an idea of the experiment, some video frames of the *Task 2* performed by Subj.8 are shown in Fig. 5. As a general outcome, we can report for 100% success rate in controlling the UB Hand. Indeed, all of the participants correctly grasped the object with minimal cognitive effort.

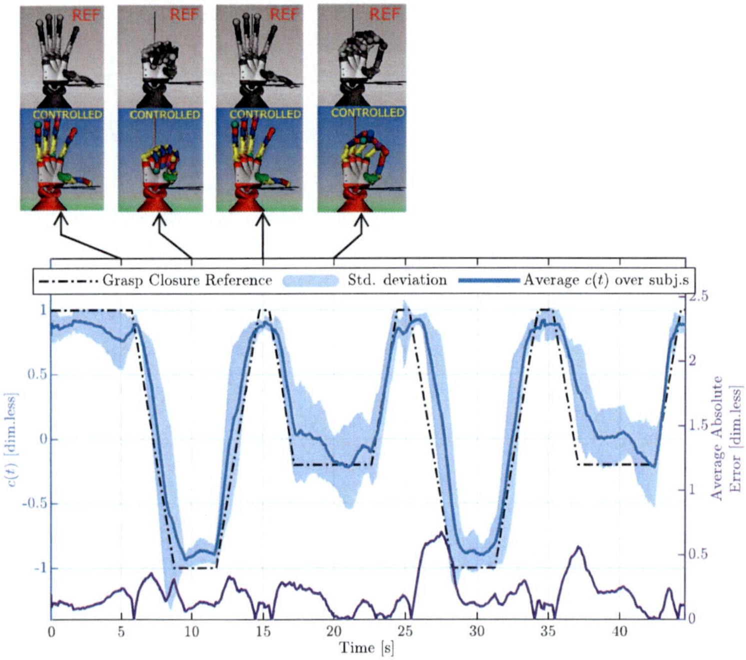

Fig. 4. Results of *Task 1*: average controlled grasp closure level $c(t)$, with standard deviation and average absolute error with respect to the grasp closure reference

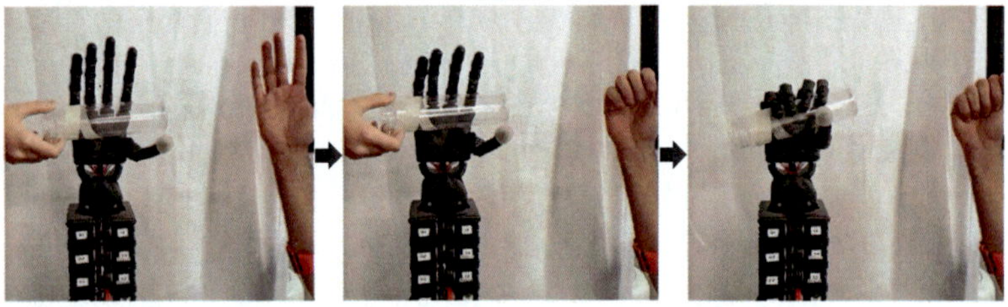

Fig. 5. Video frames of *Task 2*: cylindrical object grasping executed by Subj.8 with the real UB Hand

5 Conclusions

This study presented a soft-DTW neural network for minimally-supervised regression of sEMG signals into continuous grasping control commands. The absence of point-by-point labelling of the training dataset is the key advantage of the proposed architecture with respect to classical supervised regression approaches, because it simplifies considerably the training procedure without degrading performance. Results of both simulation and experimental studies have been reported. Specifically, once the effectiveness of the network has been tested in a simulative scenario, an offline study demonstrated its robustness in compensating for labelling misalignments. Finally, the online experimental session provided positive outcomes in controlling both the simulated and the real UB Hand robotic hand. Future developments concern the exploitation of the non-linear fitting capabilities of supervised regression approaches, in order to extract from the sEMG signals control commands to regulate different grasping motions.

References

1. Ingraham, K.A., Smith, L.H., Simon, A.M., Hargrove, L.J.: Nonlinear mappings between discrete and simultaneous motions to decrease training burden of simultaneous pattern recognition myoelectric control. In: 2015 37th Annual International Conference of the IEEE Engineering in Medicine and Biology Society (EMBC), pp. 1675–1678. IEEE (2015)
2. Meattini, R., Nowak, M., Melchiorri, C., Castellini, C.: Automated instability detection for interactive myocontrol of prosthetic hands. Front. Neurorobot. **13**, 68 (2019)
3. Hermens, H., Stramigioli, S., Rietman, H., Veltink, P., Misra, S.: Myoelectric forearm prostheses: state of the art from a user-centered perspective. J. Rehabil. Res. Dev. **48**(6), 719 (2011)
4. Lin, C., Wang, B., Jiang, N., Farina, D.: Robust extraction of basis functions for simultaneous and proportional myoelectric control via sparse non-negative matrix factorization. J. Neural Eng. **15**(2), 026017 (2018)
5. Meattini, R., De Gregorio, D., Palli, G., Melchiorri, C.: Design and evaluation of a factorization-based grasp myoelectric control founded on synergies. In: 2019 12th International Workshop on Robot Motion and Control (RoMoCo), pp. 252–257. IEEE (2019)
6. Vujaklija, I., Shalchyan, V., Kamavuako, E.N., Jiang, N., Marateb, H.R., Farina, D.: Online mapping of EMG signals into kinematics by autoencoding. J. Neuroeng. Rehabil. **15**(1), 1–9 (2018)
7. Müller, M.: Dynamic time warping. Information retrieval for music and motion, pp. 69–84 (2007)
8. Cuturi, M., Blondel, M.: Soft-dtw: a differentiable loss function for time-series. In: International Conference on Machine Learning, pp. 894–903. PMLR (2017)
9. Blondel, M., Mensch, A., Vert, J.P.: Differentiable divergences between time series. In: International Conference on Artificial Intelligence and Statistics, pp. 3853–3861. PMLR (2021)
10. Jiang, N., Englehart, K.B., Parker, P.A.: Extracting simultaneous and proportional neural control information for multiple-DOF prostheses from the surface electromyographic signal. IEEE Trans. Biomed. Eng. **56**(4), 1070–1080 (2009)
11. Dimitrov, G.V., Dimitrova, N.A.: Precise and fast calculation of the motor unit potentials detected by a point and rectangular plate electrode. Med. Eng. Phys. **20**(5), 374–381 (1998)
12. Fuglevand, A.J., Winter, D.A., Patla, A.E.: Models of recruitment and rate coding organization in motor-unit pools. J. Neurophysiol. **70**(6), 2470–2488 (1993)
13. Bayly, E.J.: Spectral analysis of pulse frequency modulation in the nervous systems. IEEE Trans. Biomed. Eng. BME **15**(4), 257–265 (1968)
14. Melchiorri, C., Palli, G., Berselli, G., Vassura, G.: Development of the UB hand IV: overview of design solutions and enabling technologies. IEEE Robot. Autom. Mag. **20**(3), 72–81 (2013)

Toward Human-Robot Cooperation: Unsupervised Domain Adaptation for Egocentric Action Recognition

Mirco Planamente[1,2,3], Gabriele Goletto[1(✉)], Gabriele Trivigno[1], Giuseppe Averta[1], and Barbara Caputo[1,3]

[1] Politecnico di Torino, Corso Duca degli Abruzzi, 24, 10129 Torino, Italy
mirco.planamente@polito.it, mirco.planamente@iit.it, gabriele.goletto@polito.it, gabriele.trivigno@polito.it, giuseppe.averta@polito.it, barbara.caputo@polito.it
[2] Istituto Italiano di Tecnologia, Via Morego, 30, 16163 Genova, Italy
[3] Consortium Cini, Roma, Italy

Abstract. With the advent of collaborative manipulators, the community is pushing the limits of human-robot interaction with novel control, planning, and task allocation strategies. For a purposeful interaction, however, the robot is also required to understand and predict the action of the human not only at a kinematic level (i.e. motion estimation), but also at an higher level of abstraction (i.e. action recognition), ideally from the human own perspective. Dealing with egocentric videos comes with the benefit that the data source already embeds an intrinsic attention mechanism, driven by the focus of the user. However, the deployment of such technology in realistic use-cases cannot ignore the large variability of background characteristics when changing environment, resulting in a domain shift in features space not learnable from labels at training time. In this paper, we discuss a method to perform Domain Adaptation with no external supervision, which we test on the EPIC-Kitchens-100 UDA Challenge in Action Recognition. More specifically, we move from our previous work on Relative Norm Alignment and extend the approach to unlabelled target data, enabling a simpler adaptation of the model to the target distribution in an unsupervised fashion. To this purpose, we enhanced our framework with multi-level adversarial alignment and with a set of losses aimed at reducing the classifier's uncertainty on the target data. Extensive experiments demonstrate how our approach is capable to perform Multi-Source Multi-Target Domain Adaptation, thus minimising both temporal (i.e. different recording times) and environmental (i.e. different kitchens) biases.

Keywords: Human–robot cooperation · First person action recognition · Unsupervised domain adaptation

1 Introduction

Current robotics research demonstrated a significant trend in the development of technologies to support the physical interaction between humans and machines, ranging

P. Borja et al. (Eds.): HFR 2022, 26, pp. 218–232, 2023.
https://doi.org/10.1007/978-3-031-22731-8_16

from the planning and control [2], up to their social impact [26]. However, the deployment of such technology in the real world, e.g. in household or industrial environments, requires an extension of the human intention retrieval capabilities of robots, from a mere pose estimation and forecast, to an high level description of the action executed. As an example, considering a companion robot assisting the human in preparing a meal, the feasibility to infer from video the current action performed can enable the prediction of the next steps of the receipt, and eventually assist the cook with proper tools handover. To reach this goal, a very promising solution relies on the usage of egocentric vision, in which the human activity is recorded by wearable cameras placed on the head of the user [50]. In contrast with standard third person Computer Vision (CV) tasks, this setting comes with the benefit that source data are characterised by a rich multi-modal information, thanks to the proximity of audio/video sensors to the action scene, and by an intrinsic embedding of an attention mechanisms that stems from the human gaze direction itself. Although egocentric vision rapidly attracted the interest of the research community [21,23,30,58,60,66], this particular setup of data collection comes with several difficulties: (i) ego-motions represents a significant source of noise for the dataset, because changes in head posture cause a shift in the point-of-view and background, introducing confusion between ego-motion and the real action of the subject; (ii) model predictions tend to be strongly correlated with the surrounding environment, which represents a bias in the dataset (usually referred to as *environmental bias*) [42], thus resulting in decreased performances when the environment changes (e.g. different kitchens); (iii) video recordings of actions can change in time, e.g. as a consequence to differences in illumination (night vs. day), habits, or changes in human skills on the longer distance.

It is important to recall that the effect of this problem is not consistent across different sensing modalities. Considering, as an example, the ego-motion, the impact on the auditory channel is extremely limited, while the visual domain is strongly affected. The optical flow, instead, is more focused on the motion in the scene, rather than the appearance, and is therefore less sensitive to environmental changes [42] (see e.g. Fig. 1). Despite being more subject to environmental bias, RGB data are richer information sources, representing in detail all the objects present in the scene. As such, they play a crucial role for understanding the affordances of the scene [24]. Lastly, the domain shift of the audio signal is further distinct from the visual one (e.g., the sound of 'cut' will differ from a plastic to a wooden cutting board). These observations suggest that domain shifts are not all of the same nature, and their impact can vary significantly across modalities. As a consequence, it is of crucial relevance to develop classifiers able to assess - depending on the conditions - which modality is more informative, and therefore modulate on the flight the weights that combine different sensing inputs for the output definition. This approach demonstrated how increasing the network's multi-modal learning capability promotes model resilience across domains, allowing the model to better recognize action under diverse domain shifts.

As a step in this direction, in our previous work [46], we proposed a multi-modal framework, called Relative Norm Alignment network (RNA-Net), which aims at progressively aligning the feature norms of audio and visual (RGB) modalities among multiple sources in a Domain Generalization (DG) setting, where target data are not

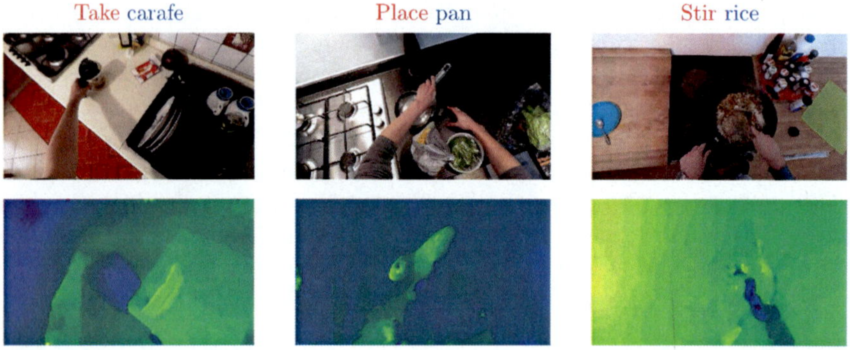

Fig. 1. Three examples of RGB frames (top line) and colorized optical flow (bottom line) with the corresponding verb and noun labels. The three frames come from different kitchens in the Epic-Kitchens-100 dataset.

available during training. Interestingly, our results demonstrated that a mere feeding of all the source domains to the network without applying any adaptive techniques leads to sub-optimal performance, while a multi-source domain alignment allows the network to promote domain-agnostic features [46]. However, it must be noted that the RNA-Net assumes a simplified (and unrealistic) scenario, in which only one single domain shift (environmental bias, i.e. different kitchens) and few different actions (8 labels) are considered. With this work, we extend and generalise the field of application of RNA-Net to a more realistic and challenging scenario, where i) most of the possible domain shifts are considered and ii) we considerably increase the number of actions and objects available for the classifier (up to 97 actions and 300 different objects).

To reach this goal, in this work we developed a method to tackle the most realistic setting possible, adopting the Epic-Kitchens-100 dataset and facing the corresponding UDA challenge. Indeed, both the presence of different environments and the fact that the source and target domains are captured in different temporal moments make it a multi-shift problem ideal for our experiments. Our basis idea is that, to increase the consistency of the predictions and close the source-to-target accuracy gap, it is necessary to tackle concurrently both the temporal shift and the environmental bias. Such a problem is an extension of a simple UDA setting and may be referred to as Multi-Source Multi-Target Domain Adaptation. The term "Multi" refers to the various environments found in both the source and target datasets.

As anticipated before, to face this new setting we exploited the capability of the RNA-Net method of close the gap between feature norms of different modalities. Our method was complemented with several domain adaptation techniques which tackle other aspects of the domain shift. In particular, we integrated several branches at distinct levels of abstraction, namely frame-level and video-level, into our framework to use the adversarial alignment technique for feature adaptation as in [9]. All of the techniques presented attempt to use the target data by acting simply on the model's features; differently, we introduce in this framework a set of losses aimed at minimising the classifier's uncertainty on the target data, thereby boosting the model's adaptability. Lastly, to deal with the significant challenges posed by this dataset, we adopted different models as an ensemble to obtain the final prediction and we introduce also a set of ensemble UDA losses.

To summarize, this paper advances the state of the art with the development of a strong UDA pipeline that we tested on a competitive international challenge (Epic-Kitchens-100).[1] Our contributions are (see also Fig. 2):

1. RNA-Net was extended to the Flow modality, obtaining remarkable results without accessing target data;
2. with further modifications, RNA-Net was adapted to work with unlabelled target data under the standard Unsupervised Domain Adaptation (UDA) setting;
3. the challenge's setting was revisited by identifying a new concurrent shift denominated "environmental bias". Our framework was modified accordingly to perform Multi-Source Multi-Target Domain Adaptation;
4. the final results were obtained by combining different model streams by means of DA-based losses, namely Min-Entropy Consistency (MEC) and Complement Entropy (CENT).

2 Related Works

First Person Action Recognition So far, most of the research effort has been focused on data provided by a specific view of the camera (often fixed), i.e., third person view [5,54,64]. With the recent release of a large-scale dataset of first-person actions [14], the community started to investigate the potential and the challenges of videos recorded from an egocentric viewpoint. As we anticipated in the Introduction, first person action recognition suffers from sudden changes of view caused by the motion of the camera. To tackle this problem, the main approaches proposed so far are based on multi-stream architectures [5,6,30,37,40,41,54], many of which are inherited from the third-person action recognition literature. The networks used to extract spatial-temporal information from egocentric videos can be divided into two main groups. The first exploits Long Short-Term Memory and variants [21,58–60] to generate an embedding representation based on the temporal relations between the features frames. The second [29,55,62,66] leverages 3D convolutional kernels, which jointly generate spatial-temporal features by sliding along the spatial and temporal dimensions. Recent works also exploit an attention mechanism at frame or clip level [39,40,44,58,60] to re-weight the spatial or temporal features, obtaining interesting results. By observing the importance of multi-stream approaches in this context, several papers investigate alternative methods to fuse streams w.r.t. the standard late fusion approach, creating a more compact multi-modal representation [57,65,66,69]. The most popular technique in this context is the multi-modal approach [5,21,42,58,64], especially in EPIC-Kitchens competitions [14,15]. Indeed, RGB data is frequently combined with motion data, such as optical flow and audio information [31].

Unsupervised Domain Adaptation (UDA). The goal of UDA is to bridge the domain gap between a labeled source domain and an unlabeled target one, which often

[1] https://epic-kitchens.github.io/2022.

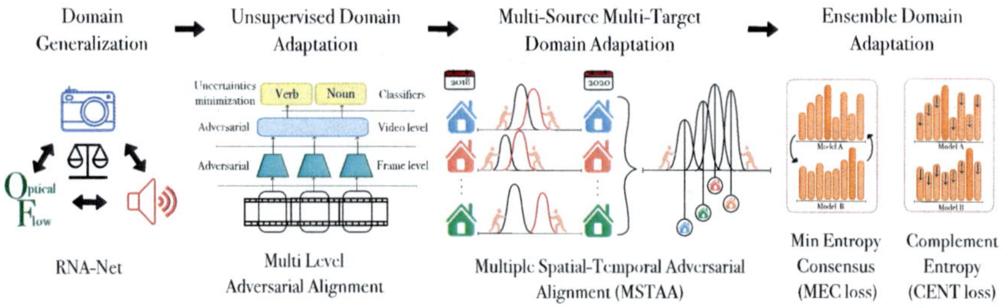

Fig. 2. An overview of the proposed approach. It can be summarized in four main aspects: **1.** Domain Generalization through RNA-Net [46], **2.** Unsupervised Domain Adaptation via Multi-Level Adversarial Alignment and entropy minimization, **3.** Multi-Source Multi-Target Domain Adaptation extension and **4.** Ensemble Domain Adaptation losses.

are drawn from different data distributions. We can divide unsupervised domain adaptation approaches into *discrepancy-based* methods, which explicitly minimise a distance metric between source and target distributions [38,53,67], and *adversarial-based* methods [19,61], which target the same goal using a gradient reversal layer (GRL) [22]. Other works, instead, exploit batch normalisation layers to normalise source and target statistics [7,35,36]. The approaches described above have been designed for standard image classification tasks. Considering methods developed specifically for video tasks, instead, it is worth mentioning UDA for action detection [1], segmentation [10] and classification [9,12,28,42,43,56]. To align the temporal dynamics of feature space for videos, several works on video domain adaptation use an adversarial learning framework like DAAA [28], also at multi-level such as in TA3N [9], in conjunction with an attention mechanism. TCoN [43] exploits a cross-domain co-attention mechanism to match the feature distributions between source and target domains, for temporal alignment. Other methods, instead, exploit jointly with the adversarial approach also auxiliary self-supervised tasks. As an example, in Munro et al. [42] the authors propose a synchronisation task to learn the multi-modal correspondence of RGB and optical flow. SAVA [13], instead, proposes a self-supervised predictive method for video domain adaptation, which aims to predict the clip order. Instead in [32,52,56], the authors propose a self-supervised contrastive learning approach for video domain adaptation.

Domain Generalization (DG). The DG scenario is a particular setting in which no target data is available at all, and the model is expected to learn to generalise using inputs from a single or multiple source domains, as it may happen in realistic scenarios. Previous approaches in DG are mostly designed for image data [3,4,20,33,34,63] and are divided in *feature-based* and *data-based* methods. The former focus on extracting invariant information which are shared across-domains [33,34], while the latter exploit data-augmentation strategies to augment source data with adversarial samples to get closer to the target [63]. Interestingly, using a self-supervised pretext task is an efficient solution for the extraction of a more robust data representation [3,4]. Recently, in [46] we proposed a feature-level solution for Domain Generalization problem in first person action recognition by leveraging audio-visual correlations.

3 Problem Definition

Epic-Kitchens Action Recognition Challenge, is based on a dataset of video consisting of trimmed actions, where the start and end times of each action are given. The objective of the challenge is to understand the activity executed in each sample, uniquely defined by the couple 'verb' (i.e. the actual action) and the main interacting object ('noun'). In the official dashboard[2] the authors report the performance linked to all those 3 categories separately. This is done because objects and actions represent the two main pillars that describe the high-level activity implemented, but since they are encoded differently in the input data (mainly RGB for objecs, motion for actions), in some cases the model could be more accurate in classifying only one of the two. The overall activity classification is assumed to be correct only if both the object and the action are classified properly.

Given $ks \geq 1$ source domains $\{\mathcal{S}_1, \ldots, \mathcal{S}_{ks}\}$, where each $\mathcal{S} = \{(x_i^s, y_i^s)\}_{i=1}^{N_s}$ is composed of N_s source samples with label space Y^s known, our goal is to learn a model representation able to perform well on $kt \geq 1$ target domains $\{\mathcal{T}_1, \ldots, \mathcal{T}_{kt}\}$, where each $\mathcal{T} = \{x_i^t\}_{i=1}^{N_t}$ of N_t target samples whose label space Y^t is unknown. If we consider $\mathcal{D}_{s,i}, \mathcal{D}_{s,j}$ distributions of the i-th and the j-th source domain and $\mathcal{D}_{t,w}, \mathcal{D}_{t,z}$ distributions of the w-th and the z-th target domain. Our two main assumptions are that the distributions of all the domains are different, i.e., $\mathcal{D}_{s,i} \neq \mathcal{D}_{t,w} \wedge \mathcal{D}_{s,i} \neq \mathcal{D}_{s,j} \wedge \mathcal{D}_{t,w} \neq \mathcal{D}_{t,z}$, with $i \neq j$ and with $w \neq z$, $i, j = 1, \ldots, ks$ and $w, z = 1, \ldots, kt$, and that the union of all label spaces is shared, $\mathcal{C}_s = \mathcal{C}_{s,1} \cup \cdots \cup \mathcal{C}_{s,ks} = \mathcal{C}_{t,1} \cup \cdots \cup \mathcal{C}_{t,kt} = \mathcal{C}_t$. In this work we consider three different scenarios:

Domain Generalization (DG), where at training time the model can access one or more fully labeled source datasets $\mathcal{S}_1, \ldots, \mathcal{S}_{ks}$, but no information is available about the target domains $\mathcal{T}_1, \ldots, \mathcal{T}_{kt}$. The objective is to train a model able to predict an action of the target domain without having access to target data at training time, thus exploiting the knowledge from multiple source domains to improve generalization. The literature refers to this setting considering the number of target domains kt equal to 1.

Unsupervised Domain Adaptation (UDA), where at training time it is possible to access a set of unlabeled target samples belonging to the target domains $\mathcal{T}_1, \ldots, \mathcal{T}_{kt}$, jointly with one fully labeled source domain $\mathcal{S}_1, \ldots, \mathcal{S}_{ks}$. Usually the literature refers to this setting considering both the number of target domains kt and the number of source domains ks equal to 1.

Multi-Sources Multi-Target Unsupervised Domain Adaptation, an extension of the previous setting, with the only difference that ks and kt are larger than one.

4 Our Approach

In this section, we first describe the DG approach used. Then, we show our UDA framework and its extension for Multi-Source Multi-Target Domain Adaptation. Finally, we demonstrate how to re-define existing DA-based losses to induce consistency between different architectures.

[2] https://codalab.lisn.upsaclay.fr/competitions/1241#results.

4.1 Domain Generalization

The multi-source nature of the proposed challenge setting makes it perfect to deal with the domain shift using DG techniques. Thus, we first exploited a method which has been recently proposed to operate in this context, called Relative Norm Alignment (RNA) [46]. This method consists of an *audio-visual domain alignment* at feature-level through the minimization of a cross-modal loss function ($\mathcal{L}_{RNA}$). The latter aims at minimizing the *mean-feature-norm distance* between the audio and visual features norms among all the source domains, and it is defined as

$$\mathcal{L}_{RNA} = \left(\frac{\mathbb{E}[h(X^v)]}{\mathbb{E}[h(X^a)]} - 1 \right)^2, \tag{1}$$

where $h(x_i^m) = (\|\cdot\|_2 \circ f^m)(x_i^m)$ indicates the L_2-norm of the features f^m of the m-th modality, $\mathbb{E}[h(X^m)] = \frac{1}{N}\sum_{x_i^m \in \mathcal{X}^m} h(x_i^m)$ for the m-th modality and N denotes the number of samples of the set $\mathcal{X}^m = \{x_1^m, \ldots, x_N^m\}$.

Authors of [46] proved that the norm unbalance between different modalities might cause the model to be biased towards the source domain that generate features with greater norm, thus causing wrong predictions. Contrarily, by simultaneously solving the problem of classification and relative norm alignment on different domains, the network extracts a shared knowledge between the different sources, resulting in a domain-agnostic model.

In this work, we extended the RNA-Net framework to the optical flow modality, in order to exploit the multiple sources available from the official training splits while showing the effectiveness of RNA loss in a multi-source DG setting.

4.2 Domain Adaptation

The UDA techniques embedded into our pipeline can be divided in two main groups: *feature*-level and *classifier*-level. The first aims at aligning the distribution of source and target, and works at different levels of representation (frames- and video-level); the latter, instead, reduces the classifier's uncertainty on target data.

Multi-Level Adversarial Alignment. Following popular unsupervised domain adaption techniques for videos, we integrate into our framework an adversarial approach [9,42], consisting of an extension of the DANN [22] typical UDA image-based method. We apply it at two different feature levels; frame- and video-level. It entails the introduction of two separate branches in our framework. Down-stream of said branches there are discriminators that try to distinguish the two domains (source and target). Contrarily, by maximising the corresponding discriminator losses, the network learns feature representations invariant to both domains.

Attentive Entropy. In order to reduce the uncertainty of the classifier on the target data, we minimize the attentive entropy loss proposed in [9] as in [48]. This action minimizes the entropy, resulting in a refinement of the classifier adaptation. The term "attentive" refers to a loss re-weighting approach that prioritizes videos with low domain discrepancy by focusing on minimizing entropy for these videos.

4.3 Multi-source Multi-target Domain Adaptation

The previous Epic Kitchen challenges [16, 17], as well as the literature on unsupervised domain adaptation for first person action recognition [42,45–47,49], reveal a strong dependency of the models on the environment where the actions are recorded. This problem, known as "environmental bias", causes a decrease in performance in occurrence of environment switches. As regards past action recognition challenges, we see this behavior by comparing performances of the models when tested on S1 (seen) and S2 (unseen). In the setting proposed in [42], similar behavior is observed, demonstrating the model's low generalization ability when tested on different kitchens.

The above considerations allow us to identify a secondary shift in this challenge, that occurs along with the temporal shift. Indeed, the training data are collected from different environments i.e. kitchens, thus introducing an environmental shift. As a result, we may rename the challenge setting *Multi-Source Multi-Target Unsupervised Domain Adaptation*. To deal with this new setting, we propose a novel framework - which we call Multiple Spatio-Temporal Adversarial Alignment (MSTAA) - combining Multiple Temporal Adversarial Alignment (MTAA) and Multiple Spatial Adversarial Alignment (MSAA). MTAA is obtained by adopting 2K domain adversarial branches (where K indicates the number of kitchens), aligning the source and the target distribution both at video- and frame-level for each kitchen. Instead, MSAA consists in adding another adversarial branch with a k-dimension discriminator in order to align the distribution of different kitchens and alleviate the environmental bias issue.

4.4 Ensemble UDA Losses

For our final testing, different models have been used in order to fully exploit the potentiality of popular video architectures. However, training individually each backbone with standard UDA protocols would result in independently adapted feature representations, which consequently vary between different streams. Our intuition is that this aspect could impact negatively the training process and the performance on target data. Indeed, since the domain adaption process acts on each architecture independently, naively training the backbones separately would yield mismatching prediction logits on target data, which, when combined, could increase the level of uncertainty of the model. For this reason, we use the Min Entropy Consensus (MEC) loss, to impose a consistency constraint between feature representations from various models. Then, repurposing the existing Complement Entropy (CENT) loss, we attempt to exploit the target data samples based on the assumption that there are some conditions in which it is easier to answer the question *"Which classes does this action not belong to?"* rather than *"Which class does this action belong to?"*.

Min Entropy Consensus (MEC loss). We extended the loss proposed in [51] to encourage coherent predictions between different models. The resulting loss is defined as:

$$\mathcal{L}_{MEC} = -\frac{1}{m}\sum_{i=1}^{m}\frac{1}{b}\max_{y\in\mathcal{Y}}\sum_{b}\log p_b(y|x_i^t) \qquad (2)$$

where m is the cardinality of the batch size of the target set, y is the predicted class, and $\log p_b(y|x_i^t)$ is the prediction probability of the b-th backbone network. The intuitive idea behind the proposed approach is to encourage different backbones to have a similar predictions.

Complement Entropy (CENT). The Complement Entropy (CENT) loss aims at neutralizing the negative effects on the final prediction of clips whose logits present high degrees of uncertainty. It accomplishes this by "flattening" the predicted probabilities of "complement classes", i.e., all classes except the predicted one. As a result, when predictions are ensembled, the noise due to uncertainty on complement classes is reduced. We refer to this loss as "complement entropy" objective, as it consists in maximizing the entropy for low-confident classes rather than minimizing it for the most confident one, as standard entropy minimization does. It is defined as:

$$\begin{aligned}\mathcal{L}_{CENT} &= \frac{1}{N}\sum_{i=1}^{N}\mathcal{H}(\hat{y}_{i\bar{c}}) \\ &= -\frac{1}{N}\sum_{i=1}^{N}\sum_{j=1,j\neq p}^{C}\left(\frac{\hat{y}_{ij}}{1-\hat{y}_{ip}}\log\frac{\hat{y}_{ij}}{1-\hat{y}_{ip}}\right)\end{aligned} \tag{3}$$

where N is the total number of samples in the batch, $\hat{y}_{ip} = max_j(\hat{y}_{ij})$ represents the predicted probability of the class p with the higher score for the i-th sample, and $\mathcal{H}(\cdot)$ is the entropy function computed on the prediction of complement classes $\hat{y}_{i\bar{c}}$ ($\bar{c} \neq p$). The formulation is similar to the one in [8], and we extend it to operate in an unsupervised fashion.

5 Framework

In this section, we describe the architectures of the feature extractors used to produce suitable multi-modal video embeddings, and the fusion stategies adopted to combine them. Finally, we deepen the analysis describing the hyper-parameters used.

Backbone. For our submission, we adopted three different network configurations. In the first one, corresponding to the RNA-Net framework in [46], we used the Inflated 3D ConvNet (I3D), pre-trained on Kinetics [5], for RGB and Flow streams, and a BN-Inception model [27] pre-trained on ImageNet [18] for the auditory information. Each feature extractor produces a 1024-dimensional representation which is fed to an action classifier. In the second configuration, we used BN-Inception models for all the three streams, using pre-extracted features from a Temporal-Binding-Network (TBN) [42] model trained on EPIC-Kitchens-55. In the last configurations, we used standard ResNet-50 architectures [25] equipped with the Temporal Shift Module (TSM) [37] pre-trained on EPIC-Kitchens-55.[3]

Multi-modal fusion strategies. In all the above mentioned configurations, each modality is processed by its own backbone, and the corresponding extracted representations are then fused following different strategies. For RNA-Net, we followed a standard late fusion strategy, consisting in averaging the final score predictions obtained from two different fully-connected layers (verb, noun) from each modality. In the other configurations, we adopted the recent mid-fusion strategy, called Semantic Mutual Refinement

[3] https://github.com/epic-kitchens/epic-kitchens-55-action-models.

sub-module (SMR), proposed in [68], to generate a common frame-embedding among the modalities. Then, using temporal pooling, we obtain a final video-embedding that is sent to the verb and noun classifiers.

6 Experimental Setting

Dataset. EPIC-KITCHENS-100 [15] is the dataset utilized in our experiments; it was recorded by 16 individuals from diverse nations (in North America and Europe). The dataset is divided into two major groups: *Source* and *Target* data, which include 16.115 and 32.024 samples, respectively. The source data consists of labelled videos recorded in 2018, whereas the target data consists of unlabeled videos collected in 2020. Both categories are further subdivided into train, validation, and test sets. The dataset includes a total of 3369 possible actions, each of which corresponds to a combination of "verb" and "noun". The total classes for verb and noun are respectively 97 and 300.
Input. For RNA-Net, we use 16 continuous frames (segment) randomly sampled for each modality during training, while at test time 5 equidistant segments spanning across all clips are fed to the network. At training time, we apply random crops, scale jitters and horizontal flips for data augmentation, while at test time only center crops are applied. Regarding aural information, we follow [30] and convert the audio track into a 256 $\times$ 256 matrix representing the log-spectrogram of the signal. The audio clip is first extracted from the video, sampled at 24kHz and then the Short-Time Fourier Transform (STFT) is calculated of a window length of 10ms, hop size of 5ms and 256 frequency bands. Hence, the x and y axis represent time and frequency, respectively. As regards the other two architectures, TSM and TBN, we use respectively 8 and 25 frames uniformly samples along all the videos.
Implementation Details. We trained I3D and BN-Inception models with SGD optimizer, with an initial learning rate of 0.001, dropout 0.7, and using a batch size of 128, following [46]. Instead, when using pre-extracted features from ResNet50 or BNInception, we trained the SMR modules on top of them for 45 epochs with an initial learning rate of 0.03, decayed after epochs 25 and 35 by a factor of 0.1. We used a batch size of 128 with SGD optimizer. We weighted RNA, CENT and MEC losses $\lambda_{RNA} = 1$, $\lambda_{CENT} = 0.31$ and $\lambda_{MEC} = 0.22$ respectively. In addition, we report the values used to weight the attentive entropy loss, $\gamma = 0.003$, and the domain losses at different levels for MSTAA, $\beta = (0.75, 0.75, 0.75)$.

7 Results

In Table 1 we report our best performing model on the target test, achieving the **2nd** position on 'verb', and the **3rd** on 'noun' and 'action'. The **3rd** place on 'action' obtained in the challenge is a demonstration of the robustness of the pipeline developed. Indeed, unlike the first two positions [11], our approach focuses on developing an unsupervised domain adaptation strategy that is independent of the backbone used. Indeed, the disparity between our results and the other two approaches is justified by the fact that they either used large models for action recognition (**2nd** place) or introduced hand detection as a secondary branch (**1st** place [11]). Our results, instead, demonstrate that our UDA pipeline is competitive even without adopting state-of-the-art models or auxiliary tasks (such as hand detection). Adding the techniques reported above is complementary to our work and is another step toward solving the multi-source multi-target domain adaptation realistic setting. Additionally, in Table 2 (left and right) we show an ablation of the proposed UDA and DG methods described in Sect. 4.

Table 1. Leaderboard results of EPIC-Kitchens Unsupervised Domain Adaptation Challenge. The results obtained by the top-3 participants and the provided baseline methods are reported. **Bold:** highest result Underline: second highest result; Green: our final submission.

UNSUPERVISED DOMAIN ADAPTATION LEADERBOARD							
	Rank	Verb Top-1	Noun Top-1	Action Top-1	Verb Top-5	Noun Top-5	Action Top-5
VI-I2R	1	**57.89**	40.07	**30.12**	**83.48**	64.19	**48.10**
Audio-Adaptive-CVPR2022	2	52.95	**42.26**	28.06	80.03	**67.51**	44.03
plnet	3	55.51	35.86	25.25	82.77	60.65	40.09
CVPR2021-chengyi	4	53.16	34.86	25.00	80.74	59.30	40.75
CVPR2021-M3EM	5	53.29	35.64	24.76	81.64	59.89	40.73
CVPR2021-plnet	6	55.22	34.83	24.71	81.93	60.48	41.41
EPIC_TA3N [15]	8	46.91	27.69	18.95	72.70	50.72	30.53
EPIC_TA3N_SOURCE_ONLY [15]	9	44.39	25.30	16.79	69.69	48.40	29.06

Table 2. Left. Results on the EPIC-Kitchen validation set with different ensembling UDA losses. **Right.** Results on EPIC-Kitchen test set under the DG setting. **Bold** highest result.

UNSUPERVISED DOMAIN ADAPTATION			
	Verb	Noun	Action
Ensemble (E) *Source Only*	53.64	32.65	22.98
E-UDA	53.88	33.10	23.22
E+MEC	53.67	34.32	23.91
E+MEC+CENT	54.20	33.92	23.99
E-SMR+MEC+CENT	**54.55**	34.72	24.22
E-SMR+MEC+CENT+MTAA	54.09	33.72	23.77
E-SMR+MEC+CENT+MSTAA	54.01	**34.82**	**24.24**

DOMAIN GENERALIZATION			
	Target	Verb Top-1	Verb Top-5
Source Only	✗	44.39	69.69
EPIC_TA3N [15]	✓	46.91	72.70
RNA-Net [46]	✗	47.96	79.54
EPIC_TA3N+RNA-Net	✓	**50.40**	**80.47**

How well do DG approaches perform? The results in Table 2 (right) are obtained under the multi-source DG setting, when target data are not available during training. Noticeably, RNA outperforms the baseline Source Only by up to 3% on Top-1 and 10% on Top-5, highlighting the importance of using ad-hoc alignment techniques to deal with multiple sources in order to effectively extract a domain-agnostic model. Moreover, it outperforms the recent UDA technique TA^3N [9] without accessing target data. Interestingly, when combined with EPIC_TA3N, it further improves performance, proving the complementarity of RNA to other existing UDA approaches.

In Table 2(left) it can be seen how the proposed UDA approaches improve Top-1 accuracy on all categories by up to 1%. Although using an additional adversarial branch for each kitchen does not appear to provide a significant improvement on the validation set, it increases the top-1 action accuracy on the test set, allowing us to obtain the third position in the challenge. Without MSTAA, the accuracy on the action top-1 reaches just 24.83%. This outcome was predictable given that the validation set is populated with a different set of kitchens than the test set, whereas the kitchens in the test set are the same as those used for the target and source training. This aspect confirms the *Multi-Source Multi-Target Unsupervised Domain Adaptation* setting and the presence of two different shifts, the *temporal* shift (2018–2020) and the *environmental* shift (among the kitchens).

8 Conclusions

In this paper, we introduced and discussed the potentiality of egocentric vision for human-robot cooperation. Indeed, this source of information may come with the interesting benefit of an intrinsic attention mechanism associated to the user head posture and gaze direction. The wearability of the sensing setup makes this technology particularly suitable to be deployed in unstructured scenarios, where robots and humans are co-existing in a daily-life environment. While from one side this is a key enabling factor for the actual deployment of human-robot cooperating frameworks, it also brings several challenges, such as the severe noise superimposition caused by ego-motion, and the strong domain shifts associated to the particular unstructured tasks (i.e. changing environments, subjects, habits). For this reason, to really unlock the potential of egocentric vision, it is important to develop models able to generalize efficiently across domains. In this paper, we presented an unsupervised approach for Multi-Source Multi-Target Domain Adaptation for egocentric action recognition as a strong solution to the problem of temporal and spatial biases. Moving from these promising results, we plan to integrate our framework in a more general human-robot cooperation framework, in which the manipulator will be able to identify the action performed by the human, and eventually plan a consequent action to support human tasks.

Acknowledgements. This work was supported both by the CINI Consortium through the VIDESEC project and by the Italian Ministry of University and Research under the DM1061. The research herein was carried out using the IIT HPC infrastructure.

References

1. Agarwal, N., Chen, Y.T., Dariush, B., Yang, M.H.: Unsupervised domain adaptation for spatio-temporal action localization. arXiv preprint arXiv:2010.09211 (2020)
2. Ajoudani, A., Zanchettin, A.M., Ivaldi, S., Albu-Schäffer, A., Kosuge, K., Khatib, O.: Progress and prospects of the human-robot collaboration. Autonom. Rob. **42**(5), 957–975 (2018)
3. Bucci, S., D'Innocente, A., Liao, Y., Carlucci, F.M., Caputo, B., Tommasi, T.: Self-supervised learning across domains (2020)

4. Carlucci, F.M., D'Innocente, A., Bucci, S., Caputo, B., Tommasi, T.: Domain generalization by solving jigsaw puzzles. In: CVPR, pp. 2229–2238 (2019)
5. Carreira, J., Zisserman, A.: Quo vadis, action recognition? a new model and the kinetics dataset. In: CVPR, pp. 6299–6308 (2017)
6. Cartas, A., Luque, J., Radeva, P., Segura, C., Dimiccoli, M.: Seeing and hearing egocentric actions: how much can we learn? In: ICCV Workshops (2019)
7. Chang, W.G., You, T., Seo, S., Kwak, S., Han, B.: Domain-specific batch normalization for unsupervised domain adaptation. In: CVPR, pp. 7354–7362 (2019)
8. Chen, H.Y., Wang, P.H., Liu, C.H., Chang, S.C., Pan, J.Y., Chen, Y.T., Wei, W., Juan, D.C.: Complement objective training. arXiv preprint arXiv:1903.01182 (2019)
9. Chen, M.H., Kira, Z., AlRegib, G., Yoo, J., Chen, R., Zheng, J.: Temporal attentive alignment for large-scale video domain adaptation. In: ICCV, pp. 6321–6330 (2019)
10. Chen, M.H., Li, B., Bao, Y., AlRegib, G., Kira, Z.: Action segmentation with joint self-supervised temporal domain adaptation. In: CVPR, pp. 9454–9463 (2020)
11. Cheng, Y., Fang, F., Sun, Y.: Team vi-i2r technical report on epic-kitchens-100 unsupervised domain adaptation challenge for action recognition 2021. arXiv preprint arXiv:2206.02573 (2022)
12. Choi, J., Sharma, G., Chandraker, M., Huang, J.B.: Unsupervised and semi-supervised domain adaptation for action recognition from drones. In: WACV, pp. 1717–1726 (2020)
13. Choi, J., Sharma, G., Schulter, S., Huang, J.B.: Shuffle and attend: Video domain adaptation. In: ECCV, pp. 678–695. Springer (2020)
14. Damen, D., Doughty, H., Farinella, G.M., Fidler, S., Furnari, A., Kazakos, E., Moltisanti, D., Munro, J., Perrett, T., Price, W., Wray, M.: Scaling egocentric vision: the epic-kitchens dataset (2018)
15. Damen, D., Doughty, H., Farinella, G.M., Furnari, A., Kazakos, E., Ma, J., Moltisanti, D., Munro, J., Perrett, T., Price, W., et al.: Rescaling egocentric vision. arXiv preprint arXiv:2006.13256 (2020)
16. Damen, D., Kazakos, E., Price, W., Ma, J., Doughty, H.: Epic-kitchens-55—2020 challenges report (2020)
17. Damen, D., Price, W., Kazakos, E., Furnari, A., Farinella, G.M.: Epic-kitchens—2019 challenges report (2019)
18. Deng, J., Dong, W., Socher, R., Li, L., Kai Li, Li Fei-Fei: Imagenet: a large-scale hierarchical image database. In: CVPR, pp. 248–255 (2009). https://doi.org/10.1109/CVPR.2009.5206848
19. Deng, Z., Luo, Y., Zhu, J.: Cluster alignment with a teacher for unsupervised domain adaptation. In: ICCV, pp. 9944–9953 (2019)
20. Dou, Q., Coelho de Castro, D., Kamnitsas, K., Glocker, B.: Domain generalization via model-agnostic learning of semantic features. NeurIPS **32**, 6450–6461 (2019)
21. Furnari, A., Farinella, G.: Rolling-unrolling LSTMS for action anticipation from first-person video. T-PAMI (2020)
22. Ganin, Y., Lempitsky, V.: Unsupervised domain adaptation by backpropagation. In: ICML, pp. 1180–1189. PMLR (2015)
23. Ghadiyaram, D., Tran, D., Mahajan, D.: Large-scale weakly-supervised pre-training for video action recognition. In: CVPR, pp. 12,046–12,055 (2019)
24. Gibson, J.J.: The theory of affordances. Hilldale, USA **1**(2), 67–82 (1977)
25. He, K., Zhang, X., Ren, S., Sun, J.: Deep residual learning for image recognition. In: CVPR, pp. 770–778 (2016)
26. Henschel, A., Hortensius, R., Cross, E.S.: Social cognition in the age of human-robot interaction. Trends Neurosci **43**(6), 373–384 (2020)

27. Ioffe, S., Szegedy, C.: Batch normalization: accelerating deep network training by reducing internal covariate shift. In: F. Bach, D. Blei (eds.) ICML, Proceedings of Machine Learning Research, vol. 37, pp. 448–456. PMLR (2015). URL http://proceedings.mlr.press/v37/ioffe15.html
28. Jamal, A., Namboodiri, V.P., Deodhare, D., Venkatesh, K.: Deep domain adaptation in action space. In: BMVC, vol. 2, p. 5 (2018)
29. Kapidis, G., Poppe, R., van Dam, E., Noldus, L., Veltkamp, R.: Multitask learning to improve egocentric action recognition. In: ICCV Workshops, pp. 0–0 (2019)
30. Kazakos, E., Nagrani, A., Zisserman, A., Damen, D.: Epic-fusion: audio-visual temporal binding for egocentric action recognition. In: ICCV (2019)
31. Kazakos, E., Nagrani, A., Zisserman, A., Damen, D.: Slow-fast auditory streams for audio recognition. In: ICASSP 2021-2021 IEEE International Conference on Acoustics, Speech and Signal Processing (ICASSP), pp. 855–859. IEEE (2021)
32. Kim, D., Tsai, Y.H., Zhuang, B., Yu, X., Sclaroff, S., Saenko, K., Chandraker, M.: Learning cross-modal contrastive features for video domain adaptation. In: ICCV, pp. 13,618–13,627 (2021)
33. Li, H., Jialin Pan, S., Wang, S., Kot, A.C.: Domain generalization with adversarial feature learning. In: CVPR, pp. 5400–5409 (2018)
34. Li, Y., Tian, X., Gong, M., Liu, Y., Liu, T., Zhang, K., Tao, D.: Deep domain generalization via conditional invariant adversarial networks. In: ECCV, pp. 624–639 (2018)
35. Li, Y., Wang, N., Shi, J., Hou, X., Liu, J.: Adaptive batch normalization for practical domain adaptation. Pattern Recogn. **80**, 109–117 (2018)
36. Li, Y., Wang, N., Shi, J., Liu, J., Hou, X.: Revisiting batch normalization for practical domain adaptation. In: ICLR. OpenReview.net (2017). URL https://openreview.net/forum?id=Hk6dkJQFx
37. Lin, J., Gan, C., Han, S.: Tsm: Temporal shift module for efficient video understanding. In: ICCV, pp. 7083–7093 (2019)
38. Long, M., Cao, Y., Wang, J., Jordan, M.: Learning transferable features with deep adaptation networks. In: ICML, pp. 97–105. PMLR (2015)
39. Lu, M., Li, Z., Wang, Y., Pan, G.: Deep attention network for egocentric action recognition. IEEE Trans. Image Process. **28**(8), 3703–3713 (2019)
40. Lu, M., Liao, D., Li, Z.N.: Learning spatiotemporal attention for egocentric action recognition. In: ICCV Workshops (2019)
41. Ma, M., Fan, H., Kitani, K.M.: Going deeper into first-person activity recognition. In: CVPR, pp. 1894–1903 (2016)
42. Munro, J., Damen, D.: Multi-modal domain adaptation for fine-grained action recognition. In: CVPR, pp. 122–132 (2020)
43. Pan, B., Cao, Z., Adeli, E., Niebles, J.C.: Adversarial cross-domain action recognition with co-attention. In: AAAI, pp. 11815–11822 (2020)
44. Perez-Rua, J.M., Martinez, B., Zhu, X., Toisoul, A., Escorcia, V., Xiang, T.: Knowing what, where and when to look: efficient video action modeling with attention (2020)
45. Planamente, M., Bottino, A., Caputo, B.: Self-supervised joint encoding of motion and appearance for first person action recognition. In: 2020 25th International Conference on Pattern Recognition (ICPR), pp. 8751–8758. IEEE (2021)
46. Planamente, M., Plizzari, C., Alberti, E., Caputo, B.: Domain generalization through audio-visual relative norm alignment in first person action recognition. In: WACV, pp. 1807–1818 (2022)
47. Plananamente, M., Plizzari, C., Caputo, B.: Test-time adaptation for egocentric action recognition. In: International Conference on Image Analysis and Processing, pp. 206–218. Springer (2022)

48. Plizzari, C., Planamente, M., Alberti, E., Caputo, B.: Polito-iit submission to the epic-kitchens-100 unsupervised domain adaptation challenge for action recognition. arXiv preprint arXiv:2107.00337 (2021)
49. Plizzari, C., Planamente, M., Goletto, G., Cannici, M., Gusso, E., Matteucci, M., Caputo, B.: E^2 (go) motion: Motion augmented event stream for egocentric action recognition. arXiv preprint arXiv:2112.03596 (2021)
50. Rodin, I., Furnari, A., Mavroeidis, D., Farinella, G.M.: Predicting the future from first person (egocentric) vision: a survey. Comput. Vis. Image Understand. **211**, 103252 (2021)
51. Roy, S., Siarohin, A., Sangineto, E., Bulo, S.R., Sebe, N., Ricci, E.: Unsupervised domain adaptation using feature-whitening and consensus loss. In: CVPR, pp. 9471–9480 (2019)
52. Sahoo, A., Shah, R., Panda, R., Saenko, K., Das, A.: Contrast and mix: temporal contrastive video domain adaptation with background mixing. NeurIPS **34** (2021)
53. Saito, K., Watanabe, K., Ushiku, Y., Harada, T.: Maximum classifier discrepancy for unsupervised domain adaptation. In: CVPR, pp. 3723–3732 (2018)
54. Simonyan, K., Zisserman, A.: Two-stream convolutional networks for action recognition in videos. In: NeurIPS (NIPS'14), pp. 568–576. MIT Press, Cambridge, MA, USA (2014)
55. Singh, S., Arora, C., Jawahar, C.: First person action recognition using deep learned descriptors. In: CVPR, pp. 2620–2628 (2016)
56. Song, X., Zhao, S., Yang, J., Yue, H., Xu, P., Hu, R., Chai, H.: Spatio-temporal contrastive domain adaptation for action recognition. In: CVPR, pp. 9787–9795 (2021)
57. Sudhakaran, S., Escalera, S., Lanz, O.: Hierarchical feature aggregation networks for video action recognition. arXiv preprint arXiv:1905.12462 (2019)
58. Sudhakaran, S., Escalera, S., Lanz, O.: Lsta: Long short-term attention for egocentric action recognition. In: CVPR, pp. 9954–9963 (2019)
59. Sudhakaran, S., Lanz, O.: Convolutional long short-term memory networks for recognizing first person interactions. In: ICCV Workshops (2017)
60. Sudhakaran, S., Lanz, O.: Attention is all we need: nailing down object-centric attention for egocentric activity recognition. arXiv preprint arXiv:1807.11794 (2018)
61. Tang, H., Jia, K.: Discriminative adversarial domain adaptation. In: AAAI, pp. 5940–5947 (2020)
62. Tran, D., Bourdev, L., Fergus, R., Torresani, L., Paluri, M.: Learning spatiotemporal features with 3d convolutional networks. In: ICCV, pp. 4489–4497 (2015)
63. Volpi, R., Namkoong, H., Sener, O., Duchi, J.C., Murino, V., Savarese, S.: Generalizing to unseen domains via adversarial data augmentation. In: NeurIPS, pp. 5334–5344 (2018)
64. Wang, L., Xiong, Y., Wang, Z., Qiao, Y., Lin, D., Tang, X., Van Gool, L.: Temporal segment networks: Towards good practices for deep action recognition. In: ECCV, pp. 20–36. Springer, Berlin (2016)
65. Wang, X., Wu, Y., Zhu, L., Yang, Y., Zhuang, Y.: Symbiotic attention: Uts-baidu submission to the epic-kitchens 2020 action recognition challenge
66. Wu, C.Y., Feichtenhofer, C., Fan, H., He, K., Krahenbuhl, P., Girshick, R.: Long-term feature banks for detailed video understanding. In: CVPR, pp. 1426–1435 (2019)
67. Xu, R., Li, G., Yang, J., Lin, L.: Larger norm more transferable: an adaptive feature norm approach for unsupervised domain adaptation. In: ICCV, pp. 1426–1435 (2019)
68. Yang, L., Huang, Y., Sugano, Y., Sato, Y.: Epic-kitchens-100 unsupervised domain adaptation challenge for action recognition 2021: Team m3em technical report. arXiv preprint arXiv:2106.10026 (2021)
69. Zhou, B., Andonian, A., Oliva, A., Torralba, A.: Temporal relational reasoning in videos. In: ECCV, pp. 803–818 (2018)

Use of EEG Signals for Mental Workload Assessment in Human-Robot Collaboration

Andrea Ruo(✉), Valeria Villani, and Lorenzo Sabattini

University of Modena and Reggio Emilia, Reggio Emilia, RE 42122, Italy
{andrea.ruo,valeria.villani,lorenzo.sabattini}@unimore.it

Abstract. The present study aims to investigate the effect of human-robot collaboration on brain activity. To this end, we analyze differences in the electroencephalogram (EEG) power spectrum (Theta, Alpha and Beta frequency bands), and state-of-the-art indices, between some participants who performed different experiments, including a experiment of human-robot collaboration. In particular, tests included low cognitive load tasks, such as listening to classical music and watching a relaxing video; tasks with medium cognitive load through a collaborative robotics experiment, developed using the Niryo Ned robot, structured in three subtasks with increasing difficulty; and a high cognitive load task consisting of a sudoku game with tight time constraint. The EEG was recorded using the Neuroelectrics Enobio20 helmet. In addition to recording EEG signals, electrodermal activity (EDA) was recorded by means of a BITalino (R)Evolution Board in order to compare the results obtained from the two biosignals and evaluate the indices deriving from the spectral powers of the rhythms. Most of the workload indices, present in the literature, used for the analysis of this work have proved, on the basis of the analysis carried out, to be very good indices of the cognitive load.

Keywords: Human-robot collaboration · EEG power spectrum · EEG · EDA · Mental workload · Biosignals

1 Introduction

Mental workload (MWL) is defined as the amount of mental or cognitive resources needed to satisfy the current demands of a given task; it is also called psychological load and can be understood as the amount of brain activity in a unit of time, the occupancy rate of brain resources, the pressure or psychological information processing capacity of a person at work [1]. When perceived by a threatening stimulus, a cascade of physiological processes occurs that mobilize the body and the system to address the impending threat and ensure effective adaptation. Biosignals [2] that can be reliably measured in relation to such stressors include physiological (electroencephalogram (EEG), electrocardiogram

P. Borja et al. (Eds.): HFR 2022, 26, pp. 233–247, 2023.
https://doi.org/10.1007/978-3-031-22731-8_17

(ECG), electrodermal activity (EDA), electromyogram (EMG)) and physical (respiratory rate, speech, skin temperature, pupil size, eye activity) measurements. A key goal in psychological and physiological research is to establish reliable biosignal indices that reveal the physiological mechanisms underlying the stress response. Studies have shown that an excessive mental workload can cause quickness, reduced flexibility, increase in errors and frustrated emotions and results in errors in the analysis of information acquisition and in decision-making [3]. Additionally, a too low mental workload can cause a waste of human resources and lead to a decline in job performance. Therefore, it is very important to analyze this disciplinary area in order to define human mental load states accurately.

MWL is traditionally assessed with questionnaires such as NASA Task Load Index (NASA-TLX) [4] or the Subjective Workload Assessment (SWAT) technique [5]. Since these methods only provide a subjective assessment of an operator's workload, the current trend is to supplement these assessments with physiological measurements using devices that measure bio-signals such as EEG. To properly evaluate MWL with such devices, it is necessary to be able to recognize the workload level of the input signal, and this can be achieved with the use of various machine learning techniques. About the EEG signals, with the development of technology, some studies have shown the effects of the Delta, Theta, Alpha and Gamma bands on mental workload [14]. It is generally believed that psychological stress, active thinking and attention cause EEG activity to shift to higher frequency bands and suppress Alpha wave activity [6]. Obviously, this topic offers many extremely interesting research areas, which include the analysis of a user's cognitive load, sleep disorders, epilepsy and brain-computer interfaces (BCI). Specifically, a BCI allows a subject to communicate and control the outside world without using the brain's normal activity through peripheral nerves and muscles. Messages are conveyed by spontaneous EEG activity or evoked by muscle contractions that are otherwise used for communication through speech and writing. People with severe neuromuscular disorders, or sometimes those completely paralyzed, benefit greatly from a BCI that offers them basic communication skills through which they can express themselves, for example, by checking a spelling program or operating a neuroprosthesis [7].

An equally important role in the analyzes is played by the analysis of EDA activity [8] that represents the measure of skin conductivity and is the physiological measurement of the flow of electricity through the skin. It is used to provide information regarding changes in the sympathetic nervous system, an indicator of cognitive fatigue and emotions. EDA recordings can be obtained using a bipolar montage from the palmar sites of the hand (such as two fingers) or from the feet where the highest density of sweat glands is observed.

The contribution of this paper lies in a study that compares two types of biosignals, EEG and EDA, with the aim of measuring the workload of participants during different types of tests, including a collaborative robotics experiment using the Nyrio Ned robot. The EEG signals were recorded using the Enobio20 helmet and were processed with an automatic algorithm that allows pre-processing operations (data removal, downsampling, rereference, filter application, bad channels removal) and application of independent component analysis (ICA) algorithm for artifact removal. In particular, to facilitate the identification of ocular artifacts, two Sticktrode electrodes were used for the acquisition of electroculographic (EOG) signals. EDA activity was recorded by means of a BITalino (R)Evolution using a bipolar montage from the palmar sites of the hand. In this way it is possible to evaluate the degree of accuracy of the indices found in the literature for the assessment of mental fatigue during human-robot collaboration tasks.

2 Materials and Methods

2.1 Partecipants

A total of three male university students (age = 22 ± 1 years old) were analyzed. Participants gave written and informed consent to participate in the study. The only exclusion criterion from the following study relates to the fact that users were not to be subjected to diseases or drugs affecting the nervous system.

2.2 Experimental Procedure

Before starting the study, participants were instructed on the procedures and protocol requirements during the different trials. All participants underwent a period of familiarization with the equipment required for the test as they had no experience with robots. The experimental room was calm and the light and temperature were continuously regulated. Participants conducted four different types of experiments: (A) listening to music, (B) watching a video, (C) performing human-robot collaborative assembly, and (D) solving sudoku game. In the first test, the user had to remain relaxed with his eyes closed listening to classical music with earphones, as shown in Fig. 1. In the second test, the user had to remain relaxed while watching a video on a computer. The third test, which consists of a collaborative robotics experiment, was organized in three tasks (C.1, C.2, C.3) of increasing difficulty. During these tests, the users had to assemble different figures by using "Duplo" bricks, supplied to them in a dedicated unloading box located inside the workstation through pick & place operations. Bricks were provided to the user in sequential or random order, depending on the task, by a robot, namely Niryo Ned robot [9], and a conveyor equipped with a proximity sensor. In particular: regarding C.1, the user has to assemble one figure, the blocks of which are passed to him in order of construction and can see the photo of the figure at any time. About C.2, the user has to assemble two figures, whose blocks are passed to him in order of construction and in random order, moreover he can see the photo of the figures once only. Regarding C.3, the user must assemble three figures, whose blocks are passed in order of construction and in random order and by means of a conveyor, moreover they can constantly see a drawing with different types of figures inside. The setup for this test is shown in Fig. 2. In particular, the six constructions shown in Fig. 3 were used during these tests, for a total of 44 pieces. In the fourth test, the user had to stay focused on playing a Sudoku puzzle of medium difficulty with a time limit. Between each experimental test (A, B, C, D) there was a pause of 3 min in order to prepare the experimental setup for the following test. This does not apply to the sub-tasks C.1, C.2, C.3 which have been carried out in sequence.

Considering successive improvements due to pilot testing, the enrolled participants performed the following tests: Subject 1 did A, C, D; Subject 2 executed A, B, C; Subject 3 performed C.

Fig. 1. Experiment setup when subjects were listening to music.

During all the tests the participants were subjected to the acquisition of the EDA activity. At the end of the experiment, the participants had to fill in a questionnaire to collect their personal evaluation regarding their state of relaxation and concentration during the various tests.

2.3 Instruments

The EEG data was evaluated using the Neuroelectrics Enobio20 device [10], a wireless electrode system, using a referential montage. The device is shown in the middle panel in Fig. 4. EEG was recorded from 19 scalp locations, according to the international 10–20 system: 17 EEG electrodes were Enobio NG Geltrode using electrode gel and 2 EOG electrodes were Enobio Stricktrodes. With reference to Fig. 5, the considered electrodes included frontal (Fz, F3, F4, F7, and F8), central (Cz, C3, and C4), temporal (T3, T4, T5, and T6), parietal (Pz, P3 and P4) and occipital (O1 and O2) locations, and two EOG electrodes (EOG1 and EOG2). Data were collected through the software Neuroelectrics® Instrument Controller (NIC2) [11], which allows a computer to interact with Neuroelectrics

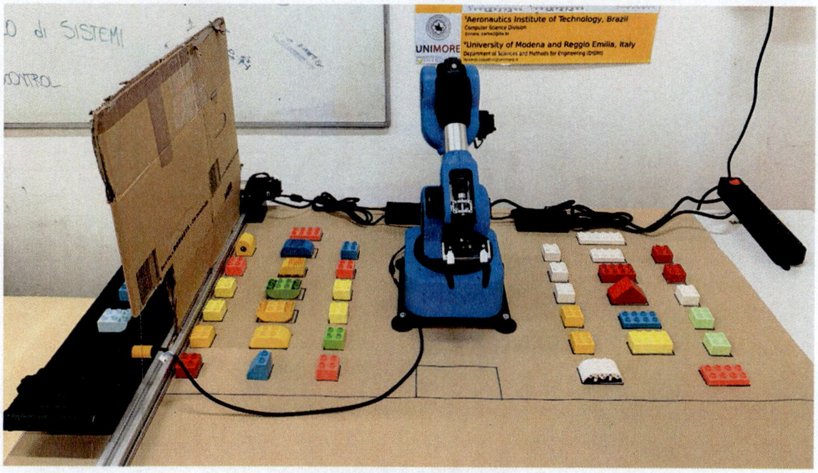

Fig. 2. Setup for the human-robot collaboration test.

Fig. 3. Figures to be assembled in the human-robot collaboration test.

devices. The electrodes placed in the mastoids served as a reference while recording on NIC2. A sampling rate of 500 Hz was used. The EEGLAB 2021.1 toolbox (MATLAB 2021.a) was used for preprocessing and data analysis. Raw data were filtered with a 50 Hz notch filter and a band pass filter with bandwidth 0.1–30 Hz. Artifacts were corrected using independent component analysis (ICA). Spectral analyses were performed for the difficulty level of each test using MATLAB. The data were grouped into Theta (4–8 Hz), Alpha (8–12 Hz), and Beta (13–29 Hz) frequency bands.

As regards EDA recordings, we used BITalino (R)Evolution device, which is shown in Fig. 6. It is a kit of components for the detection of physiological data and is composed of a series of sensors designed to collect a large number of

Fig. 4. Enobio20 device.

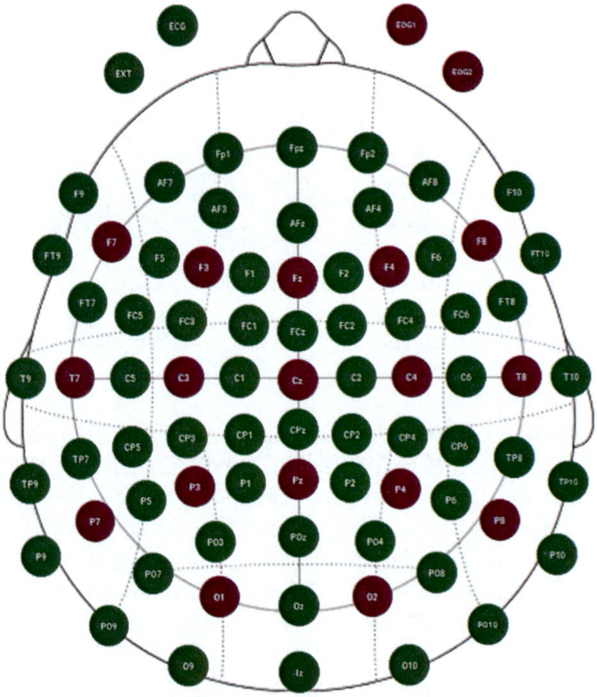

Fig. 5. Channel locations according to the international 10–20 system.

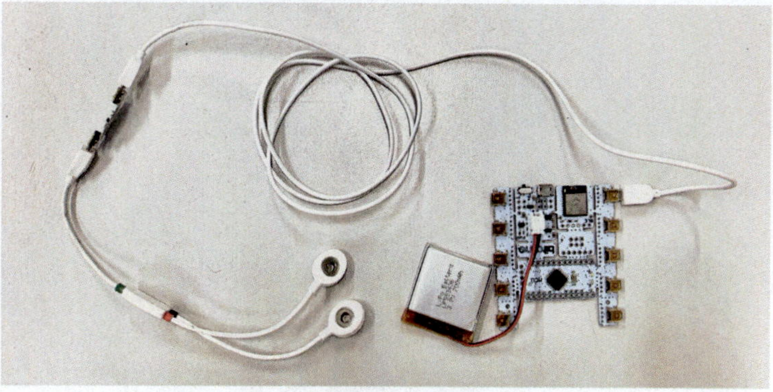

Fig. 6. BITalino (R)Evolution kit.

biological signals, from the measurement of muscle to heart activity, from small electrical signals coming from the skin to the detection of body movements. A microcontroller receives these analog signals and converts them into digital signals. EDA recordings were processed considering the approach in [13], using the software Weka [12].

Regarding the collaborative robotics experiment, the robot Niryo Ned, shown in Fig. 7, was used. It has the following characteristics: (a) Number of axis: 6; (b) Weight: 6.5 kg; (c) Payload: 300 g; (d) Max Reach: 440 mm; (e) Precision: 0.5 mm; (f) Communications: Ethernet 1 Gb/s, WIFI 2.4 GHz & 5 GHz - 802.11 g/g/n/ac, Bluetooth 5.0 BLE, USB.

2.4 Statistical Analysis

After carrying out ICA, the dataset was divided into epochs of 5 s, for a trade off between computational burden and reliability of results. In each epoch the spectral powers (μV/Hz) of the Alpha, Beta, Theta bands of the channels were calculated, considering the channels reported in Table 1. Moreover, we computed the state-of-the-art indices reported in Table 2 [14–17]. In the literature it has been found that: i) Theta variation determines fatigue and mental workload; ii) Alpha variation decreases with arithmetic tasks and increases with creative thinking; iii) Beta variation refers to visual attention and short-term memory; iv) BAT determines involvement in tasks and mental effort; v) TA refers to increased mental load; vi) TA ASS provides the same measurement of TA, with the difference that it is obtained from the ratio of two absolute measurements that are obtained from certain channels on the scalp.

3 Results

After data processing, plots of each recording were created for each subject. Figs. 8, 9 and 10 show spectral powers and synthetic indices for Subject 1 in the

Fig. 7. Niryo Ned robot.

Table 1. Spectral powers for the different frequency bands.

Rhythm power	Cortical area	Channels
Theta variation	Frontal e Temporal	F7, F3, F4, F8, Fz, T7, T8
Absolute theta variation	Frontal	Fz
Alpha variation	Parietal e Occipital	P7, P3, P4, P8, Pz, O1, O2
Absolute alpha variation	Parietal	Pz
Beta variation	Frontal	F7, F3, F4, F8, Fz

Table 2. Indices for comparison among frequency bands.

Indices	Definition
BAT	$\frac{Beta}{(Alpha+Theta)}$
TA	$\frac{Theta}{Alpha}$
TA ASS	$\frac{ThetaAss}{AlphaAss}$

different phases of the test (with reference to Subsec. 2.2, music, three collaborative tasks with the robot and Sudoku test).

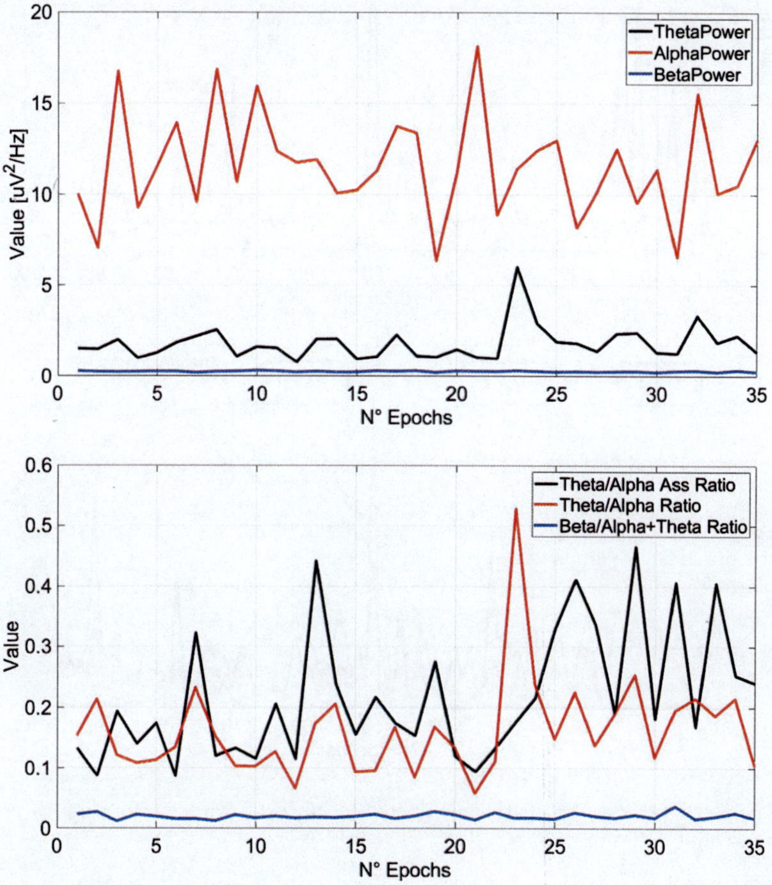

Fig. 8. Spectral powers and synthetic indices for Subject 1 while listening to relaxing music (A).

In addition, minimum, average and maximum values of each parameter of interest (see Tables 1 and 2) were computed. These are reported in Fig. 11.

Finally, as regards EDA activity, Fig. 12 plots recordings of Subject 1. It is noteworthy that an increase in the amplitude of the processed signal indicates an increase in stress and cognitive load on the user. Hence, Fig. 12 shows that Subject 1, according to the EDA activity, was more stressed during the tests at high workload (D, C.1, C.3) and less concentrated while listening to music (A).

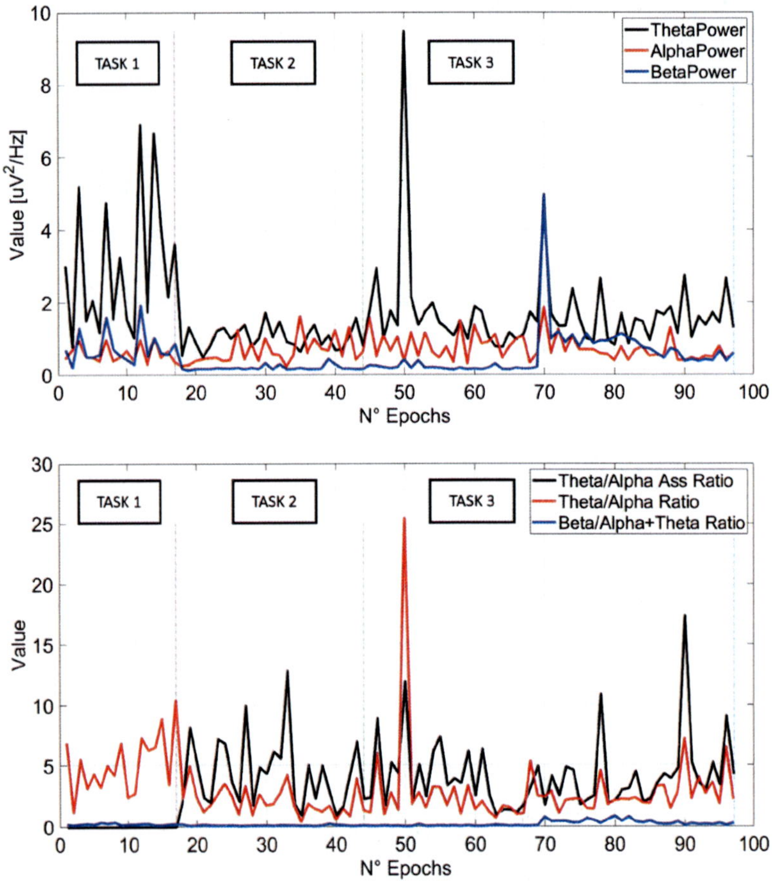

Fig. 9. Spectral powers and synthetic indices for Subject 1 while performing three tasks of human-robot collaboration (C).

4 Discussion

Comparing the results obtained from the comparison of the two biosignals with the replies to questionnaire, it was possible to carry out an analysis of the indicators present in the literature for the analysis of EEG signals. In particular, the green tag in Fig. 11 has been attributed to the very good indices and the yellow tag to the sufficiently good indices. With reference to Subsect. 2.4, it was found that: (a) MAX THETA represents a very very good index for mental fatigue assessment. It is present when there is mental fatigue and high concentration and cognitive load. (b) MAX ALPHA represents a very good index. It is present with high values when subjects had closed eyes, otherwise high values were found during creative thinking due to mental effort. (c) MAX BETA is a sufficiently good index. It is related to visual awareness and is present when there is an

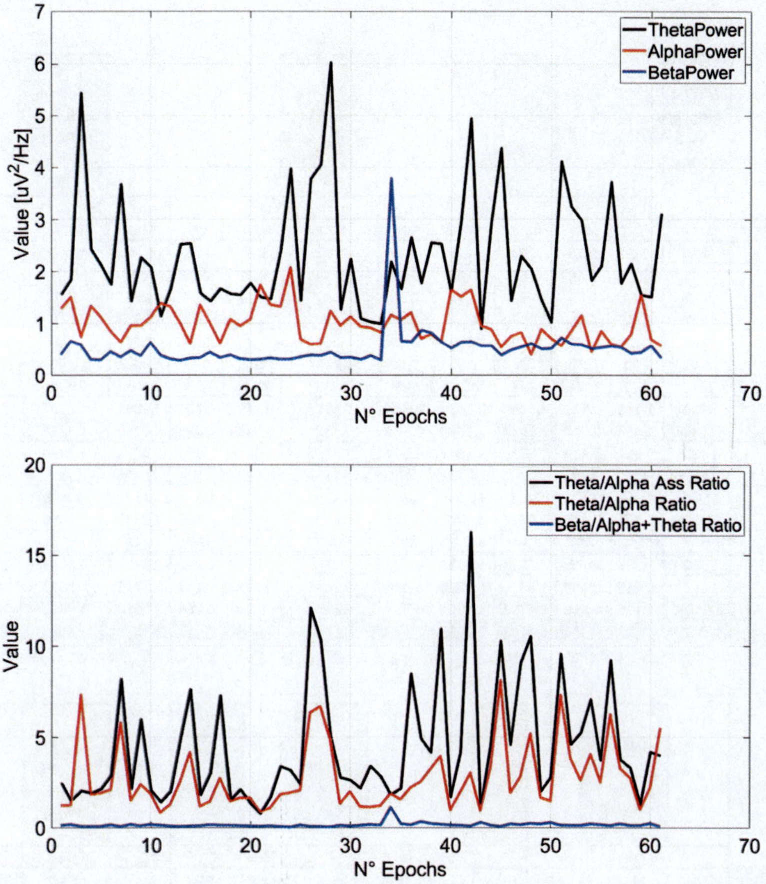

Fig. 10. Spectral powers and synthetic indices for Subject 1 while performing a Sudoku test (D).

increase in working memory. There were discrepancies in all participants with respect to the questionnaires. (d) MEAN THETA ASS is a very good index. It is related to the average concentration of the tests. (e) MEAN ALPHA ASS is a very good index. It is present with high values when subjects had closed eyes or during creative thinking due to mental effort. (f) MAX BAT is a sufficiently good index. It is related to mental effort. It has a discrepancy in Subject 2 as it should have a higher value in Task 2. (g) MAX TA is a very good index. It is related to the high cognitive load. (h) MIN TA is a very good index. It is related to low cognitive load. (i) MEAN TA ASS represents a sufficiently good index. It is related to the average stress during the tests, but did not prove reliable in our tests

On the basis of the considerations set out above, as regards Subject 1, it is possible to conclude that: (a) From the spectral power of the Theta rhythm,

	Mental fatigue and workload			More creative thinking			High concentration				Medium task difficulty (ABSOLUTE)	
	THETA			ALPHA			BETA			THETA ASS		
SUBJECT 1	MAX	MEAN	MIN	MAX	MEAN	MIN	MAX	MEAN	MIN	MAX	MEAN	MIN
TASK1	6.90	3.00	0.78	1.03	0.59	0.28	1.92	0.75	0.20	0.00	0.00	0.00
TASK2	1.71	1.03	0.49	1.61	0.65	0.22	0.44	0.19	0.12	4.06	1.94	0.67
TASK3	9.47	1.71	0.76	1.86	0.74	0.30	4.97	0.58	0.14	4.83	2.19	0.66
TASK123	9.47	1.75	0.49	1.86	0.69	0.22	4.97	0.50	0.12	4.83	1.74	0.00
MUSIC	6.01	1.81	0.77	18.15	11.56	6.33	0.33	0.26	0.19	4.66	2.09	0.80
SUDOKU	6.02	2.31	0.90	2.07	0.99	0.40	3.79	0.52	0.29	14.66	3.21	1.00
SUBJECT 2	MAX	MEAN	MIN	MAX	MEAN	MIN	MAX	MEAN	MIN	MAX	MEAN	MIN
TASK1	1.8242	1.131	0.522	1.3884	0.7634	0.386	0.9196	0.4807	0.206	2.9883	1.3327	0.587
TASK2	2.6496	1.433	0.712	2.5318	1.1048	0.623	0.7421	0.3409	0.205	3.0356	1.7228	0.793
TASK3	1.4935	0.913	0.439	2.55	1.1647	0.564	1.3896	0.3797	0.191	2.4004	1.3759	0.612
TASK123	2.6496	1.1	0.439	2.55	1.0737	0.386	1.3896	0.3874	0.191	3.0356	1.466	0.587
MUSIC	1.9564	1.096	0.723	28.7	11.595	0.263	1.004	0.5157	0.171	3.2526	1.892	1.059
VIDEO	1.4855	0.871	0.592	2.0472	0.962	0.422	0.6258	0.3213	0.186	2.8322	1.3636	0.826
SUBJECT 3	MAX	MEAN	MIN	MAX	MEAN	MIN	MAX	MEAN	MIN	MAX	MEAN	MIN
TASK1	3.2023	1.348	0.71	3.3934	1.0153	0.592	0.704	0.3957	0.296	2.4398	1.3092	0.556
TASK2	0.9663	0.665	0.439	1.8075	0.6709	0.427	0.2845	0.1792	0.113	1.5418	0.8522	0.406
TASK3	12.1753	1.176	0.596	6.6732	1.1503	0.516	0.6696	0.34	0.164	11.518	1.4691	0.477
TASK123	12.1753	1.075	0.439	6.6732	1.0117	0.427	0.704	0.3084	0.113	11.518	1.2928	0.406

(a)

		Medium creative thinking (ABSOLUTE)		Mental effort			Higher Workload		Lower Workload		Brain Beat: Stress	
	ALPHA ASS			BAT			TA			TA ASS		
SUBJECT 1	MAX	MEAN	MIN	MAX	MEAN	MIN	MAX	MEAN	MIN	MAX	MEAN	MIN
TASK1	1.19	0.57	0.22	0.36	0.22	0.13	10.39	5.15	1.15	0.00	0.00	0.00
TASK2	1.66	0.57	0.13	0.25	0.12	0.07	4.90	2.05	0.40	12.83	4.43	0.93
TASK3	1.55	0.62	0.19	0.86	0.25	0.04	25.38	2.90	0.70	17.36	4.21	0.94
TASK123	1.66	0.60	0.13	0.86	0.21	0.04	25.38	3.06	0.40	17.36	3.53	0.00
MUSIC	24.07	10.87	4.34	0.04	0.02	0.01	0.53	0.16	0.06	0.47	0.22	0.09
SUDOKU	2.57	0.93	0.27	1.13	0.17	0.05	8.10	2.72	0.83	16.25	4.47	0.79
SUBJECT 2	MAX	MEAN	MIN	MAX	MEAN	MIN	MAX	MEAN	MIN	MAX	MEAN	MIN
TASK1	1.3064	0.6118	0.288	0.415	0.256	0.142	2.5937	1.585	0.6013	4.617	2.4597	0.751
TASK2	1.7578	0.8921	0.347	0.247	0.136	0.077	2.1635	1.348	0.4177	6.5246	2.2677	0.703
TASK3	1.4037	0.6733	0.184	0.765	0.192	0.082	1.7298	0.869	0.3501	11.318	2.4055	0.9
TASK123	1.7578	0.7238	0.184	0.765	0.188	0.077	2.5937	1.137	0.3501	11.318	2.3766	0.703
MUSIC	23.866	8.6462	0.147	0.188	0.067	0.019	4.523	0.491	0.031	18.314	1.5754	0.075
VIDEO	2.2531	0.7548	0.409	0.298	0.183	0.094	1.9023	1.032	0.3595	4.667	2.1183	0.371
SUBJECT 3	MAX	MEAN	MIN	MAX	MEAN	MIN	MAX	MEAN	MIN	MAX	MEAN	MIN
TASK1	0.8507	0.4449	0.225	0.259	0.181	0.107	2.7498	1.438	0.8976	4.5749	3.1417	1.025
TASK2	0.936	0.4787	0.265	0.223	0.138	0.078	1.5557	1.059	0.4713	3.3055	1.9576	0.758
TASK3	2.6105	0.6165	0.192	0.33	0.169	0.017	2.0393	1.029	0.4857	8.4294	2.6829	0.664
TASK123	2.6105	0.5572	0.192	0.33	0.163	0.017	2.7498	1.097	0.4713	8.4294	2.5706	0.664

(b)

Fig. 11. Minimum, mean and maximum value of spectral powers and synthetic indices for each test participant, in each phase of the experimental procedure.

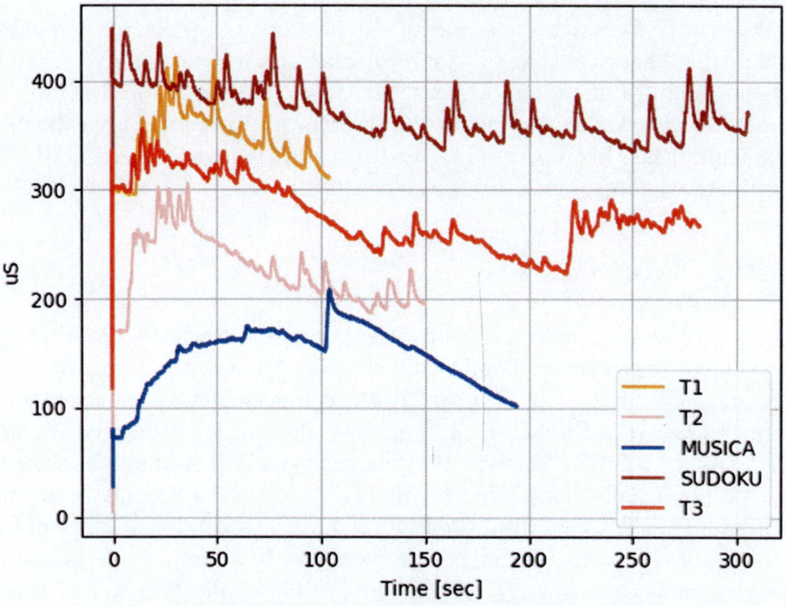

Fig. 12. EDA activity for Subject 1.

calculated in F7 - F3 - F4 - F8 - Fz - T7 - T8, it can be deduced that C.2 was by far the test that required the least mental effort; (b) From the spectral power of the Alpha rhythm, calculated in P7 - P3 - P4 - P8 - Pz - O1 - O2, it can be deduced that, during C.3 and D, an increase in creative thinking was found, probably due to to a mental effort; (c) The BAT index confirms a state of high concentration during D; (d) The TA index confirms a high workload during C.1, C.3, D and a reduced workload during A as the user is particularly relaxed. These results can be considered in agreement with those obtained from the analysis of the electrodermal activity, where C.1, C.3, D were found to be the experiments with greater amplitude value than the C.2, A.

5 Conclusion

In conclusion, within this paper, two different types of biosignals, EEG signals and EDA activity, were analyzed by means of experiments designed specifically to cause different cognitive loads to the participants. In particular, a collaborative robotics experiment was developed to analyze in more detail the degree of stress and concentration on test subjects during engagement tasks with a robot. Recorded data were processed by means of EEGLAB for the EEG signals and Weka for the EDA activity and analyzed to in terms of spectral powers of the rhythms (Alpha, Beta, Theta) and the indices of workload present in the literature. Results have shown that spectral powers and state-of-the-art synthetic indices can be reliable indicators of workload during activities of everyday life and collaborative robotics experiments. In particular, in the participants, an increase in the Theta rhythm was found during the more complex tasks in the frontal and temporal area, and a greater creative thinking. Moreover, we found an increase in the Alpha rhythm in the parietal and occipital area during the more difficult tasks, likely caused by creative ideation and divergent thinking, as the subject was looking for alternative solutions.

6 Future Works

This study represents a preliminary investigation and the achieved results need to be consolidated over a larger number of subjects. Additionally, a possible future development may be to measure cognitive load by calculating the average blink rate using EEGLAB's Blinker Tool. This tool allows the automated extraction of ocular indices from EEG, thus allowing large-scale [18] analysis. Therefore, it could be potentially interesting to analyze the users workload by means of another parameter deriving from the EOG signals. A second possibility of future development of this work may be to use the TPC/IP socket, made available by NIC2, to transmit data to any TCP/IP compatible application in real time.

Acknowledgement. The authors would like to thank Marianna Turrà for providing support in experiment design and analyzing EDA signals.

References

1. Hao, G., Zhong, Y.: MentalWorkload assessment based on EEG and a hybrid ensemble classifier. Softw. Guide **18**, 1–4 (2019)
2. Giorgos Giannakakis Member, IEEE, Dimitris Grigoriadis, Katerina Giannakaki, Olympia Simantiraki, Alexandros Roniotis and Manolis Tsiknakis, Member, IEEE: Review on psychological stress detection using biosignals
3. Salomao, T.; Alberto, L.: Operator functional state modelling and adaptive control of automation in human-machine systems. Ph.D. Thesis, University of Sheffield, Sheffield, UK (2016)
4. Hart, S.G., Staveland, L.E.: Development of NASA-TLX (task load index): results of empirical and theoretical research. Adv. Psychol. **52**, 139–183 (1988)
5. Reid, G.B., Nygren, T.E.: The subjective workload assessment technique: a scaling procedure for measuring mental workload. Adv. Psychol. **52**, 185–218 (1988)
6. Li, Z.; Dong, Y.; Guo, X.: Hybrid brain-computer interface system based on alpha wave and motor imagery. J. Electron. Meas. Instrum. **28**, 625–629 (2014).https://link.springer.com/article/10.1557/jmr.2014.218
7. Sornmo, L., Laguna, P.: Bioelectrical signal processing in cardiac and neurological application
8. Cacioppo, J., Tassinary, L.G., Berntson G.G.: The handbook of psychophysiology
9. Nyrio Ned. https://niryo.com/product/ned-education-research-cobot/
10. Enobio 20 Neuroelectrics. https://www.neuroelectrics.com/solutions/enobio/20
11. Neuroelectrics® Instrument Controller (NIC2). https://www.neuroelectrics.com/solution/software-integrations/nic2

12. Weka 3: Machine Learning Software in Java. https://www.cs.waikato.ac.nz/ml/weka/
13. Martinez, R., Salazar-Ramirez, A., Arruti, A., Irigoyen, E., Martin, J.I., Muguerza, J.: A self-paced relaxation response detection system based on Galvanic skin response analysis. https://ieeexplore.ieee.org/document/8680625
14. Electroencephalographic workload indicators during teleoperation of an unmanned aerial vehicle shepherding a swarm of unmanned ground vehicles in contested environments. https://www.frontiersin.org/articles/10.3389/fnins.2020.00040/full
15. Estimating brain load from the EEG. https://www.researchgate.net/publication/26681847_Estimating_Brain_Load_from_the_EEG
16. Int, J.: Environ. Res. Public Health (2020). https://doi.org/10.3390/ijerph17010046
17. Bujar, R., Longo, L.: An evaluation of the EEG alpha-to-theta and theta-to-alpha band ratios as indexes of mental workload. https://www.researchgate.net/publication/358919573_An_Evaluation_of_the_EEG_alpha-to-theta_and_theta-to-alpha_band_Ratios_as_Indexes_of_Mental_Workload
18. Github Repository. http://vislab.github.io/EEG-Blinks/

Author Index

P. Borja et al. (Eds.): HFR 2022, 26, pp. 248–249, 2023.
https://doi.org/10.1007/978-3-031-22731-8